THE
ANCIENT
LANGUAGE
OF
SACRED SOUND

———————

"This book is a touchstone for rediscovering some of the deepest mysteries of antiquity. Mystical sciences behind all religions are a revelation that will cause a seismic shift in our perception of the wisdom of the Ancients."

PAUL BROADHURST, COAUTHOR OF
THE SUN AND THE SERPENT

"David Elkington's theories could be a major development in our understanding of the Ancient Mysteries."

ADRIAN GILBERT, COAUTHOR
OF *THE ORION MYSTERY* AND
THE MAYAN PROPHECIES

THE
ANCIENT
LANGUAGE
OF
SACRED SOUND

The Acoustic Science
of the Divine

- - - - - - - - - - - - - - - ✸ - - - - - - - - - - - - - - -

David Elkington

Inner Traditions
Rochester, Vermont

Inner Traditions
One Park Street
Rochester, Vermont 05767
www.InnerTraditions.com

Text stock is SFI certified

Originally published in the United Kingdom in 2001 by Green Man Press under the title *In the Name of the Gods*

Cataloging-in-Publication Data for this title is available from the Library of Congress

ISBN 978-1-64411-165-9 (print)
ISBN 978-1-64411-166-6 (ebook)

Printed and bound in the United States by Lake Book Manufacturing, Inc. The text stock is SFI certified. The Sustainable Forestry Initiative® program promotes sustainable forest management.

10 9 8 7 6 5 4 3 2 1

Text design and layout by Debbie Glogover
This book was typeset in Garamond Premier Pro with Majesty, Trenda, and Futura Std used as display fonts

Illustrations by Amanda Patten

To send correspondence to the author of this book, mail a first-class letter to the author c/o Inner Traditions • Bear & Company, One Park Street, Rochester, VT 05767, and we will forward the communication.

For Jennifer and for Eloise

---- ✳ ----

ASET—A dedication

Within the sheer grasp of beauty
I shivered
as if to meet your smile
and trembling thoughts
brought to book
the wild audacity of nature

And then
all the vibrancy of your gift
and this, our world
stirred deep within me
like the last son of the lark
or the blackbird at dawn

For within my heart
have stirred illusions
as if willing myself
to keep this world at bay,
and ever the tide lapped at my feet

But solace betook me
and having broken through,
the wonder of it all
is that all the time
you were there.

Contents

---- ✳ ----

Foreword to the
Second Edition (2021)

By Robert Watts

During the mid-1970s when I was working as a producer on *Stars Wars,* neither George Lucas nor I had any idea how popular the films would prove to be. Later, I had the same experience working on the *Indiana Jones* movies along with George and Steven Spielberg.

The secret to the success of these films was the magic of myth—it is to be seen directly at the center of them all. I suspect that without myth there would have been no story, no success, and no fascination for the general public about the issues they raised. The impact of myth and legend upon our work, so taken for granted nowadays, was immense. It would be safe to say that both franchises have, in the forty-odd years since, become cultural phenomena.

And at the heart of this phenomenon is the myth of the hero.

George Lucas had read in detail the work of the late Joseph Campbell, whose groundbreaking life's work gave tremendous new insight into world myth and legend. It was, more or less, a continuation of the insights of Carl Jung, the great pioneer of the new discipline called *psychology:* from *psyche,* the human soul or spirit and also a figure of Greek legend, representing the experience of love, or *eros.*

Beyond the effect of the films, I was on my own quest for understanding, and the question remained: What is myth? Why has it had, and remains to have, such a significant impact on our culture—and not just in the world of entertainment?

I have had much experience of life both in and out of the film business, and my personal experience of myth and its interpretation of what we see around us is amazing in that it has lasted so long: scorched as it were into our cultural ethos. And then, in 2007, I encountered the work of David Elkington and read through his highly original thesis with delight and not a little wonder. It was when I got to the end and read about the Djed pillar that I put down the book and, amid utter astonishment, realized that myth is much more deeply ingrained in our consciousness than I or anyone seems to have previously appreciated.

George had called his chivalric order in *Star Wars* "Jedi knights." How could we know how accurate a description this was? It was as if, in calling the relevant section of this work "The Return of the Djed and I," it was written specifically for us to understand, a belated appreciation of the path that I as producer and George as director were on.

How remarkable therefore that the work should end up in my hands. Myth was working its extraordinary way with us all. David provides the answers to the practical mechanism of myth, the ancient and innate power of place and its manipulation by us as an act of self-awareness.

And, of course, all of this is expressed by the greatest gift of all to humanity—the gift of the written script. Not, as David here reveals, an invention by us but rather an expression of the power of place, of an inherent feeling within us that takes form within our brains and emerges, poetically, to give us expression, understanding, and self-consciousness: all themes within the *Stars Wars* and *Indiana Jones* films.

In this sense the cinema is but the latest kind of temple to these powerful forces: a temple of entertainment, yes, but the effect as we walk out from within them, having sat within the close darkness under the spell of manipulated light, is that we have witnessed mysteries indeed and we have affirmed our pathway into the light of day, and sometimes enlightenment to all that is around and within us.

For many people *Stars Wars* was and is a spiritual experience, one that set us on the path to wisdom and the quest for knowledge. *The Ancient Language of Sacred Sound: The Acoustic Science of the Divine* will leave you wanting more and will give you the courage to continue the journey.

May the force be with you.

ROBERT WATTS is one the foremost film producers of the past fifty years. As well as working on the James Bond film *You Only Live Twice* and *2001: A Space Odyssey,* he produced the *Stars Wars* sequels *The Empire Strikes Back* and *Return of the Jedi* and the Indiana Jones films *Raiders of the Lost Ark, Indiana Jones and the Temple of Doom,* and *Indiana Jones and the Last Crusade.* He also produced the Emmy Award–winning *The Mystery of the Sphinx* and followed it up with *The Mysterious Origins of Man.*

Foreword to the
First Edition (2001)

By Jean Houston

In this astonishing book the author, David Elkington, offers us a life work. With the energy of a true polymath, he weaves together archaeology, mythology, brain research, literature, history, geology, biblical criticism, linguistics, acoustics, physics, metaphysics, and a spate of other disciplines to explore his remarkable thesis: the ways in which the resonance of earth energies informs the names of our gods, sacred spaces, and their functions in our own spiritual progression.

In a foreword one can give only a taste of the feast to follow, but let me here offer some of my own reflections on three of the themes to be found in this work: sacred space, sacred sound, and the passion of Jesus.

It is not uncommon for people who travel to ancient sacred sites to come away with a sense that they have changed, their hearts opened, their spirits rejuvenated, the possibilities extended. How can we explain such miracles, except that the sacred temples—even the earth beneath one's feet—still reverberate with ancient power, charged by hundreds and even thousands of years of conscious invocation, prayer, incantation, and meditation. It is the author's contention that ancient peoples, especially the Egyptians, were well aware in their construction of temples and pyramids of the nature of sonic frequencies and of geometrical progressions, seeing these as cosmic or secret powers and employing them as secret science to affect consciousness. These were people for whom every animal, plant, and stone embodied the divine, a people for whom daily life vibrated with spiritual meaning, for whom the mere act of waking from sleep was a resurrection from death, akin to Ra being reborn each morning.

When we view the world as sacred—when every beast, grain of sand, molecule of air, tree, seed, and river is viewed as being empowered with gods, and we know the name (that is, the frequency) of these gods—then the world views us as sacred, too, and we rise in consciousness accordingly. For the most part, the temples of Egypt were erected in the center of the village, with the dwellings of the townspeople built around each temple. Spirit was the keystone of life. The temples were laid down during the dawn of Egyptian history, according to plans designed for them by the mysterious and godlike Companions of Horus. Each new temple was erected on the site of a previous one, with careful observance to all details ranging from architectural structure to astrological significance, from symbolic meaning to functional aspects. Each temple construction began with an invocation to the divine architects of the universe. The invocations called forth the divine powers of universal harmony, mathematics, myth, and symbolism—powers that were then incorporated into the temple design so that the temple on Earth would reflect the same design principles used by the divine architects of the universe. The function of a temple—whether that temple be an actual architectural structure or a group of initiates who form a template, an intellectual structure—is to maintain and perpetuate a vast body of knowledge intact. The establishment and maintenance of any temple requires extraordinary skill by knowledgeable and dedicated individuals who believe that adherence to divine creative principles is the sacred task of all, who view the structure of their temples as well as their daily lives, as artists and craftsmen imitating the creative acts of the divine. They are a fully conscious and deliberate organization who have come together to keep a body of wisdom intact.

In Egypt, it is true that the colossus of Memnon no longer sings, that the Nile no longer overflows its banks, that the paint is now flaking from the tomb ceilings, that ancient cliffs crumble every day to sand, and that these losses are deeply felt. And still the power thrums beneath the sand and under blue sky, the light of heaven pierces the hearts of those who come to revere this place. "The gods want you here," a fellah once told me. "They need someone to speak their names."

In this book David Elkington does indeed speak their names and shows how these names the world over are not only similar in pattern but also have the same expressive power in form and function.

In this book we have the coming together of world thought, new science, and social artistry, giving us perspectives on what the world and reality may be. It presents us with the beginnings of a new natural philosophy. This new philosophy has the potential to invoke and shape our future sciences, arts, economy, politics, technology, psychology, and spirituality. It is a view of the world as living within a larger living life—a complex webbing of interdependent relationships—not unlike Indra's net. Within this net, everything is energy, vibration, frequency, resonance. Even the most solid of material objects is ultimately a dance of constantly changing energy patterns. Ultimately it is all rhythm, all music; the world is sound—Nada Brahma—as the classical Sanskrit philosophers put it.

Whereas our eyes can rarely penetrate beyond the three dimensions of space, our ears have the capacity for multidimensionality. In Indian music there is the concept of *anahata* and *ahata,* the unplayed and the played note; both are equally important, and the unplayed is often thought to be more important than the played since it provides the foundation for what is actually played. Heraclitus once said that "the hidden harmony is mightier than what is revealed." Not only is the world sound and vibration, but in physical terms there are trillions of possible vibrations as well. There is a great music master behind it all. Research on harmonics has shown that the cosmos, the universe, and nature have a tendency to select from these innumerable possibilities the few thousand that give rise to proportions of vibration that give us music, architecture, the forms of animal and human bodies, the DNA relationships containing our genetic code, earth magnetism, states of consciousness, and access perhaps to higher realities.

The vantage point of this potent book gives us new insight into the Western world's central faith as well. As we move into religious syncretism, many are coming to a new appreciation of the meaning of their Christian roots. As the author shows so brilliantly, Christianity's seminal story, the death and resurrection of Jesus Christ, recapitulates the myth of the dying and rising god venerated in the millennia that proceeded his birth, most particularly in Greece and the ancient Near East. So Isis searches for the scattered parts of her husband, Osiris, binds them together, and animates him to produce new life. Demeter calls forth her daughter, Persephone, from her dwelling place in the Kingdom of the Dead. Tammuz, Adonis, and Dionysius are all destroyed, and are remade. The Christ story is the culminating expression of this cycle, the apotheosis of an instant pattern celebrated in the ecstatic and highly ritualized piety of the mystery religions. Now, two thousand years later, a new mystery arises and with it a new spiritual myth. The story of Jesus differs from that of the traditional mystery cult figures in that he was a historical person. Because he lived in space and time, Jesus brought the dimension of human experience to the transpersonal and archetypal field of God identity. But traditional Christian thought held Jesus to be unique, the only son of God. The new mythos of all time democratizes divinity. "That art Thou," as the Hindus put it. Many of us have come to believe that we are all God in hiding, seeded with Christic identity or Buddha nature, the literal capacity for cosmic consciousness. The single melody under the cacophony of twenty-first-century spiritual seeking is our collective awakening to the felt and live awareness that the divine is not somewhere "out there" to be supplicated but somewhere "in here" to be discovered.

In Jesus, moreover, as Elkington has shown, two causalities meet: one individual and national, the other collective and universal. On the horizontal level, the Christ story is a Jewish political tragedy. On the vertical, it is a myth that speaks to humankind's yearning for transformation to a higher level of being. Today, all over the world, the same story is being enacted. Our horizontal planetwide crisis of war and mayhem cries for the vertical solution of universal species transformation. Now the task of transformation does not belong to Christ or Buddha or Quetzalcoatl or any of the Saviors alone. We will be equal to the requirements and the responsibilities of the twenty-first century only if we have nurtured the innate seed of our own divinity. The spiritual paradigm has shifted from the One to the many, from the journey that was Christ's Passion, all the passion of the many rising and dying heroes and heroines of myth and legend, to the many existential journeys along the personal and collective mystic path. Like the Passion, the drama of the world's unfolding is a mystery play with several acts. Jesus the Nazarene had his death and resurrection as Christos; our version is the extinction of parts of our limited local self and our resurrection into a unitive reality that is both spiritual consciousness and global spirit, and that is ultimately spiritual resonance, what can be called mysticism or the art of union with Reality. As you read this book, you enter into a once and future world, and you become the music of the new song . . .

JEAN HOUSTON, scholar, philosopher, and researcher in human capacities, is one of the principal founders of the Human Potential Movement. She founded the Foundation for Mind Research alongside her husband, Robert Masters. She holds Ph.D.s in psychology and religion and is the author of more than twenty-five books. John Lennon called her book *Mind Games* "one of the two most important books of our time." She was also an adviser to the White House at the invitation of Hillary Clinton. Through her work with the United Nations Development Program and as an adviser to UNICEF, she has worked in more than 100 countries training leaders in human and cultural development.

Acknowledgments

An especial thanks to John and Caitlín Matthews and Claire Palmer, who between them got this edition off the ground and into the ionosphere. To Paul Broadhurst, Gabrielle Tyrso, Julie Burt, and Chris and Caroline Woodard for their encouragement, friendship, and support. Dwina Gibb and all at the Prebendal for their laughter. Geraldine Beskin at Atlantis for her common sense and good suggestions. To Cherie Lunghi for excellent advice. Lynn Picknett and Clive Prince for pointing out the "bleedin" obvious, and to Robert Watts for his friendship and observations about immortality.

I would also like to thank the following people whose friendship and support have seen me make it into the sunlit uplands: David McIntyre, Keith and Araxia Hearne, Professor Roger Webb and all at the Ion Beam Centre, University of Surrey, Allison Powell, Jon Legalloudec, Audrey Renton, John Day, Brian Mathew, Neville Farmer, Ralph Whittaker, Matthew Hood, Boyd Anderson, Anders Kvist, Sam Kandiyali, Willy Thornton, Peter Hiscock, Isam Bannani (thanks for the time in Den Hague!), Eamon Everall, Ramon Goose, John Reid, Robin Heath, and Jean Houston.

But most of all for Jennifer, Alex, and Eloise, to whom this book is dedicated.

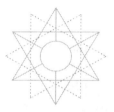

Introduction

Humanity today stands on the brink of space and its mysteries; however, little has been done to understand and appreciate the connection we have with our home planet—an intimate connection, via our brain wave patterns, to the Earth itself.

If we are one day destined to sever ourselves from Mother Earth then the wisdom contained in the myth of the hero must be examined in a new light; the spirituality within the confines of religion must emerge as a science that goes beyond the sacred, as a science of consciousness with all of its implications.

THE EXPERIENCE

What is religion and why has it played such an important role in the rise of humanity? This was a question I asked myself at the outset of this personal quest many years ago.

To answer this question, it is necessary that we see the situation as it now stands to appreciate how we got here—and perhaps to see where we are going.

In an increasingly secular world, religion is no longer seen as the answer to all problems and fears. Rather, in 2020, it has become a Pandora's box from which has poured many of the world's ills. Beset by serious internal scandal and encompassed by a moral vacuum, organized religion is increasingly seen as the abode of extremists and extremism.

Consequently, people in the West are looking elsewhere for an answer to their spiritual questions.

In the 1950s in Great Britain more than 15 million people attended regular Sunday church services. However, a recent poll demonstrated a vertiginous decline: less than 400,000 people now in attendance; most of these are over seventy-five years of age. The Church of England is facing its demise—and this is not just happening in Great Britain. In the United States the rise of secularism has also seen a steep decline in churchgoing activity. Yet, despite this, alternative modes of spirituality have never been so popular—into the breach other paths to spiritual fulfillment have stepped. This demonstrates a clear need: but one detached from the idea of a collective dogma. The flavor of these alternative paths is most often decidedly pre-Christian—with an emphasis on the connection to Mother Earth.

Religion has been political since the acceptance of Christianity by Emperor Constantine in 313 CE, when it became the official religion of the Roman Empire—and those who had been its true originators, the Nasoreans, were cast out as heretics. At this point another great shift took place: all other belief systems within the empire were gradually forbidden and their sanctuaries either converted to the new faith or destroyed. If god was at the heart of religion, then religion was the heart of the empire. Christianity was declared the victor in the battle of beliefs.

However, other pagan beliefs never quite went away; if anything, they reshaped themselves and retained their pagan flavor in all but name. Brigid, Apollo, Helios, and a host of other gods became Christian saints: an act of survival in an increasingly political world born of the empire.

Religion was its manifesto: dogma became the be-all and end-all. The result of this shift was that *spirituality* in all areas went underground and, like all fecund streams, surfaced where there was a weak spot in the rock upon which the church was built. Mystical approaches to the sacred texts began to appear, though cautiously, but were soon driven underground again by the Inquisition and dogma. Christianity began its long journey toward an inevitable decline.

Nor is this surprising when all the facts are examined. Christianity is, in reality, the last vestige of the long, slow burnout of the old Roman Empire. For all of the official Christian faiths today stem from the Council of Nicea presided over by the Emperor Constantine in 325 AD: a meeting of church elders that configured what the church was to be—and was not to be. The result, the Nicene Creed, is at the heart of all Christianity. So, if we want to look for the original faith then we have to see beyond and behind it. It was at this Council that the original Christian groups were declared to be *heretics:* no better than pagans. The church fathers stated that the new Christianity of Rome was unique, that only Jesus could save, that he was the Savior.

However, when we begin to look at the familiar patterns of Christianity, its ritual, liturgy, and the story of Jesus, there are remarkable similarities to world myth: so remarkable in fact that the Jesus Seminar, a group of very eminent scholars, long since declared that Jesus could not have existed, since everything about him appears to have been borrowed from elsewhere. This is the increasingly official view of the new secular history that is arising from the ruin of the empire.

In the face of this my question is: What if the church fathers were correct? What if Jesus truly was the Savior, but he was a savior who predated the historical figure of Jesus by many thousands of years, and who is almost entirely wrapped up in the science of the sacred? If Christianity truly does comprise all the other faiths, what and where is the evidence?

This is the theme of this work.

It is also a question raised by both Saint Paul and Saint Augustine. In Galatians 3:6 and Romans 4:3, Paul drops a large hint about the preexistence of Jesus by speaking of Abraham, whose son Isaac is an early echo of the legend all too familiar to us in the Gospels.[1]

Saint Augustine of Hippo (354 to 430 CE) wrote, "That very thing which is now designated the Christian religion was in existence among the ancients, nor was it absent even from the commencement of the human race up to the time when Christ entered into the flesh, after which true religion, which already existed, began to be called Christian." However, we rarely look to the sacred site itself for an answer. It has all too often been overlooked.

In the late 1980s, I experienced a profound moment of change while undertaking a visit to a sacred site. It was Christmastime, and I was at Wells Cathedral in Somerset, England. It had been a difficult and traumatic year, and I was there to begin a quest—though I did not know it at the time.

I have never been religiously inclined and describe myself as, at best, an agnostic, although a mystically inclined one in the best tradition of Blake, Vaughan-Williams, and others. It is a question to which I have sought an answer for a very long time. On that day I caught a glimpse of it.

Walking up the well-worn steps to the Chapter House, I was suddenly taken out of myself by a convergence of two things: the sound of a choir singing in the Chapter House—and the place itself. I felt myself drawn to stand tall, with my back straight, walking uprightly, as my senses seemed, momentarily, to leave my body and to float on high. I was physically rooted to the spot, while spiritually I was elsewhere—and everywhere.

It was momentary and it was timeless. And when it was over, my heightened emotional state saw me crying tears at the sheer beauty of it all. Sound and place were, I realized, connected at a level never before suspected.

I vowed, at that moment, to pursue a quest for an answer, never to stop until I had found it.

THE THESIS

Human beings, it seems, have a natural urge toward spirituality and the need to grow inward and toward it as an expression of wisdom. We have a need to know where we came from and why: and this is curious—the word *religion* means "to bind back" (to the source).

Could it be that *religion* is an ancient language of scientific expression? What if a thing that we take for granted, so much so that we rarely give it a second thought, was not a human invention at all but an environmental response mechanism? I refer, of course, to writing, to script.

My theory is this: In emerging from the cavern environment at the dawn of history the first urge of humanity was to stay connected to Mother Earth. So began the construction of sacred sites—recognized from the Paleolithic to more recent eras. From these came civilization as we understand it.

We are inclined to see temples and pyramids very much in the past tense and fail to see that they were, and still are, living, breathing places of creation: of the outpouring of creation that came from them. What we do not and cannot see is what we cannot now hear—these were places of music and dance. Britain had Perpetual Choirs, there were Choral Guilds in the Temple of Jerusalem, the pyramids rang out to the music and hymns of the gods: wherever there is a sacred place it was a place of music and chant. The purpose of music and chant was to alter our brain wave patterns, to take us away from the mundane and into the heavenly.

Over the course of many years I was witness to and partook in a series of experiments at these places—experiments that charted their individual resonance. But there was one clear distinguishing aspect to them all that was nothing short of astonishing. When sound wave patterns from these sites were manifested in pictorial form, using a process known as cymatics, they demonstrated a clear linguistic form, exhibiting script types unique to those cultures. We saw Celtic knot work at Celtic sites, Buddhist and Hindu yantras at Buddhist and Hindu sites; Paleo-Hebrew script; the *Om* of Hinduism—and, most astonishingly, hieroglyphs deep inside the Great Pyramid of Giza. The list is by no means definitive, but the implication is clear: script, in its earliest form, emerged from these sites in a form created by sound. However, it didn't just appear there: it was a gift of the gods given to humanity via the hero.

It is the hero who is the fulcrum point at these places, whose myth informs us that he was the meeting point of humanity and god—the perfect arbiter between both. It is the hero who, in the mythologies of the world, brings language and the skill of writing to humanity. It is the hero who draws down knowledge and brings the civilizing arts to the world. It is also the hero who is born in the temple—and who dies there: at the center of the world.

As I began to look deeper into the legends it became increasingly obvious to me that what we are looking at is not just a metaphor for an experience but also an expression of what I can only describe as spiritual technology.

But who is the hero really?

The hero is a state of mind, almost the calm after the storm of not knowing. He brings self-awareness and the ability to grow, to ascend, to climb to heaven by learning wisdom. In effect, we are witnessing a technology of the Soul, the science of our intellectual and conscious ascent—and in this scheme of things the sacred site becomes the umbilicus connecting us to an unseen world.

These places were constructed with a clear and resonant purpose: they are places of communion—between humanity and the divine. But what exactly are the gods? What do they represent?

The general answer is that they are forces of nature, expressions of the different aspects of the world around us. However, somewhere in all of this is a meeting point, where man meets god, where man even becomes god. (I say *man* not because of any inherent sexism but because the feminine is already seen as an aspect of the divine in the form of the Earth Mother, whereas a man has to demonstrate his latent divinity.) That place is the temple.

There was no denying the remarkable similarities of worldwide hero myths—the story, with variations based on geography and culture, is largely the same, even to the degree that the names are the same and have the same meaning. In this scheme the sun as the Father-god shines upon the Earth Mother, adding to the life she gives. The earth aspect is wisdom: our ability to listen to it is *our* wisdom, the beauty of which is reflected in the shades of the moon. The child of these deities is the hero, whose task is to unite heaven and earth by teaching us language and how to sing it, and in his quest to bring us self-awareness, dies to become a Judge of the Dead: an ancestral memory. Humanity is brought to consciousness by the Divine Child—and in all of this the sacred place plays a central role.

THE RENEWAL

In December 1999 it was announced that the very beautiful white marble baptistery at Pisa, Italy, completed in 1363 CE, is in essence the largest musical instrument in Christendom.[2] Even more extraordinary is the fact that the circular structure, surmounted by a 250-foot cupola, was in all probability designed by its architects to mimic the pipes of a church organ. The acoustics beneath the cupola are so perfect, scholars have realized, that it must be either an incredible coincidence or the work of genius. Other experts have called it the Italian Stonehenge, believing it to have been built to record the winter and summer solstices.

According to Professor Silvano Burgalassi the cathedral and the baptistery are on a perfect east–west axis, while the famous leaning tower is 23.5 degrees off the axis (taking into account the angle of the lean). "The explanation is that the Sun strikes the tower later this month (December), about four days before Christmas, coinciding with the winter solstice. At Midsummer, the Sun streams through a south facing window in the baptistery on the feast day of Saint John the Baptist, who is Pisa's patron saint, striking the statue representing him on the font."[3]

As this work will demonstrate, all sacred places were carefully designed to maximize the most beneficial acoustical effects, for the precise days to which they were aimed, more often than not the summer and winter solstices, though there are plenty of exceptions to the rule. While this analysis is extraordinary it leaves out one central point: that *we are the instrument of measurement in the new science—the technology is ourselves.*

To measure the sacred place, its esoteric value, and its appearance in history, as well as the fact that it was an emergent science—emergent from within us—we need to own our tools and use the one tool we are not yet good at relying on: ourselves. In a way this is what religion is all about.

It is the tragedy of Western religion that it was separated from science during and after the Renaissance, each going their separate way. Perhaps it was at this point that religion became the perceived political entity it developed into, while spirituality trod a lonely path upon the fringes of science and religion. Each has derided the other over the centuries, and, while there have been fine examples of scientists as churchmen and vice versa, they have had to tread a cautious path.

The persecution of science by the church in the run up to and during the Age of Reason has seen the triumph of scientific values in our own world—and the diminishment of the church. Yet both are, in their idiosyncratic way, trying to answer the ultimate question of the meaning of human existence; and, scientifically speaking, this was never more so than in the growing scientific inquiry into the nature of consciousness.

In the mythologies associated with these sites it is the hero who brings consciousness, who brings self-awareness by his heroic actions on behalf of humanity: because the hero is humanity, he is humanity as self-aware. His is the journey to the dark heart of wisdom—dark in the sense that wisdom is about seeing both sides of the argument to reach the higher good.

And indeed, on the morning of the winter solstice the sacred places are equally dark as their resonant frequencies take effect. In this way they acted as chambers of initiation wherein the hierophant would undergo spiritual death to become the hero. Quite often this hierophant was the king—and the duty of the king was his duty to his people. Thus the beauty of the sacred space was as a touchpaper whose fires were lit by the presence of the inquiring mind: *ars sine scientia nihil est*—the practice of an art without proper knowledge accomplishes nothing.

It is my belief that it was at the sacred site where that knowledge first emerged. This gives us a new perspective on the story of the historical Jesus and the rise of Christianity at the outset of the first century CE; for what the Gospels do not tell us and what has been written out of them is the centrality of the temple to the peoples of the period. That this was so is borne out by the rise of Christian architecture from the earliest period after the resurrection of Jesus.

The historical Jesus was unique in this period because he actively sought to fulfill the prophecies by living them, by undergoing spiritual death to achieve them. However, in terms of the presence of the hero the name of *Jesus* appears to be unique to a vast majority of them, *from the outset of civilization*. He is the original hero bringing language, naming the articles of Creation, a vessel for their identity. This is precisely the role of the shaman in world mythology and rite: the shaman helps his or her peoples to celebrate consciousness by the naming of things. The Aboriginal peoples of Australia call this the Dreamtime. In Hebrew mystical tradition from the historical Jesus's time it was called Merkavah mysticism and was preserved in the later kabbalistic teachings. It is curious that the early Christian saints tell us that Jesus preexisted the first century, that he was with Moses in the Ark, that he existed even before that.

According to the Gospel of Saint John, Jesus was *the Word*. Jesus, in becoming Christ "draws" down the Christ spirit from the highest heaven.[4] In the Word is meaning: What can be a greater mystery than the pulling down of language to the dimension of humanity that it might push its way back up to a higher consciousness, starting with self-consciousness? For if language is not an invention but a response to the sacred environment, then who is communicating to us and why? The "Word" of God was the defining script given to the hero-king in his shamanic journey—it was the development of sound, of resonance: the rhythms of sound into the rhythms of poetry, meter, and liturgy soon to become prose.

Today we use and abuse the Word, almost devoid of any awareness of its inherent rhythmic meaning and power. This was the reason why the priests held on to the meaning of script as an elite teaching for so long—it was soon to become sundered from the place of its emergence, of its birth, as science and religion were equally sundered from each other.

The prologue of Saint John's gospel is the great "I am"—the author is saying in the prologue that the hero *is* the Word, that it is the greatest of all modes of self-expression, that to be it is to be at once both human and God. This is a statement to the effect that the very name of the hero is a word of healing and marks a continuing act of Creation, the high point of human consciousness. Saint Paul, in speaking of the name of Jesus, says as much in his Letter to the Romans 8:18–25: "The awareness within us!"

Sound becomes ritual.

However, another letter, Ephesians 1:22–23, describes the church as the Body of Christ; it is also to be found in the catechism of the Roman Catholic Church: Is this a metaphorical statement or one of fact? It is, thanks to world mythology, not an original statement but something far more profound. Paul is making an architectural as well as a theological point.

Once you have reached the culmination of the journey you will be pleasantly surprised by the answer. It is truly revelatory. For the hero in his birth pose is still present to offer us insight in these troubled times.

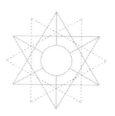

1
Secrets of Life

I sing the body electric . . .

WALT WHITMAN, *LEAVES OF GRASS*

The singing of words reveals their true meaning directly to the soul through bodily vibration.

HILDEGARD OF BINGEN, FROM
HILDEGARD VON BINGEN'S PHYSICA,
TRANSLATED BY PRISCILLA THROOP

Hurtling through the cosmos at great speed is a crystalline, but living, rock, in orbit about the sun. Since the formation of the solar system approximately 4,600 million years ago, our home, the Earth, has developed in a profoundly different way from all of the other planets in our solar system. By virtue of being the third planet from the sun, with a unique axial tilt and a twenty-four-hour spin, Earth is subject to precisely the right conditions for carbon-based life-forms*—life as we know it—to thrive, away from the dangers that lurk in space. By a curious twist of fate, the planet is shielded from a majority of incoming meteors and other potential hazards by the gravitational pull of the great planet Jupiter.

However, we have created hazards of our own. For years, the indigenous peoples of the world, from the Kogi of Colombia to the

---

*Silicon-based life-forms, too, have also been discovered in deep cave systems in Romania.

Aborigines of Australia, have warned the Western peoples about their ignorant, wanton ways, justifying their comments with the threat that the Earth Mother won't like it, that she will fight back. They seem to mean this literally. Can they possibly be correct? Is she really a being in her own right?

At the 1995 Rio Summit, the world's industrial nations agreed that they were going to have to clean up their act. Agenda 21, as it was called, was the long-awaited document that would set the environmental themes for the twenty-first century, with all countries agreeing to cut their industrial emissions by at least a third within the decade. The warnings were, we thought, being heeded at last. The conference, though not achieving much in substance, had broken new ground in admitting the folly of our ways. The world's indigenous peoples were, however, not only correct in their observations about worldwide pollution, conclusions that have been accepted across the planet and that are now the prime focus of scientific endeavor, but they are also gaining support for a deeper, more profound observation—that the Earth lives and that we live within her embrace.

GOOD VIBRATIONS

As we begin this new millennium, humanity's age-old need to find a reason for our existence is pulling the realms of science and religion closer together: though not without resistance.

The question of What is God? has, in the past eighty years, been replaced largely by What is Consciousness? The views of the great Jesuit scientist Teilhard de Chardin now seem commonplace. Teilhard de Chardin, while sojourning in England in the early 1900s, had a mystical experience in which he became aware of the universe as no longer an abstract notion or a machine but as energy transformed into presence, of which he was a crucial part.[1] Ideas such as this, once considered to be foolish, now seem to have the support of scientific data. Laboratories are now abuzz with speculations about the nature of consciousness,* and

---

*For example, the Institute of Noetic Sciences.

many organizations are dedicated to learning more about what is, in effect, super nature.

Beyond the laboratory, James Lovelock's Gaia hypothesis, that the Earth is a single, self-regulating organism, has caught the popular imagination, and this feeling has been carried forward to maintain that planet Earth is sentient. Many are becoming more and more convinced that we inhabit a living, breathing world and that, rather than being a dead piece of rock floating through space, our planet may well be a being in itself. Though this is not yet scientifically proven, there is no doubt that we live in Earth's embrace, her atmosphere, her vibration. The term *vibration* is a truism in every sense of the word; hence its popularity with New Agers and others. Moreover, it is a remarkably astute way in which to describe the state of matter within the entire universe.*

Matter, in order to exist, must vibrate, and anything that vibrates has a frequency of vibration. When a string is plucked, its rate of vibration is measured, and if it should happen to vibrate ten times a second, then it may be said to have a frequency of ten cycles per second, or 10 Hertz (Hz) after Heinrich Hertz (1857–1894). More familiar to radio users and astronomers are the terms *kilohertz* (kHz) and *megahertz* (mHz): these are thousands and millions of Hertz. We can see therefore that the wavelength of 1 Hz must be very long (the higher or faster the frequency, the shorter the wavelength).

Within the known universe there are literally billions of frequencies, as matter vibrates not only from within but also by being struck from without. The Earth resonates, or vibrates, at an extremely low frequency, or ELF for short. This resonance is produced from a variety of sources, but the most influential are from space: electromagnetic radiation from the sun and cosmic rays, which are actually high-energy particles coming in from all over the cosmos. The origins of these particles are many and varied, from supernovas to something far more violent—gamma ray bursts, possibly the most violent events in the universe. Another source could be the background radiation of the big bang itself, resonating

---

*See also Greg Taylor, "Beyond Time and Space: Are Synchronicities Evidence That Our Universe Is Made of Information?" March 7, 2019. Accessed at DailyGrail.com (website).

from one end of the universe to the other, like the ripples in a pond.

The best way to describe earth resonance is to picture the planet as a gigantic bell. We know that the Earth is not actually hollow, but it is much less solid—fluid even—beneath the crust. Indeed, Mother Earth was described by the Ancients as an orb, meaning literally "hollow sphere," not unlike a bell. When a church bell is struck by a hammer it vibrates, and no matter how loud it might seem, the frequency of vibration always remains the same. There may, however, be overtones, or undertones, which are variations on the resonant frequency. These are dependent upon the shape of the bell, its material, and any underlying cracks or anomalies. Similarly, our Earth is struck by the incoming energies and therefore vibrates at a particular frequency along with its overtones. Thus it is that everything resonates, beats to its own rhythm, and in the process gives off energy that remained, until fairly recently, imperceptible. Now there are ways of measuring it, even at incredibly low frequencies.

While the seismic (underground) vibration of the Earth is 32 Hz, the primary aboveground vibration of the planet resonates at the ELFs of between 7 and 10 Hz. These frequencies are measured as fluctuations in the planet's magnetic field. ELF wavelengths are extremely long: in earth resonance they approximate the circumference of the planet. This is known as Schumann resonance, after W. O. Schumann, who first demonstrated in 1952 that the atmospheric cavity, known as the ionosphere, was in fact a gigantic electrodynamic resonator. It is this region that, in resonating at about 8 Hz, dominates all the aboveground frequencies—and all life itself.

Ions are atoms that, by virtue of gaining or losing an electron, have become negatively or positively electrically charged. As the name "ionosphere" implies, it is a highly electrically charged area of the atmosphere, caused mainly by the incoming ultraviolet radiation from the sun. Incidentally, the layers within the ionosphere, the magnetosphere and the troposphere, are also responsible for the reflection of radio waves and are thus very important to long-distance communication.

By a curious turn of nature, certain frequencies emitted by the brain fall within the band of earth resonance. The brain has varying bands of frequency, reflecting different activities at different times of the day. The lowest of these bands is the delta rhythm 0.5 to 4 Hz. The association here is with deep sleep, and there is also a link to the onset of paranormal experiences. Theta waves occur between 4 and 7 Hz, when we are only half awake, dreaming, or musing meditatively. Alpha rhythms, at 7 to 13 Hz, are states of passive alertness, wherein the mind is relaxed to the point of emptiness. Intriguingly, alpha is the wavelength associated with altered states of consciousness, the frequency of expanded awareness. The fourth level of brain wave activity is beta, at 13 to 30 Hz. This is the frequency of normal everyday wakefulness, of active thinking and interaction with the physical world.[2]

So, the brain contains electrical frequencies that occur naturally in the energy fields of the planet itself. This state of affairs may have been the natural result of life evolving to respond rhythmically to the pulses of Mother Earth. The fact that certain frequencies could coincide with earth resonance means that, in effect, the Ancients were correct in calling this planet the Earth Mother. Indeed, this connection between earth frequencies and our own alpha brain wave frequencies is like an unseen umbilicus. Coincidentally, to a degree, all animal life seems to share alpha frequencies of about 10 Hz, significant evidence that "every creature is hooked up to the earth electromagnetically through its DC system."[3]

Much of what is spoken of as animal intuition or instinct is in fact an acute sensitivity to electromagnetic fields both within the body and without. It is well known that whales and dolphins navigate the world's oceans by detecting magnetic stripes on the sea floor. What is less known is the acute sensitivity of land animals to changes in the pulse of the earth. Days before an earthquake is due to hit a locality, animals will be seen vacating it—slight variations in frequency having warned them of impending danger. The signature of an oncoming earth tremor or quake falls within the ELF band 1.6 to 3.2 Hz, very long wavelengths indeed.[4] Animal sense must be very highly evolved to be able to detect them.

Experimentation has shown that humans, too, are sensitive. Researchers have found that a 10 Hz band can restore normal metabolic rhythms in people denied access to the natural fields of earth, sun, and moon, and that electrical stimulation of the region of the brain known as the hippocampus, again to about 10 Hz, can lead to alterations in the perception of time and space, leading also to apparitions. Alpha states can be induced in all sorts of ways; the very act of staring into space, of observing nature and natural sounds can lead the brain into the alpha/theta frequency range. Measuring brain wave patterns via electroencephalograms, researchers have discovered that simple meditative states coincide with earth resonance, particularly the alpha frequencies. In fact, there are frequencies to be found at points all over the brain that relate to those that occur naturally within the energy fields of the wider planet. In short, the 0.4 to 30 Hz wave band, particularly the alpha wave band within it (8–13 Hz) remains supremely important for all life on Earth.

A CRYSTAL CALLED HOME

The Earth is essentially crystalline in nature and, in response to incoming electromagnetic waves—namely, the very spectrum of light—it exudes a natural radiation. There is also evidence to suggest that this incoming energy interacts with the Earth's own energy fields, leading to heightened sensitivities at locations wherein the interaction of the two forms of energy is particularly lively. But not only is the Earth crystalline in nature, it also seems that her offspring are. Within the cells of living organisms, a certain amount of self-luminosity has been located, and this offers a clear indication of the crystalline makeup of organic structures.[5]

Crystals have the ability to convert certain frequencies of light and sound—vibrational energy—into electromagnetic and electric energy. They have the ability to absorb, transduce, amplify, and transmit these energies. The primary component of the Earth's crystalline makeup is silicon dioxide (SiO_2), better known to us as quartz. In the world of microchip technology, it is the wonderful ability of quartz crystals to store energy that is used to great effect. Quartz is electrostatic—rub two pieces together in a darkened room and you will soon see a warm glow. The same effect comes from rubbing amber, or even ebony, with some other material. Significantly, the Greeks called amber *elektron*. This electric charge can be transmitted through the human body, water, or even certain metals, a property that makes them natural conductors. Silicon is a natural semiconductor; electricity flows through it with moderate resistance.

Sand, silicon, or quartz is found in nearly all stones, but it is for a specific quality that quartz is put to good use in the world's kitchens: the spark that comes from pressure exerted upon a tiny crystal is used to ignite gas. The effect of vibration upon a crystal is one that alternately compresses and decompresses it as the wave form strikes the crystalline surface in a series of rhythmic pulses. The result of this energy transference process is electrical output: piezoelectricity. The spark that arises in the world's kitchens is due to the piezoelectric effect.

On a much larger scale, the tectonic level, where huge continental landmasses rub together, friction is produced in the form of earthquakes, and, every so often, strange phenomena called earthlights are witnessed. These earthlights have recently become the subject of much research and have even been captured on film.[6] It has been suggested, and with good cause, that the strange lights are in fact balls of plasma or superhot gas, sometimes familiar to us as ball lightning. They are, in effect, gas ionized by electricity coming from quartz-bearing rock, mainly granite. These extraordinary effects would therefore show that quartz is also a transducer: the property of transferring one form of energy, electromagnetism, into another—plasma in this case—with spectacular results.

The human body is itself largely crystalline in its structure and organic makeup. The blood flowing through our veins is really liquid crystal. This is easily demonstrated when the skin is broken. Blood seeps out of the wound and separates into blood plasma that forms a lattice, interlinking all areas of the wound and sealing off any outward flow. In response to outside exposure to the air, iron crystals, which color the blood blue inside the body, oxidize and turn red in the process of crystallizing and forming a hard shell over the wound. Another example of the crystallization of certain body parts occurs around puberty when the pineal gland, located near the forehead, begins a process of calcification. The reason for this remains unknown, but it is as if softer parts of the body tissue suddenly turn to bone. Bone is primarily calcium, as can be seen by looking at the teeth, the only bones directly accessible to the outside world.

The bone structure of the body has long been recognized as crystalline in nature; it is a solid crystal structure complete with piezoelectric energy transference properties. Recent research has shown that energy transference in crystals is to a large extent ruled by the geometric structure of the individual crystal. Indeed, crystals are organized structures that, when not in a liquid state, may take one of several different shapes, according to the atomic makeup of the crystalline substance. Whether cubic, tetragonal, orthorhombic, triclinic, monoclinic, hexagonal, or trigonal, crystals seem to arise spontaneously, always in the same pattern. Furthermore, when a piezoelectric crystal such a quartz is exposed to an electromagnetic field (EMF), the crystal will change shape and generate an EMF of its own.

Other crystals with this property of piezoelectricity and shape-shifting are tourmaline and bone. Teeth and the cartilage around bone and muscle have these properties, too. The forces that produce the piezoelectricity in these areas are particular forms of muscle movement and the rhythmic beat of the cardiovascular system; the environment, too, plays its part. EMFs promote crystal growth and types of bonds between crystals and therefore have the ability to effect a greater rate of growth in bone structure. Crystals, like very finely tuned violins, are constantly murmuring to themselves, an environmental response and one easily detected. It is this very murmuring that heightens the magnetic forces of attraction, thus aiding the healing process by mending bone fractures through growth. When subjected to a greater intensity of magnetism, bone fractures mend at a quicker rate.[7]

A liquid crystal may be defined as having form but in the shape of stored information; in fact, it may act simultaneously as a liquid or as a crystal. Fatty tissues, nerve tissues and muscle, the lymphatic system, white blood cells, and pleural linings are all liquid-crystal systems, all held in place by an overall solid crystal structure—the skeleton. All of this is then held in shape by the skin, which itself exudes, on occasions, copious amounts of liquid crystal through the pores—perspiration.

Further to all this, all of the bodily cell structures are considered as being liquid crystal in form. Cellular membranes operate as typical liquid crystals, as do plasma membranes, mitochondrial membranes, and the nuclear membranes. As Gabriel Cousens, M.D., points out:

> Bodily fluids also have crystal qualities. The water molecule contains in itself the potential forms of all crystals in its primary form of a tetrahedron. Water can bring all different forms of ions into a crystalline state and hold them in solution. In addition, the more structured the water is, the higher concentration of ions it can hold. One of the most important of these ion solutions is the dissolved cell salts.[8]

In terms of the pattern of life, water has its own structure and, as has been discovered and acknowledged, is able to store data.[9] This data is then organized around particular ions and once it is located within particular cells is able to attract other ions or even cell salts into the same cell.

It is held that life adapts to physical circumstance via a process known as natural selection. Over the course of millions of years, environmental changes have produced variations within life as its forms have adapted, allowing the profusion of different life-forms scattered across the entire surface of the planet. However, in the beginning, the very device of adaptation, DNA, if it existed at all, was far too fragile to have been solely, if at all, responsible for all adaptations. Ultraviolet radiation was very profuse upon early Earth, and DNA is all too easily damaged by it. The Earth's protective ozone filter had yet to become formed. Something else must have existed in its stead, something that was adaptable, something that would not simply wear away but would replicate with almost precise accuracy and yet contain the capacity to evolve. Only one state of matter in the early world contained all of the necessary ingredients: the matter of the earth itself, its very substance—dust, otherwise described as sand, silicon, or crystal. Crystals are self-organizing lumps of matter. They are regular geometric forms and are able to replicate in a fairly stable manner. Nevertheless, it is *what* they contain that provides deeper answers to the question of life.

THE LIFE FORCE

As we have observed, crystals are continuously murmuring; that is, they are resonating in response to incoming frequency pulses of electromagnetic radiation. The fact that they respond by giving off their own magnetic energy tells us that they are also surrounded by an electromagnetic field.

Einstein was at particular pains to point out (as in his unified field theory) that it is precisely this kind of energy field that creates the form. In other words, as matter is created out of nothing, a zero-point, it becomes defined within the limits of space and time. Space-time gives what was nonlinear a set of coordinates: direction, mass, breadth, and velocity. No-thing becomes a thing. The Russian scientist and pioneer in the study of organisms and geomagnetism, A. P. Dubrov, underlined Einstein's observation when he wrote, "life became into being and has evolved in the presence of the geomagnetic field."[10] Only recently have we been able to measure the extraordinary subtlety of this EMF.

Regarding organisms, it was the pioneering work of Harold Saxton Burr in the 1930s and '40s that first brought to the attention of the world the concept, and the initial evidence, for what he called "L fields." Burr found that where there is life there is a field. He found fields on a wide range of organisms and made meticulous observations of them for a period of more than twenty years. As professor of anatomy at Yale University, he mounted an extensive campaign of research to investigate this phenomenon. Using extremely sensitive equipment he located minute EMFs on every living thing that came his way, whether it was newt eggs, leaves, or human beings.

It was not until the 1960s and '70s that two Russian scientists, Semyon and Valentina Kirlian, developed the means to take actual photographs of these fields. Kirlian photography is dramatic in its support of Saxton Burr's work. It even goes so far as to suggest that if tissue is removed from a living organism its field persists for quite a time after removal.

Saxton Burr saw these EMFs as organizing fields that were just as valuable a part of the living makeup as the genetic code within it, and it is here that I wish to make a rather obvious point: we can see that living things are, to a significant degree, crystalline in nature. In response to incoming electromagnetism, crystals give off their own fields; en masse, crystals produce an EMF—Saxton Burr's L-field. However, Mother Earth, too, is crystalline, and so it follows that she, too, has a field—the geomagnetic field. A huge EMF, this field is just as organizing as Saxton Burr's original L-field, thus giving rise to life on Earth. Unfortunately, Western science has been slow, at times reluctant, to imbibe the extraordinary implications of this research, whereas, by contrast, a lot of work has been done from within the old Soviet bloc and what is now Russia. Now, however, in the West, the work of a number of researchers is bearing fruit, even in the face of academic intransigence.

The new work has shown that all organisms are surrounded by electrical fields that guide the direction of shape and form of growth. Furthermore, such work has promoted a greater awareness that electrical fields in the environment have an effect upon growth, health, and vitality. Distortions within these fields, leading to a breakdown of order within them, may arise from time to time. Modern technology uses overwhelming doses of electrical power that may create such distortions. This, in turn, might lead to overstrong fields whose influence upon life and subtle organizing fields is massively out of balance, leading to biological damage on a cellular level, wherein tissue grows out of control and beyond the control of other factors. It is an uncomfortable fact that cancer is endemic within the Western industrial world whereas in nonindustrialized areas, beyond the reach of pollution and technology, it is much less in evidence. Furthermore, it has been shown that exposure to strong energy fields, for instance power lines, can produce a wide range of unpleasant effects, from headaches and nausea, to blackouts, genetic damage, epilepsy, and leukemia.

Exposure to the lower energy fields for long periods at a time, such as watching television or working on a computer, can reduce blood sugar levels and lead to nervous disorders—chronic fatigue syndrome (or ME) and also epilepsy. Such fields reach well beyond the appliance. In 1987 a major report revealed that computer systems create a field that extends even through and beyond the walls of the building in which they are housed. Thirty-three years later, there is no change.

In 1996 a report published by the University of Bristol, England, caused widespread consternation among politicians and scientists when it was suggested that large concentrations of radon gas, having seeped from the ground, were to be found in close proximity to electricity pylons, with detrimental effect upon human health. Two influential fields were coming together and bringing imbalance. Radon gas is a natural form of radiation. It is one of the noble gases (gases that rarely react with other elements) and is found in quantity at the Earth's fault lines. Other places of seepage are water wells and underground streams. Moreover, where there is radon there is quartz. Natural radiation is a fundamental force of the universe, radiation being, in one sense, a type of resonance on the ultrascale (higher scale) of frequency. In addition to these radiations, we must always remember that the Earth is similar to a giant bar magnet and that together, electricity and magnetism permeate all matter and thus all life on Earth as an electromagnetic field.

Having drawn attention to the dangers of field imbalance, I should reiterate that electromagnetism is, on the whole, natural, balanced, and life supportive. Electromagnetism has the ability to heal. We have already seen how it can promote bone growth in cases of fracture or breakage. Magnetic fields can also help to treat various forms of cancer and, indeed, rhythmic pulses of electromagnetism have been shown to generally encourage the body's own defenses to heal and grow stronger.

Our scientific age, with its increasingly sensitive technology, has led us to these subtle realms of vibrating energies. One might think that with our rising knowledge of the Earth's fields and frequencies, our own fields and brain wave frequencies, and investigations into heightened sensitivity, humankind stands on the edge of a breakthrough in the study of earth science and perhaps an understanding of consciousness itself. However, recent experiments at sacred sites tell us that this knowledge is not being discovered for the first time.

REVELATION AT NEWGRANGE

In 1994, in association with British researchers, Professor Robert Jahn of the Engineering Anomalies Department, Princeton University, carried out a series of experiments at selected ancient sacred sites in Britain and Ireland. Bob Jahn had been to Ireland the year before and while on a visit to the famous Neolithic cairn at Newgrange had come across a much-noticed but little-investigated effect. There, in tandem with a still and heavy atmosphere, weird and wonderful things happen when you light up a cigarette. The smoke seems to respond in a resonent fashion. Acting on a hunch, Jahn returned a year later, this time with a large array of equipment in tow.[11]

Newgrange dates to about 3500 BCE and, as a cairn, is much like an underground church. The interior is cruciform—indeed, Newgrange is the largest structure of its sort in Europe, a Neolithic cathedral. Carved into both the interior and exterior of the monument—famously in front of the entrance to the passage grave—are a series of concentric spirals.

Many theories have been put forward to explain these vivid but enigmatic symbols. Some researchers believe that they are astronomical markers with perhaps an agricultural purpose in mind, others pass them off as mere ornamentation. Such theories are in many cases accommodating, but nonetheless inadequate.

The structure is oriented so that at dawn on the midwinter solstice, the rising sun shines beyond the doors and into an aperture, small in size but covered in spiral motifs; the result is that a shaft of sunlight penetrates deep into the passage, to the very heart of the chamber.

Inside Newgrange there are large water basins, rather like the baptismal fonts inside churches. In the days when Newgrange was in full

Fig. 1.1. Spiral-etched stone from Newgrange entrance

use, these would have been filled with water to induce the still and heavy atmosphere mentioned earlier. This is a very damp setting in which water vapor plays a key role. As I have said, smoke, in this circumstance, does some very strange things, and it has been noted that deep inside the cairn it can take on a spiral form that seems to respond to the frequency of sound.

Jahn and his team found that the resonant frequencies at Newgrange were well defined, the chamber behaving like an acoustically designed building. When they compared the newly gained data with the chamber's carved decorations, they were intrigued. Such is the accuracy of these carvings regarding the chamber's acoustic value that they even reflect critical niceties such as nodes and antinodes—the actual construction of a standing-wave form or wave pattern (see figure 1.2).

One can see the way that the wave form works by using a piece of string. If the string is attached to two fixed ends and plucked, waves are sent along it in both directions. They are reflected at both ends and, as they return along the string, they pass through each other and combine, producing a standing wave.

Acoustically speaking, a standing-wave form is a wave that is in harmony with the interior, rather like bubbles of vibration within a fixed space. Once a pattern is established, sound reverberates from wall to wall, and as it does so its frequency diminishes by a harmonic with each reverberation. Normally, due to poor acoustic design and an inappropriate volume, sound will dissipate in a building, or the attempt to achieve

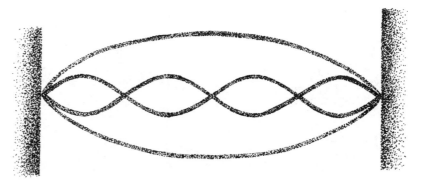

Fig. 1.2. Example of a standing-wave pattern between two walls

a standing wave will result in a cacophony of noise. However, in the right conditions, one can be formed easily.

Besides Newgrange, tests were made at Loughcrew, also in Ireland, and at Carn Euny in Cornwall, and Wayland's Smithy (Berkshire). The Princeton team's conclusions were dramatic and certain features were clearly established as described below.

1. Despite the substantial small- and large-scale irregularities in the boundary walls of these structures, their resonant frequencies were well defined. After very little practice, the experiments could "blind tune" the source frequency to a clearly audible resonance with a reproducibility of +/-1 or 2 Hz with little ambiguity.
2. Although many shapes and sizes of cavity were presented, the resonant frequencies of *all* of them lay in the range between 95 and 120 Hz (my italics).
3. In all cases, principal antinodes of resonant standing-wave patterns were established at the outer walls, as would be expected theoretically. The number, configuration, and relative magnitudes of the other antinodes and nodes leading back to the source depended on the particular chamber configuration.
4. In some cases, rock art on the chamber walls bore some similarity to the observed standing-wave patterns.[12]

Given that Newgrange, Loughcrew, and the other monuments were constructed 5,500 years ago, this is extraordinary stuff, especially the latter point. But the Princeton report goes further.

The Newgrange and Loughcrew sites present extraordinary and well-known examples of diagrammatic rock etchings conventionally regarded as astronomical, seasonal, or environmental representations. In several cases, however, the experimenters were also struck by the similarities of certain sketches with the resonant sound patterns characterizing these chambers. For example, a number of the sketches feature concentric circles, ellipses, or spirals that are not unlike the plan views of the acoustical mappings. In other sketches, sinusoidal or zigzag patterns resemble the alternative nodes and antinodes. . . . Note especially that the two zigzag trains etched on the corbel at the left side of the west subchamber of Newgrange have precisely the same number of "nodes" and "antinodes" as the resonant standing-wave pattern we mapped from the chamber center along the passage. . . . Conceivably, the triple spiral configurations sketched on the magnificent entrance stone and elsewhere could be somewhat more metaphysical representations of the interactive resonances of the three subchambers.[13]

I wondered, how could the Ancients have known about such properties of acoustics and architecture? And why was it important to them to create such places and to mark out acoustic qualities as if to define a building's attributes? These questions sprang to mind as I heard of the Newgrange experiments firsthand from one of the team. I had just finished a public lecture in which I had suggested that the various sacred names of the heroes in myth were linked by a common thread—and that these names were somehow names of great potency, uttered to powerful effect—and for an express purpose. The member of the team introduced himself, saying that he felt there was some link between the theme of my lecture and his work.

Since the late 1970s I have been researching the subject of the hero in this and in the world's oral traditions. What particularly intrigued me were the actual names of the hero and the god with whom he was associated. In many of the world's great mythic and religious traditions, God has a secret holy name, a name that is ineffable; that is to say, it was unspeakable except upon specific days or festivals, and even then it was limited to the high priest, who would utter it under his breath. According to these traditions, the name was a name of power, a dangerous thing. I had often wondered if there was more to this than meets the eye, and, from the time of my meeting with the Princeton team member, the acoustic link led me further along the path of inquiry. I proceeded to look at the names, languages, and places involved. Were they in some way connected, and could it be that the vital clue for which I have been searching was resonance? The time had come for a closer look.

RESONANCE

One of the most interesting observations made by the Princeton team was the comparison of Newgrange to some gigantic musical wind instrument.[14] The most significant factor of all of the sites described was that they were constructed with a view to human interaction with the numinous. Newgrange, in acting like a giant flute, was aiding this process— the sixty-two-foot-long entrance passage accentuating any sound that entered. In fact, the whole edifice was constructed to a pattern, with everything from the giant stones to the huge water basins placed with a particular purpose in mind.[15] The builders were very careful to maintain a strict acoustical balance, as revealed by the experimental results, and as the interior was constructed with a view to acoustic interactions, it is significant that the whole was built to within the human male voice range, within 95 to 110 Hz.

It is now necessary to understand more deeply the concept of resonance and how we as humans interact with it.

All things in the known universe have their resonant frequencies; a resonant frequency is a frequency of vibration that interacts with other frequencies of vibration. If we pluck a guitar string, it will be seen to vibrate up and down; one complete cycle, from up to down and up again, gives us a wavelength. Should the guitar string move up and down 10 times a second its cycle is described as 10 Hz, if 100 times a second then 100 Hz, and so on. There are billions of different wavelengths in the known universe, from billions or even trillions of cycles a second down to sub-Hertzian wavelengths. Thus it is that every chemical element, every crystal and every living thing has a signatory vibration, its individual wavelength.

On a quantum level, the structure of the wave pattern is seen not as a movement by an actual physical wave—like the rushing sea, for instance—but as the passage of energy through a series of individual oscillating (vibrating) particles, moving at a fixed point. What is being described is called kinetic energy (from the Greek *kineein*, "to move"). The same effect is better demonstrated using a line of pool balls. When the cue ball hits one end, its energy is transferred down the line with the result that it is the last ball, at the end of the line, that is ejected (hopefully into the pocket!).

Sinusoidal waves (all of the acoustic waves propagated and studied during the Princeton University survey at Newgrange were of this kind) are waves that oscillate regularly at a fixed frequency. It is in this way that radio waves are transferred via radio antenna in a form known as longitudinal waves.

These are waves that we, as humans, sense as sound. Sound waves occur within the limited frequency band—fixed, of course, by the human inability to hear above or below a certain measure. Transverse waves are of the sort found in water. They are transverse because individual particles move perpendicular to the direction of travel—in other words, left and right—while the direction is ahead. All electromagnetic energy travels in this way, and it is in this form that they can be seen to give off fields that are themselves resonant, as well as being the result of other resonating waves, a sort of harmony meets harmony to create yet more harmony. Waves can be reflected when they meet an obstacle, such as a wall; upon reflection they will follow the original path from which they came. Thus incoming waves will meet outgoing waves, creating a standing-wave pattern; hence the illusion of one's voice being "thrown" (echoing) in certain types of landscape. "Resonating frequencies are primary physical bonds in nature. For every frequency or frequency bond, there exists natural or created resonators. In other words, a field's frequency pattern at a given time is a resonating structure that determines the energy it will absorb or by which it will be affected."[16]

Basically, it can be seen that resonance is implicit in the nature of matter. Indeed, recent research and experimental data have shown that all matter may be nothing other than an interlocking pattern of standing-wave forms.[17]

Described in the *Oxford English Dictionary* as "sympathetic vibration," *resonance* may be seen as the reinforcement or prolongation of sound by reflection, "as from the walls of a hollow space." An example of sympathetic vibration can be seen in the phenomenon of tuning forks singing to each other: if one tuning fork calibrated to 440 cycles a second is sounded anywhere near another fork with the same frequency, the second will begin to vibrate in sympathy with the first; without touching, energy has been transferred from one to the other.

In precisely the same way, a cosmic event—for example, a supernova (the explosion of a giant star)—sets off a string of vibrations, electromagnetic waves, that travel across space and, upon striking the Earth, set up a sympathetic vibration within her: that of her natural frequency. This frequency will be a harmonic of the original incoming frequency. In music these harmonics are often called undertones or overtones, and they are in proportion to the original incoming frequency of vibration. These harmonics may be multiples of the basic frequency or divisions; for instance, a frequency of 32 Hz will respond to an incoming frequency of 16 Hz, and so on. In this sense harmonics can be seen as colors on a paint chart: there is the primary color and, surrounding it, the various tones, lighter and darker, of the basic color scheme.

Perhaps the best demonstration of the sheer power of low-frequency resonance comes from an experiment performed by Professor Gavraud, an engineer from Marseille, France. Fascinated by the phenomena of low-frequency vibration, Gavraud set out to construct a machine that could produce infrasound. Infrasound is sound that occurs below the normal limits of human hearing. Gavraud took as his model the pea-whistle used by all French policemen of the time. He found that the pea in the whistle helps to produce a wide range of low-frequency sounds, infrasounds; it was all a matter of building a larger whistle: one exactly six feet in length and powered by compressed air. What happened next is the stuff of modern horror movies.

A technician was instructed to give the whistle its first trial blast, and upon doing so he collapsed dead on the spot. A postmortem examination followed and revealed that the cause of the unfortunate man's death was resonance: it had very effectively turned all his internal organs into jelly. Death was instantaneous. Undeterred, Gavraud went on to develop machines that can aim low-frequency sounds sufficient in power to demolish entire buildings.

As we have noted, everything in the universe resonates; for every frequency of vibration there is something that vibrates in sympathy with it. So, to call something specific "a resonator" is perhaps misleading;

however, some things connect to our frequency ranges more than others, and so, literally, to us they are more resonant. This sympathetic connection can bring attention to very subtle phenomena. We sometimes call this the supernatural. "Life may respond to the stimuli directly, but more often it reacts by resonating in sympathy with part of its immediate environment. . . . A very weak electrical or magnetic field becomes noticeable because it resonates on the same frequency as the life field of the organism reacting to it. . . . The supernatural becomes a part of natural history."[18]

Organisms, then, can become aware of very subtle fields through a process of sympathetic resonance. We humans are used to sensing life's rhythmic vibrations through smell, taste, touch, sight, and sound. However, not only do each of these senses pick up wide bands of vibrations, but they also can interact across their distinctive boundaries. Certain sensitives, often indigenous peoples, particularly Australian Aborigines, have the ability to "smell" color, "see" sound, and "hear" sight. This indicates that the human body is a good versatile resonator with excellent potential for sensory refinement and adaptation.

The Soviet Academy of Sciences has researched these realms of subtlety. Dermochromatics is a prime example—the ability to feel the different vibrations of light known to us as color. Colors absorb and refract different levels of electromagnetic energy; the sensation of color is the different oscillatory reactions of matter (atoms) to incoming frequencies of energy. The more sympathetic the frequency, the more sympathetic the reaction. Different colors have different levels of energy refraction; for example, black totally absorbs and white reflects. Dermochromatics is the sensory ability to observe the energy refraction of different colors by touch, thus realizing what those colors are. Thus, red burns, yellow is warm, blue is cold, and so on (exactly the way an artist uses the palate). Sensing color in this way is achieved by the use of the fingertips. In a further, striking example of sensorial refinement, certain students, completely blind, could read newspapers and books in prodigious feats of eyeless vision.[19]

Earlier, I mentioned that some sensitives have the ability to see sound. Now, in the West, recent research has shown that young children also have the ability to see sound as color and vice versa.[20] According to the *Oxford Companion to Music,* sound is not a "thing," it is a sensation that occurs as a series of vibrations that cause a tingling of the ear. Sound is created in the brain as a result of the movement of air pressure as it enters the ear. We are then able to tune ourselves to various rhythms within sound. These rhythms act upon the brain as stimuli, causing it to adjust frequency, just as a radio might be attuned to specific wave bands. Along the same lines, certain "complementary" health care systems encourage patients with heart defects to stroke cats. The idea is that the cat's purr has the ability to induce alpha states in the brain, thus relaxing the patient. A further example is the ability of some musical works to pacify overactive states of mind. In an effort to calm a class of unruly children in 1998, a school-teacher in Wales introduced them to tape-recorded music of Mozart.[21] The effect was, in her own words "awesome." Not only did the music calm their minds, but it improved their learning ability. Physiologically, music has the property of working on the brain's limbic system; thus, soothing endorphins are released into the blood. Music triggers memories, all sorts of them, and this also helps to release endorphins. Literally, in response to rhythmic vibration, the brain changes its frequency.

Anecdotal evidence of altered states while making or listening to music is rife. Many elements may come into play in creating these experiences, but my own inquiries indicate that certain types of music, in certain types of buildings, produce more consistent results than others, the effects upon choir members singing Plainsong chant in Gothic cathedrals being particularly notable. And then there is the accompanying aspect of good health: highlighted in the interesting case of the monks of a Benedictine monastery in France who had abandoned chant altogether. Soon the monks could not operate on their customary three to four hours of sleep per night. Now, however long they slept, they remained tired. Meat was introduced to their meals in case the vegetarian diet was at fault (after a tradition of 700 years, this was unlikely). Eventually, Dr. Alfred Tomatis was called in to help solve the problem. Upon examination, he found that many monks were not only tired, but their hearing had also suffered. Taking steps to improve their hearing, he also suggested that chant be reinstated at the monastery. This was done, and within a few months the majority of monks were healthy again on the short sleep, frugal diet, and hard physical work routine. Dr. Tomatis believes that, through the ear, sound can stimulate brain activity through charging the cerebral cortex with electrical potential.[22]

It seems to me that if the quality of the sound was important, then the acoustics of the space the sound was made in would also be very important.* This would make a great deal of sense regarding the Newgrange research. In the world of music, acoustics are all too important to leave out of the equation. This realization helped me to understand just what the ancient worshippers were up to in their sacred buildings: transformation! I became convinced that they used specifically chosen sounds, words of power, at specifically sited and designed locations, places of power, to attain at-one-ment—states of power! My work was evolving. I was now researching an ancient, sacred technology.

COILS AND TRANSFORMERS

The human organism resonates, and in doing so it interacts softly with the phenomenal world about it. Within that same human organism the many parts that make up the whole are also resonating, with the result that together they can form a harmonious whole. A radio, in order to receive a signal, contains a tuned coil that resonates to the various signals that come through it. These vibrations are then processed by other electronic components, resulting in sound being transmitted from the speaker. In their 1990 book *Science of the Gods,* David Ash and Peter Hewitt make the important analogy of the DNA coil to the coil of a radio.[23] Since the discovery of its double helical form by Crick and Watson, not much attention has been paid to its actual shape, only its

---

*I have heard from a number of sources that in an experiment performed in the 1980s a human guinea pig was monitored for changes in brain wave patterns as a response to Gregorian chant. As an added bonus, the experiment took place within the environs of a monastic building. It is said that the results were dramatic, displaying a lowering of brain rhythms to precisely the rhythm of the interior in which the experiment took place. I have to admit that, to date, I have not been able to trace related published material, and therefore I'm planning EEG research on a similar basis with my associates in the future.

component parts. As Ash and Hewitt point out, the DNA molecule, acting as a resonant coil, could receive vibrations from a field of super energy; in this way DNA could bridge the gap between the super-physical world and our own.[24]

Within our own finely tuned bodies, the fact that DNA might act as a resonator makes sense in that, in line with the rest of the body, DNA is largely crystalline in nature and is to be found everywhere in the makeup of our life, whether in the blood, hair, or individual skin cells. As a reso-nator, it would also be responding to other vibrant influences within the body. "Each organ system or subsystem gives off a specific measurable electromagnetic field. The EMFs are measurable, subtle, vibratory fields that can have a great effect upon the behavior of an organism."[25]

It is electromagnetism that stimulates the very process of life in all its various forms. The work of astrophysicist Michael Shallis has shown how sensitive the human being really is to the surrounding fields. Shallis's work has shown that we are naturally inclined to orient ourselves along the north–south axis, in line with the Earth's magnetic field. It has even been suggested that by orienting ourselves conversely to the Earth's field, we might compromise the information system and circuitry that is the human body.[26] Much of this circuitry is of course in the brain. Brain cells, too, act like the coils in radios and, in response to a small incoming magnetic field, produce small amounts of electricity. Of additional inter-est is their own source of magnetism: brain cells contain microscopic particles of magnetite (naturally occurring magnetic iron oxide).

In 1983, Robin Baker, a reader in zoology at Manchester University, located another human magnetic source within the ethmoid bone, which is located close to the pituitary and pineal glands in the brain. Although modern science is not certain of their role in the functioning of the body, these glands are traditionally associated with psychic activ-ity and transcendental experience.

THE VIBRATING BRAIN

Clearly, resonance plays a hugely significant part in the everyday pro-cess of life and what we perceive as material reality. As incoming waves of electromagnetism encounter our planet, a sympathetic vibration is set up—earth resonance, to which life on Earth also resonates. This "life resonance" is an effect of vibrational pressure upon the crystalline makeup of life and of matter in general. As it resonates, this energy radiates outward, causing living things to have subtle interaction with other living things and with the environment at large.* It is significant that the human head, our communication center, is surrounded by a measurable electromagnetic field—one much stronger than the field around the rest of the body. It is here that the subject of humans and life-forms in general having auras comes into its own, and I find it intriguing that, just as a computer's memory is stored in the electrical field of its crystal-based technology, to be read on command, so psychics say that they can read memories, records of interactions, from the aura surrounding an individual's crystal-based body. To do this, the psychic must be in a relaxed but alert state, a state that science associates with alpha rhythms, brain wave activity at 7 to 13 Hz.

We have already seen that there is a relationship between earth resonance and alpha rhythms of the brain. This relationship can be tested by denying an individual the requisite rhythm. ELF waves directly connect us to the Earth and to our environment, and when we have been removed from that environment, we are being excluded from the embrace of the Earth's electromagnetic fields. Within a human it is the temporal lobe section of the brain that is acutely sensitive to these fields. Deep inside the temporal lobe is the hippocampus, which is associated with memory and dreaming. Stimulation of the same, via tiny currents of electricity, can induce various altered perceptions, including distortion of time and space, the appearance of apparitions, and various effects upon the ability of hearing. The key frequencies of Schumann resonance are altered. ELFs are thus intimately associated with key areas of the brain. As mentioned earlier, neural stimulation within the hippocampus connects to earth frequencies at approximately 10 Hz.[27]

---

*Some fascinating interplays have been noted, especially with crystals. For example, it has been demonstrated that ruby and emerald vibrate in sympathy with the liquid crystal structures of the heart muscle system.

Furthermore, as Michael Persinger suggests, the amygdala, which is allied to the hippocampus (in that it is associated with emotional experiences), may use the Earth's ELF channel and interact via the temporal lobe area of the brain. It is here that we enter into a broad range of paranormal experiences and out-of-body sensations; such phenomena lead some to believe that our physical bodies may be used as channels for ever higher degrees of conscious experience.[28]

Though often derided, this view is borne out by a discovery called the "God spot" by geneticists. What is being suggested by scientists at the University of California in San Diego is that a particular region of the brain is associated with religious and mystical experiences. Crucially, this region is "particularly active in people who suffer from Temporal Lobe epilepsy, sufferers of which are often deeply religious and subject to transcendental experiences. The theory is that epileptic seizures strengthen the formation of the neural connections between the region known as the inferior temporal cortex and the region devoted to emotion—the Amygdala."[29]

However, in contrast to epilepsy, there is plenty of evidence that religious belief is linked with improved health, happiness, and a longer life.[30] Certainly, religious or not, within many there is a passionate yearning to experience the unknown and to be more aware of the subtle, ongoing, everyday experiences that most of us, quite simply, have neither the time nor the sensitivity to notice. Often when wishing to experience more sensitivity, to be inspired, or to enhance or change atmospheres, we humans take recourse to music.

Music is not an irregular mass of different wavelengths, it is a collection of tones with measured intervals between them—for sound, when mathematically proportionate, is harmonious. This is pleasing to the ear, it appeals to our innate sense of rhythm, it is interactive, and thus it inspires; the muse is upon us. We know how music can change our mood and uplift us and that harmonious sounds have been used for spiritual purposes from time immemorial. The Newgrange experiments show me that, to the Ancients, these things went beyond mere harmonious sound and into a practical knowledge of acoustics, architecture, and EMFs.

COMPONENTS OF AN
ANCIENT TECHNOLOGY

According to mythology, various religious teachings, and the evidence of archaeology, the temple is the meeting place of the human and the heavenly. The temple is a place wherein the devoted may encounter their god or goddess. But, to interact, to commune, one's mind must open beyond the daily material environment to the wider influences of the cosmos. Sacred structures were built with this very purpose in mind. We have already seen how the atmospheric and acoustic pattern of the interior of Newgrange reflects, with uncanny accuracy, the spiral representations carved into the walls inside and outside the cairn. These Neolithic artists seem to be saying, "We were here, and to great purpose."

The Ancients, by a process of ritual chant attuned to the resonant frequency of the interior chamber, plugged their brains into a sort of harmonic overdrive. This produced an altered state, wherein awareness and, consequently, psychic activity were greatly enhanced. Following on from my musings upon acoustics and a possible spiritual technology, I now began to realize a deeper relationship between the brain's mind-expanding alpha rhythms at 7 to 13 Hz and the complementary ionosphere-centered resonance of 7 to 10 Hz. The physical position of the ionosphere is in itself the natural link between the cosmos and the Earth. Here, in this spacious and refined realm, the gifts of the heavens, those messengers of the gods, the cosmic and, particularly, the dominant solar rays pour in and feed our world with power and light. I contend that, by virtue of this sympathetic resonance, literally through frequency, associated with the sky and the incoming rays, the practitioner of rites was brought to the gates of heaven.

Further evidence came to me regarding sacred sites and psychic activity. At the height of the Cold War, scientists on both sides of the Iron Curtain were involved in a program of experimentation today called remote viewing. Although it is now well known that the CIA was involved in its own series of investigations, the Russians have been a lot more forthcoming with information. Toward the end of the 1980s a Soviet scientist performed a multitude of experiments at various Neolithic sites throughout Europe. Much to his astonishment, he came to realize that telepathy can be enhanced many thousands of times at these sites.[31] The choosing of the sites and the design and the building of those ancient structures must have been based upon a profound knowledge. Diverse traditional lore that tells us that, worldwide, this knowledge was indeed resonance based.

> In both Mexico and Peru there is a legend which tells that the ancient peoples were scientists of "sound," with which skill they had no need of technological impedimenta. They could split massive stone slabs along precise harmonic lines with sound alone, and then "resonate" them into position. Thus the vast and precisely laid temples of Uxmal and Machu Picchu were raised and patterned—according to this legend—in symphonies of sound. Their religion recognised each individual as having a particular note and pitch. With the "sound knowledge" a man could be "purified" and raised by vibrationary mantras or conversely, slain by a single note. An echo of this is found in the "kiai" of the samurai warriors in medieval Japan. The biologist Dr. Lyall Watson suggests that when uttered at the correct pitch the "kiai . . . produces partial paralysis by a reaction that suddenly lowers the arterial blood pressure."[32]

In northern India, Thailand, and Tibet there is a tradition, just as insistent as in South America, wherein it is said that the largest of the monuments were raised by means of sound. Unfortunately, myth, no matter how stubborn, remains unsubstantiated. However, investigation of its constituent parts does reveal that "sound technology" is a distinct possibility.

Certain shapes and forms have the ability to reflect and radiate energy in different ways; for example, crystal structure varies, as does crystalline resonance. Some crystal forms radiate electromagnetic energy at different levels from others. A prism, for example, has the property of splitting light into its separate wavelengths of color—the spectrum. Other crystals can focus light, some being more effective than others:

the lens is a good example. On a microcosmic scale, different elements exude different properties by virtue of their atomic makeup. Water does not become water until its molecular structure is correct. If it has only five molecules or fewer it acts as a film rather than liquid. With six molecules it suddenly acts in three dimensions instead of two. Also, many radioactive elements, such as chromium, have isotopes with variations in the number of neutrons. I pondered upon these building blocks of nature, upon how certain numbers and configurations create particular effects. I wondered why not certain numbers and configurations of objects, buildings, or blocks, as at Stonehenge and Avebury, for example?

What would be at work is a concept first raised by Herbert Weaver in *Divining the Primary Sense*[33] and since expanded upon by others.[34] He suggests that by virtue of their form and location, all buildings, sacred or otherwise, give out a specific type of energy. Weaver categorizes this energy into three different types: Signaling (radiating energy), Suppressing (nonradiating), and Suppressing Locally (within observable boundaries). Indeed, if sacred sites such as temples and cathedrals can resonate internally, then this does not preclude the idea that they resonate externally. This is the basic philosophy behind the Chinese practice of feng shui—the balancing of two opposing forms or principles of energy radiation, *feng shui,* meaning literally "wind and water," the tenet being to bring about harmony between opposites.

Rectangles and squares were seen as radiating energy and therefore as bright signals, whereas six-pointed stars, pentagrams, and cruciforms were seen as widely suppressing energy. Weaver found that configurations of monuments and sacred sites had specific effects, too. In an observation that lends credence to the idea of ley lines, Weaver noted that when configurations of sites were joined by a straight line, energy radiation was confined to *within* those sites.

Perhaps the most startling confirmation of this observation comes from sites such as the Castlerigg stone circle in Cumbria and the Rollright Stones in Oxfordshire, both of which are on straight line configurations—and have remarkably different energy levels inside and outside the circles. Readings of electromagnetic energy suddenly go off the scale when the circles are entered.[35]

PURPOSEFUL POWER

At the outset there is one thing that distinguishes the secret site from any other site: sheer power. These sites are made distinct by the presence of ELFs and electromagnetic fields. Significantly, the presence of specific types of architecture at the sites suggests an awareness of these forces and of their manipulation. Pyramidal structures and domed structures, such as those at Newgrange and a host of other locations, hermetically sealed off a space, not just aurally and visually but also electromagnetically[36]—the power of place was confined to within strict boundaries. Furthermore, a majority of sites have focal points. In the majority of religious beliefs, focus is stressed: both physically and spiritually. The physical point of focus is usually the altar on which the sacrament is given. It is a point of concentration at which prayers are offered. In the same ways that energies are focused, concentration of mind is, too.

As highlighted by Russian research, absence of interference at sacred places makes the progress of thought transference many times more efficient. There, results can be quite dramatic; with the airwaves unsullied, one's link to the gods is more direct. The wearing of special costumes and decoration by indigenous peoples can also be understood in this context. Bone, feathers, horn, and hair can produce their own electrical fields; thus, Indian headdresses really did have a technological purpose: "The structure of tents, wigwams and hogans provided the American Indians with protection and suppression when needed. The horns and antlers worn by the warriors also provided protective force fields which prevented chemical emissions from attracting distant foes. Not only were animals and enemies blocked but it was possible to protect individuals from the influence of the gods."[37]

The stronger energies, electromagnetism in particular, stem from the actual site. Measurements of radon gas emissions have a tendency to be greatly increased at these locations. This is primarily because of a geological condition known as faulting. Fault lines are fractures deep within the Earth's crust and are areas of tectonic or seismic activity. At such places there is a greater amount of mineral encrustation than normal. Minerals, of course, have very magnetic and electrical properties and, as Paul Devereux points out, "different minerals have different densities, so the value of gravity can change at such places."[38] The electrical properties of areas also attract electrical currents such as lightning. In simple geological terms these are very special places. Creatures have the ability to recognize these powerful places—animals are acutely sensitive to fluctuations in electromagnetic fields—and, living more "at one" with nature, ancient humans were, too. Having recognized the spots, it was all a matter of enhancing the effect.

Besides being located near geological faulting, the sacredness of a place was marked out by other factors. Some gods were even believed to inhabit individual stones. The stones were often labeled *omphalos* stones (meaning "navel" stones). They demarcated a certain area as the world center to the Ancients, a god-meets-human interaction point. Although not a specific requirement, these places are almost always linked to sources of water—sacred springs and wells. It was at such places that oracular priests and priestesses communed with the resident deity. Interestingly, stones can "speak"—they give off varying signals. A curious confirmation of this is, again, a particular stone within the Rollright Circle in Oxfordshire. This stone was found to emit a signal of 37 Hz—but only at a certain time of the year.[39] Signals from stones elsewhere have also been picked up on sensitive equipment. These signals can measure a few feet across and be a few feet above ground level. Beyond this, some stones have even been found to exude strange haloes, affecting photographs that have been taken of them. Certainly, these are fascinating phenomena, giving some credence to the traditional lore that the monuments concerned were entrances to the mythic realms, the domain of the gods and their powers.

Quite who these gods were is to be discussed later; however, what is important to realize is that their sacred status drew upon more than just earth energies. Astronomers have discovered that many megalithic burial mounds and stone circles are either astronomically aligned or have astronomical significance apart from being aligned north–south and east–west—Newgrange, Stonehenge, Knowth, and other monuments built circa 3500 BCE are very accurate measures of solar and lunar movements. As we have seen, at the winter solstice (December 21) a beam of sunlight penetrates the inner sanctum of Newgrange, illuminating it to wonderful effect. The combination of all of these details reveals to us a unique sense of cosmology, a cosmology that was geared ultimately to mind-state enhancement. By tapping in to an array of specific energies that were made all the more numinous by the quality of the site, the style of building, and the time of year, devotees resonated and attuned their consciousness to what they considered the divine.

Divinity was believed to reside in these places, but were the energies we have described, the local electromagnetic fields and radon emissions, the actual "gods" concerned, or does the answer lie deeper than that? I have expressed my realization that earth resonance could be used as a gate to the heavens, but I was left to wonder why "the gods"? Surely, a frequency is a frequency, the ionosphere is the ionosphere. Why did the Ancients make everything so personal? It was around this time that I found myself again at Newgrange.

There is something powerful about Newgrange, something that draws you in. I have been there many times during the course of this research. On this visit, as usual, I found myself walking the site and musing upon the mystery of the place when I was stopped in my tracks by a rapid sequence of thoughts—thoughts that excited me. If the Earth was seen as a living, breathing, sentient being, why not the rest of the cosmos? After all, mythology never suggests that Mother Earth is alone in the cosmos. Therefore, could it be that we inhabit a conscious universe? Full of *conscious* frequencies? That this was the tenet of ancient man's thinking, I was certain: consciousness fills all of space!

SUPERSCIENCE, SUPRASENSE

As well as gateways to the heavens, in the world's mythologies sacred sites are often said to be gateways into the underworld. The underworld means the substratum, the subtle realm that provides all and, in doing so, links the heavenly and earthly worlds. One well-known journey to the underworld is that of Hercules, who goes there to capture and bring to Earth Cerberus, the hound of hell.

Hercules enters into the underworld via an oracular cave—an abode guarded by various forbidding priestesses. Mythology gives countless examples of the same kind of journey, but all of them are linked by the location of the gateway. Be it a temple, cairn, cave, or grotto, it is always a location chosen for its sacred properties.

A key to this gateway is hidden within language itself. The word *entrance* is also *en-trance*. It comes from the Latin *transire*, meaning "to pass over." The entry into this trance was, in all likelihood, induced by ritual chanting, as in "en-*chant*-ment," meditation, or the controlled use of drugs. In many regions of the world these factors still play a vital part in religious tradition and belief: they are associated with altered states, usually induced at sacred sites. "The temple might be held to have a relationship with the powers of the underworld. Libations were poured to them, possibly down actual drainpipes. Hence, the temple was the binding post of the underworld. The numen (presiding deity) extended into the depths like an unseen pillar. It was a vertical bond."[40] The siting, the shape, and the materials used in the creation of holy buildings were a key part of the technology of the sacred. Temples were purposefully built to enhance the state of mind by carefully nurturing its sensitivity to the resident electromagnetic universe.

As we know, everything in the universe is resonant, for the universe is made up of frequencies of rhythmic vibration; all things both radiate and absorb, transmit and receive energy in this fashion, and it is recognized that energy and information are one and the same. It is conceivable that, with the correct knowledge, anything and everything can be consciously tuned in to. I have suggested that by linking with the Earth's resonance, humankind was tapping in to a wider universal resonance: after all, Earth's own resonant rhythmic vibrations are largely a response to incoming waves of cosmic energy. Rhythmic vibration, as any radio or software specialist will admit, is the most efficient way to transfer information. As the universe is made up almost entirely of various frequencies of rhythmic vibration—in essence, information transfer systems—the prospect is raised that we may be inhabiting what theorists call a mind-state universe: For what is information without mind? A mind-state universe makes the mind far more than just an offshoot of the brain. In *The Doors of Perception,* Aldous Huxley quotes the philosopher C. D. Broad: "Each person is at each moment capable of remembering all that has ever happened to him and of perceiving everything that is happening everywhere in the universe."[41]

In the view of Huxley, consciousness extends beyond human boundaries to the universe at large. "The brain," in his words, "does not produce mind, it reduces it" and into manageable portions.[42] Keith Hearne has made the significant point that "there is no doubt that a mind-state universe is capable of accepting all the peculiar phenomena studied by parapsychologists, not to mention all the inexplicable phenomena and beliefs of the various religions. We should start to theorise more along the line that we share an existence in a mind universe. It need not be an unprovable assertion."[43]

If the mind-state universe could be proved, attitudes would change, and the ideas of religion could gain renewed currency, underpinned by the knowledge that there is consciousness beyond death, even beyond all things. Research has led Antonio Damasio of the University of Iowa to break the human self into three elements: the "proto-self," "core self," and "autobiographical self."[44] His description of the core self or core consciousness as a sort of pure awareness, uncomplicated by any autobiographical knowledge interests me, for such a description fits well with the Buddhist Nirvana, the Hindu's silent Brahman, the Chinese Tao and the impersonal transcendent of the Western mystics. Despite recent researches and scriptural reference to an all-permeating transcendental consciousness, we are still held in the overriding scientific paradigm that there is little connection between separate phenomena. I am, of course, generalizing, but the "separate" view of science in the observation of life, the universe, and everything present presents very much an *it* and *us* situation.

By contrast, we know from studies of archaeology and the evidence of history that ancient humans believed in a connected universe wherein separateness only went so far as identity. Indeed, we perceive solid objects as separate items. In this way, all things are unique; everything has its individual identity. But ultimately, on a quantum level, it is all frequency, with the frequencies giving rise to the difference. In "reality," we perceive that difference, but in this act we have our own involvement. It is an interactive moment; we relate on a subjective, personal level; we are in the process of inviting, receiving, filtering, and interpreting information. The study of language confirms that earlier civilizations recognized this. The words *objective* and *subjective* describe the sense in which we see things: both are derived from the Latin *jacere*, "to throw"; we are throwing out a line in the hope of retrieving something. To make sense of the world we throw out signals and gather them in again and again.

This gathering in means that we put our preprogrammed self into the equation, not least of which is the way in which our sensory equipment tends to work. Matter, according to physical law, is illusory; it is more than 99.9 percent space—where millions of atoms race around each other at phenomenal speeds to give the appearance of a solid. This limited appearance is what is registered by the eye, which itself makes many thousands of muscular movements per second to record the illusion. It is possible then, in keeping with the ancient tenets, that consciousness may not only fill the space within all things but also the space between all things, as a field of potential from which the illusion arises. Paul Devereux's personal viewpoint is one that I share.

> Consciousness itself does not have a skull-centred source. It is processed rather than produced by the brain, and it is that processing which gives us our unique human dimension of consciousness. Consciousness could be a potential field, nonphysical in itself, that manifests at different levels. . . . In such a view, consciousness is not restricted to the human brain, but can occur in all matter. It may seem bizarre to think of a rock as possessing consciousness, but even a rock is not solid or inert. All matter dissolves into energy fields, which dissolve into more fundamental and mysterious quantum fields.[45]

Here, consciousness is nonphysical in itself, manifesting at different levels; that is, at different frequencies of vibration or resonances. In other words, one shared consciousness is given apparent individuality through a hierarchy of frequencies. Could this be the reason why humans, apparently individual and of solid body, share the same resonance as the Earth? Teilhard de Chardin thought so. He believed that as the planet evolved and bore life, it passed on a little of its consciousness into its biosphere. He was of the opinion that the whole universe vibrated, even though "things" retained definition. Teilhard came to believe that the Earth was imbued with, and surrounded by, an evolving terrestrial field of consciousness and that consciousness was a vital component in the phenomenon of nature and of the universe, that probably the universe was conscious, but also that it was living. In this light, consciousness is an act of the ultimate creative imagination, the mind of God, God being the origin and sum total of consciousness. Is the universe a consciousness, made manifest by an enigmatic, vibrating thing that we call energy?

STARDUST MEMORIES

While the purpose of this chapter is to move my findings forward rather than be a compendium of theory, it might serve some purpose here to illustrate how science's latest understandings of the nature of the universe give us clues as to how "perceiving everything that is happening everywhere in the universe"—to reiterate Huxley—might be facilitated.

Because this universal consciousness would not be an object, it is unlikely that modern science could prove its existence by objective criteria. The current impression given to us by science is that the universe was created in chaos and that ever since, it has, by the process of physical law, been sorting itself out. However, as the physicist David Bohm pointed out in 1980 in *Wholeness and the Implicate Order,* space, rather than being chaotic, does have an underlying order, an order that has, from the very outset of creation guided physical law. Space, according to Bohm, is not a vacuum but its very opposite, a plenum, and it is full of something called zero-point energy, energy that exists in space even at temperatures of absolute zero (-373 C). Zero-point energy will be discussed much later, but its most familiar description to the layman is ex-nihilo, something out of nothing. This is also a fair description of consciousness and the finest levels of creation—something that, however inexplicable, seems to come from nowhere.

Bohm made further postulations. Until fairly recently, the swiftest movement was thought to be the speed of light. However, there is a phenomenon that does suggest that somehow light's great speed can be exceeded. The example is very simple: two photons flying apart at the speed of light will each make exactly the same movement or change in direction if one or the other is shifted. This orderly information transfer is instantaneous, even though the two particles could be many hundreds of billions of miles apart! Einstein called this "spooky action at a distance." Its discovery led Bohm to the idea that we might just inhabit a holographic universe, a hologram that arises from out of an underlying and ordered reality. From out of this underlying order may come all of the information, all of the manifestations that are known to our consciousness. Furthermore, the holographic model gives us an example as to how, as Huxley and Broad suggested, all of the information in the universe may be available to us—if we can only tune in to it.

In this light, I would suggest that our normal mode of consciousness is like a partial hologram: one that can view only certain frequencies unless an effort is made to perceive more. We seem to be like television sets, perpetually tuned in to one mundane program, while an infinite number of other interesting transmissions are available, but not yet tuned in to. These latter suggestions, which may well help to explain the mystery behind the spiritual technology that I shall continue to unveil in these pages, are based upon theoretical sides of science. However, there is a more tangible and easily understood energy that links us to all of life.

While engaged in research for the present work, I recalled a well-known but nonetheless astonishing fact: that we are made of stardust. The likely process can be summed up briefly as follows: Billions of years ago a massive star exploded in an event called a supernova, sending shock waves and stellar matter out over a wide expanse of space. Eventually, some of this matter was reconstituted in a vast cloud of dust that became the primordial solar system. Shock waves from other novae ignited the core of this dust cloud, giving rise to a star, our sun, while other clumps of dust and matter accreted to become the planets and, upon the Earth, life itself. Everything in the universe is recycled, and humans, in this context, are recycled stellar matter. Every particle that constitutes our physical selves was once fuel that could be found deep in the heart of a star. We are laden with stardust memories; we are made of the stuff of the very gods themselves, and perhaps we still share a little of their resonance.

THE MEASURE OF COMMUNION

That humankind has, in the past, used the sacred site as a place of communion with god or the gods is beyond doubt. Mythology and archaeological evidence tell us so. Furthermore, the evidence points to the act of communion being related to an altered state of mind. I find it interesting that the ancient *altar* was sited on what we now find to be the spot whose radiations were most likely to *alter* our brain wave patterns. Did this word association begin as a subtle pun long ago?

I have always felt drawn to sacred sites. At any of these places, sunrise or sunset is very special. There is something contemplative about watching the sun in the last act of the day, setting over the quiet enigma of the holy place.

There is a physical phenomenon that goes some way to explaining the sunset state of mind. In 1938 the Japanese physician Maki Takata developed a test known as the Takaton reaction. It was a test for albumin (the coagulating part of blood) in blood serum. Albumin has a propensity to curdle into small lumps. This process is known as flocculation. In men it is consistent, but not in women. However, in January 1938, researchers began to notice that the flocculation indices of both men and women were rising. Detailed analysis showed that the primary agent of influence was the sun. While being very low at night, the index showed a sudden rise at dawn, and, what is more, the changing index for both men and women precisely coincided with the appearance of sunspots; 1937 had been a vintage year for sunspot activity. Clearly, such solar activity has a hidden and yet significant effect on human physiology.

Some scriptures indicate that the Ancients saw a relationship between the sun and consciousness. Vedic texts tell us that the conscious, all-pervasive reality known as Perusha "identified with the Sun, is the Self of all beings, mobile and immobile."[46] Significantly, a majority of ancient secret buildings are geared toward specific sun-related times of the year: the solstices and equinoxes. The solstices mark out the longest day in summer (June 21) and the longest night in winter (December 21), whereas the equinoxes are the only days when day and night are equal in length. Many sites—Stonehenge, for example— have a particular stone that marks out the rising sun at the solstice. Such a stone casts a long shadow, often deep into the interior of the monument. Stonehenge is, of course, both a solar and a lunar calendar of considerable accuracy.[47] Another example is the temple of Luxor, Egypt, where the central axis of the monument is oriented to the midsummer sun. However, on a daily level, it was the rising and setting sun that was of equal importance at the same site.[48] The specifics of these sites are telling us one thing: that to ancient humans, geometry was all-encompassing, all-important; that it linked heaven and earth. The ultimate link was humankind itself, in the use of these places, wherein a deep and resonant spiritual technology was applied to unlock our stardust memories.

It is interesting that when the sun rises and sets there is much less incoming ionic radiation, and earth resonance stabilizes at 8 Hz. Perhaps the Ancients knew this and felt that this period of equilibrium was advantageous. Since the earliest of times, sunrise and sunset have been particularly designated for prayer, chant, and meditation. With hardly an exception to be found worldwide in any era, the most sacred rituals of the day have always been celebrated at these times. These rituals are to be found in the vast corpus of the world's mythology and the great body of sacred literature scattered about the world's religions. We know that the word *religion* is derived from the Latin *religio,* meaning "to bind back." If we are binding back every time we sing a hymn or give praise to the Lord or Lady, are we, in reality, seeking to resonate with creation? Communing with a deeper unknown?

As previously noted, in Newgrange's interior, smoke, when influenced by a sound frequency appropriate to the building, takes particular shapes and forms. This is an ample demonstration that form is a function of frequency, of rhythmic vibration. The fact that these forms can be seen carved into the interior and exterior of the monument shows that the Neolithic builders were well aware of this. Further demonstrations of the relationship between sound and form have been undertaken in laboratory conditions with sound equipment and smoke chambers. These are revealing. As a simple example, when the sound of the letter *O* is spoken into the microphone, it is precisely an O that is formed within the smoke.[49] As the Neolithic peoples sat in their smoke chambers and used their sound equipment (the voice), it is unlikely that such examples would pass them by.

This brings us to language. Indeed, the most common and important use of vibration for humankind is as the sounds and rhythms that make up formal language. Language comes from the most ancient roots. Mundane as daily language seems, it is from this point that we embark upon an adventure into a different world, one of profundity, a world wherein language becomes the key to unlock the door of ancient humanity's mind.

Some commentators on Sanskrit, the world's most influential and intact ancient language, speak of its origins being rooted in a science of name and form. I believe that language did not emerge haphazardly from a series of bestial grunts, that its syllables and words were chosen fittingly and wisely. We utter sound as resonances that emanate from the region of the throat, mouth, and the nasal passages. These resonances are then modified by bringing pressure to bear on them from the palate, the tongue, the teeth, and the lips. Human sounds come in two distinct forms: vowels and consonants. Vowels are sounds that have no stricture; in other words, they are not strangled into shape by the tongue, palate, or lips. Air merely escapes in a relatively unimpeded way. By contrast, consonants can be "felt" easily by the way in which they have been shaped by the mouth before utterance. Consonants are squeezed and distorted into particular patterns. Vowels, being open and sustained, range in power from 9 to 47 microwatts, while consonants are restricted to a limit of 2 microwatts, seldom reaching above that threshold.[50]

Intriguingly, the various god names, held in such awe and esteem by the Ancients, had very few consonants. The power of vowels is easily demonstrated by their use in music, where a good soprano can easily shatter a glass or rattle a chandelier. This is an extreme example, but it does demonstrate why we should not undervalue the use of sound in ritual.

Back at Newgrange, walking in the mists of an early Irish morning, I could not stop myself from thinking of the purpose of the place and of the gods and heroes associated with it. Having discovered that these profoundly complex sites were the emanation of profoundly wise minds, I realized with some amusement that our pervading view of our ancestors, as it was taught to me at school, was nothing other than a view of ourselves. These sites are now a mirror held in the face of our civilization by an ancient people about whom we know so little. As I walked around the perimeter of the site, I realized, with mounting excitement, that the most revealing clue lay within its measure. Language itself is based upon measure, upon frequency of rhythm and sound. Ancient chants were, like poetry, chanted in meters, rhythmical patterns. It is said that the meters were designed to change atmospheres, and it is significant that meteorology is the science of the atmosphere. However, the important word here is *geometry*.

Geometry is two words: *geo,* meaning "earth," and *metry,* meaning "measure." It cannot be a coincidence that the syllable *ge* makes up the word for earth in quite a few languages. *Geo* and *ge* are Greek words meaning "earth," and again in Greek we have the famous *gaia,* meaning literally "Divine Earth" or "Earth Goddess." This linguistic connection, I believe, comes directly from *Geb,* the name of the ancient Egyptian god who represented the Earth. Together these roots give us words such as *genesis, generate, gene, geology, genus,* and *general.* The Latin word *genus* is from the Greek and means "of the earth."

As far away as Papua New Guinea, the creator god is called Geb. In Scottish-Irish Gaelic, *gael* is "love," and so on. Until recently, linguists believed these roots to be unconnected, a belief that is being revised radically. In Gaelic, *gu* means "go to" in the sense of return, thus implying that one comes from the earth and returns to the earth.

As demonstrated by solar and lunar alignments, the sacred measure encapsulated within ancient monuments displays a deep and profound knowledge of geometry. In a pure sense, what I was interested in, beyond measure, was its expression. As we have just seen, language *is* its expression; geo*metry* suddenly becomes an organic principle: meter, rhythm, measure, connect the sun to the site—and the sacred atmosphere of the site is linked to language itself.

We are beginning to understand that words can have an actual power, a power that contains the creative principle and can describe it so wonderfully well. Who would have thought that not only does the word *geometry* mean "measure of the earth" but that this measure, as the design and purpose of the monuments tell us, was quite specific on a vibratory level? Suddenly, just as the rising sun dispersed the mists of the morning, the vibrating, resonating gods were becoming clear and enlivened within my mind's eye.

Tradition tells us that the god who inhabited Newgrange was known as the Dagda. He was called "the Good God," and he was involved with the use of sound, for he played upon a living harp, and as he played the seasons came and went in due order. His wife was Boann, and his son was Angus, god of love—very probably a dying and rising god; that is, of the sunrise and sunset. The linguistic route of the Dagda's name is shared with the Greek, and it is from the Greek tongue that we come across an interesting clue. *Da* in Greek means, literally, "O earth," and was used for ritual purposes. It is an invocation. The central portion of the name also means *earth* and, if my assumption is correct, the Dagda's name invokes Mother Earth thrice: "Da-g(a)-da," O Earth, Earth, O Earth. As if in confirmation, in the ancient myth of this site, she was known as the triple goddess.

The Dagda's name was a powerful one. In the context of being uttered deep inside the Newgrange cairn, its effect, as a continuous chant or invocation, must have been profound indeed. At sunrise and sunset, when landscape mists arose, the effect would be intensified. Imagine then, the further potency at the winter solstice, when the air was crisp and the atmospheric moisture had penetrated deep inside to the hidden chambers. There would be gathered together devotees of the Dagda. Through concordant ritual of rhythmic chant, the atmosphere would be potentized with the name of the god, when lo, a penetrating, clear shaft of sunlight would break through the mists and into the dim interior, highlighting curved motifs upon the rock and within the air itself, those swirling mists, now not only resonating to rhythmic chant but also to the midwinter sun. The elation felt by the congregation, as their brain waves tumbled into the lower frequencies and their awareness was opened, merging, at one with the subtler, greater environment, must have been ecstatic. Through the portals of consciousness, by way of the Dagda, they and the Earth were as one.

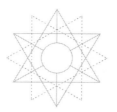

2

The Acoustic Divine

Religion is the frozen thought of men out of which they build temples.

JIDDU KRISHNAMURTI

If religion is only a garment of Christianity—and even this garment has looked very different at times—then what is religionless Christianity?

DIETRICH BONHOEFFER

A HAUNTING MELODY

In the context of religious history and phenomena, the term *ecstasy* has been given supernatural status, supernatural in the sense of being, so far, inexplicable by any known scientific hypothesis. In seeking to explain the inner workings of nature, humankind has demystified it and, some would argue, despiritualized it. We are no longer comfortable with the unknown; thus, when we are faced with super-nature our first reaction is alarm and fear. Fortunately, following on the heels of these two interlopers comes *curiosity*. Sad to say, many of us never get as far as this third impulse, preferring to ignore the mystery or to explain it away via the misuse of logical, rational argument. Logic, of course, has its limits when we starve it of fresh information, when we *ration* out data.

New knowledge has to be ridiculed before it is accepted. Alarm and fear—in the light of these typically human responses, language becomes ever more revealing, displaying to a great degree the balanced way in which our forebears thought about exactly the same phenomena that today we call supernatural, paranormal, occult, and so on.

"Ecstasy" is from the Greek *ekstasis*. The older form of the word is *exstasie* and reveals much. *Ex* is literally "out of," in this case, "out of stasis." *Stasis* is a commonly used word meaning "cessation, stoppage, stationariness." Thus, ecstasy infers un-stoppage, an unblocking. Down come the barricades and then comes change, the movement into a peak experience. This movement "out of" may not always be comfortable—after all, we are leaving our comfort zones—but the result is a joyous or blissful affair; burdens are, at least temporarily, shed. Certainly, one is taken out of normal everyday perceptions and, briefly, made aware beyond one's normal capacity. But it is clear that because of this shift, permanent changes of expanded awareness can occur.

Why is this important? Because it appears on a regular basis within religious circles and, beyond that, in an everyday context. When you win a competition or when an ordinary, everyday Joe wins the lottery, feelings can be ecstatic; this is a state out of the ordinary, wherein awareness is heightened, awareness of the internal, such as emotions and physical sensations, and awareness of the external—sometimes one's surroundings appear clearer, colors more intense. Ecstasy is often associated with music, with anthem and hymnal. But more than this, the state of ecstasy is associated with the hero of world mythology,[1] the hero who performs superhuman and supernatural deeds. Perhaps by achieving a state of ecstasy, we are communing with the heroic state of mind. Certainly, when in an ecstatic state, an illusory feeling of invincibility is common in such moments: dangerous things are often done without a hint of nervousness.

In ages past and in times of crisis, it was customary for civic leaders, kings, and potentates to consult the shades of the dead. More often than not, the shades were the souls of heroes. In Greek and Egyptian mythology there are detailed descriptions about these encounters: the meeting of Orpheus and Eurydice is among the most famous. In Hebrew myth,

King Saul consults the shade of the prophet Samuel, aided and abetted by the Witch of Endor.

These stories tell us that spirits, ghosts, and apparitions were a common feature in ages past; more so than today, where an adherence to dogma, be it scientific or religious, has replaced acceptance that there is something further, something hidden. The problem is that the hidden, or *occult,** is influential. It occurs regularly in everyday life, almost in spite of the presence of the scientist. Science, however, is at last beginning to come to terms with the idea that there may be more to such phenomena than meets the eye.

On December 23, 1998, at Belgrave Hall in Leicestershire, England, two specters decided to go for a stroll in the grounds, little realizing as they were doing so that they were being observed. In an episode not too far removed from Oscar Wilde's *The Canterville Ghost,* it was the ghosts who were being haunted, by security surveillance cameras. One of the figures captured on camera was apparently a woman in a Victorian dress with a bustle. She was identified as Charlotte Ellis, who, in her lifetime, had a deep and abiding love of the place. According to Stuart Warburton, the Hall's managing curator, the images appeared from nowhere. They make no entrance or exit but just appear and disappear. The video equipment was duly inspected and found to have no fault. Belgrave Hall already had a firm reputation for being haunted, but the appearance of the ghosts was the cherry on the cake. The film remains fascinating and unique;† hauntings are only just beginning to be understood.[2] Scientists are discovering that ghostly apparitions most probably have a logical explanation and are not, as has been too often suggested, the result of an overactive imagination.

In February 1998, Vic Tandy and Tony Lawrence of Coventry University decided to look into the case of a haunting at a local business premises: a medical manufacturing company based in the English

---

Occult literally means "hidden."

†Some doubts have subsequently been expressed about the quality of the security film or, more precisely, about the nature of what was caught on film. A logical explanation of the imagery might appear to be nothing other than oak leaves! However, the jury is still out, although in my opinion the explanation does indeed look credible.

Midlands. Tandy, an expert in computer-assisted learning at Coventry University, had been told that the building was haunted. His initial response was to dismiss it as a joke; that is, until he himself became a witness.

> As I sat at the same desk writing, I began to feel increasingly uncomfortable. I was sweating, but cold, and the feeling of depression was noticeable—but there was also something else. It was as though something was in the room with me. Then I became aware that I was being watched and a figure slowly emerged to my left. It was indistinct and on the periphery of my vision, but it moved just as I would expect a person to. It was grey and made no sound. The hair was standing up on the back of my neck—I was terrified.

Caught between the compelling nature of the apparition and sheer terror, Vic summoned the courage to look at the specter face on, only to see it fade and vanish.

> I decided I must be cracking up and went home.

Returning the morning after, he soon had an answer for the events of the night before. As a fencing enthusiast, Vic decided to make use of the vise that was set up in the room. He was adjusting one of his foils and had left the blade clamped in the vise while he went on a search for some oil. Upon his return he noticed that the free end of the blade was vibrating up and down in somewhat frantic fashion. It was possible, thought Tandy, that the blade was responding to very low-frequency sound waves filling the laboratory, sound waves so low that they could not be heard.

It was discovered that "trapped" within the laboratory was a standing-wave pattern, a pattern that just happened to reach a peak of intensity near the desk where Vic had been sitting on that hair-raising night. A standing-wave pattern, in this case an extremely low-frequency sound (ELF), caused by oscillations between a newly installed extraction fan and its mounting, was making the air vibrate at 19 cycles per second, or 19 Hz. When the fan's mounting was altered, the standing wave stopped—and so did the hauntings.[3]

Working in tandem with Tony Lawrence of the university's School of Health, Vic found that this rate of vibration, called infrasound, plays host to a whole range of physiological effects, including breathlessness, shivering, and levels of fear. More significant research by NASA had already established that if the human eyeball is led to vibrate in sympathy with infrasound at a resonance of about 17 to 19 Hz, smearing of vision would result. When a human's 19 Hz meets the world's 19 Hz, it vibrates in sympathy, rather like the effect of a vibrating tuning fork upon other tuning forks. Thus it is that humanity vibrates in sympathy with its environment and otherworldly effects begin.

Lawrence and Tandy have since come across at least two other hauntings wherein ELFs can be blamed. However, certain fundamental questions must be raised at this point, for we have the *vehicle* of the operation but not its provenance, or its origin. Is what we are looking at rather like a prerecorded message, one that in the right conditions, or frequencies, can be played back? Regarding these conditions, and given what we have already discussed in the context of Newgrange and the extraordinary findings made there, what, we might ask, are the ramifications for architecture in general and our ideas of the cosmos at large?

Tandy and Lawrence pointed out that if a building acted like a wind instrument, creating standing waves (rather like blowing over the neck of an empty bottle), apparitions could quite possibly occur, and in certain conditions—perhaps the long corridor of a tall, dark, nineteenth-century house—they are probable. We must think here of Newgrange and the host of other sites. Tandy and Lawrence's work casts new light on otherworldly experiences, but a very important factor is left unaddressed. As Professor David Fontana, former chairman of the Society for Psychic Research, pointed out: "It cannot explain those cases where there is some interaction between the person and the operation—as there is with poltergeists for example. The problem is that whenever you get a potential explanation like this, you find that there is a whole lot of things it cannot account for."[4]

VISIONS OF THE QUEEN OF HEAVEN

During times of crisis and transition, either individual or national, we are often compelled to look beyond everyday experience and are given to reflect inwardly upon our disquieted feelings. It is almost as if the outer tumultuous events of life are forcing us to look internally. This inward-looking contemplation often brings about yearnings for "something other," and in its deeper manifestations, a momentary sense of oneness, an all too brief, innocent sense of peace. Under such conditions visions can appear. Remarkably, the vast majority of these visions can take on a feminine hue and appear in all manner of guises and across a broad spectrum of cultural beliefs. Of late, these visions of transcendence have occurred with regularity. One simple observation to be made is that visions appear to be tailor made to the culture or belief in which they are set. Increasingly often, individuals of varying faiths are experiencing the presence of the Divine Feminine. To many in the West she is the Blessed Virgin Mary, the Mother of God; farther East, she is Fatima, daughter of Mohammed.*

To the Roman Lucius Apuleius, author of *The Golden Ass,* she was Isis, whereas the Greeks adored her in their many visions of Artemis. Wherever she is and whatever she is called, she is always the Queen of Heaven. Although often described as a White Lady, she sometimes appears as the Black Madonna. To some, she is seen to shed blood-red tears amid her heavenly white garb, while to others she sheds ordinary tears down her dark robe. This Black Madonna is often taken for being the Magdalene, the woman who regained her purity through Christ.

By their cultural nature, visions are linked to history; often they were associated with a sense of warning and, aside from feelings of awe and wonder, a deep sense of foreboding. The impression is given that something profound is being said, something from the depths of our archaic past.[5]

Such experiences are often deeply powerful, even completely overwhelming. Others often dismiss them as mere hysteria, though in the

---

*In pre-Islamic Arabia she was a moon goddess.

context of the visionary phenomena, hysteria has a tendency to become a mass phenomenon that occurs only after the initial "real" event. Whatever others may say, to the recipient the experience remains deeply unsettling. It is this sense of the individuality of the experience that is most likely the key to understanding the phenomena as a whole, hence the fact that they seem tailor-made to the respective cultures concerned.

Sri Aurobindo, an Indian sage who was well used to such phenomena, comments upon the mechanics involved.

The wall between consciousness and force, impersonality and personality, becomes much thinner when one goes beyond the veil of matter. If one looks at a working from the side of impersonal force, one sees a force or energy at work acting for a purpose or with a result. If one looks from the side of being one sees a being possessing, guiding and using or else representative of and used by a conscious force as its instrument of specialised action and expression. . . . In modern science it has been found that if you look at the movement of energy, it appears on one side to be a wave and act as a wave, on the other as a mass of particles and to act as a mass of particles each acting in its own way. It is somewhat the same principle here.[6]

Aurobindo's biographer, Satprem, elucidates further:

A Christian saint who has the vision of the Virgin, say, and an Indian who has the vision of the Durga may be seeing the same thing, they may have contacted the same plane of consciousness and the same forces; but quite obviously Durga would mean nothing to the Christian, and moreover, were this force to manifest in its pure state, that is, as a luminous impersonal vibration, it would not be accessible to the consciousness of either the worshipper of the Virgin or the devotee of Durga, or at any rate would not speak to their hearts.[7]

Although the apparition is usually given whatever name or form accords to the religious belief of the observer, on the wider levels of gender, purpose, and demeanor, she remains remarkably uniform in her appearances worldwide. This fact points to a common origin. This is reflected in the history of belief. Folk religions have, through the ages, calcified as history marches on. Where a new belief has not been brutally imposed, it has changed its shape to suit the cultural ethos in which it is set. Even so, imposition of a rigid faith has done little to curb the re-emergence of the older ways in one way or another, revealing the roots of the folk tradition. This was a point recognized by C. G. Jung, who wrote of it at some length. His term for it was the "collective uncon-scious." For him, the process of metamorphosis was alchemical, and the visions came to be known as archetypes, as primal manifestations of the collective unconscious. Jung was no fool and saw the manifestation of phenomena—visions, UFOs, and the like—as a continuation of a very old theme.

One of the fundaments of the past is that ancient humans saw the Earth as feminine and very much alive. This is a view that in the late twentieth century was given newfound credence by a broad diversity of writers, Teilhard de Chardin, Carl Jung, James Lovelock, and Rupert Sheldrake among them.

In 1981, in the Balkan territory of Bosnia-Herzegovina, the Virgin Mary began appearing to six young Croatians, the first of whom was Ivanka Ivanković, fifteen years old at the time. As at other times and in other places, the Blessed Virgin appeared as a young woman hovering above the ground—and she appeared to beckon Ivanka to her. Ivanka was unafraid, although days later another of the group was physically thrown into a thorn bush by an unseen force when he either refused or forgot to kneel. The description of the Virgin, as given by the young people, is particularly interesting. She was a young woman of nineteen to twenty years, extraordinarily beautiful and wearing a crown of stars, a long silver dress, and a white veil. This is a picture spoken of many years before by Dante Gabrielle Rossetti in a poem titled "The Blessed Damosel."

The blessed damosel leaned out
from the gold bar of Heaven;
Her eyes were deeper than the depth

of waters stilled at even;
She had three lilies in her hand,
And the stars in her hair were seven.[8]

A team led by Professor Henry Joyeux of Montpelier University, France, undertook scientific and medical studies on the children. They found that while there were no clinical signs of hallucination, hysteria, neuroses, or psychosis, EEG recordings clearly indicated the presence of alpha rhythms during the visions.[9]

The visions would last from ten to twenty minutes, after which the children would be blessed with the words "Go in peace with God." From here on, as with instances at Fatima, Portugal, in 1917; at Lourdes, France, in 1858; and at Zeitoun, Egypt, from 1968 to 1971, the visions took on a whole new aspect, seeming to cause considerable environmental change. For example, the sun began to dance and spin, and the word *Mir,* meaning "peace" in Croatian, appeared in the sky above a large cross that was situated on a nearby mountain.

Quite often the visions come with grave warnings. At Fatima, three momentous secrets were imparted to three young children. At Akita, Japan, in 1973 a nun was warned of the terrible punishment that awaits humanity, a punishment more terrible than the Flood. Only believers would be spared. This became one of the few modern apparitions approved by the Roman Catholic Church.

Even to a nonbeliever, the words of Conchita González from a vision of Mary at Garabandal, Spain, in the 1960s, are very forbidding. They speak of dreadful catastrophe. The third secret of Fatima has recently been revealed as a prophecy of doom, if we do not mend our ways.

Others of a more agnostic mien have shared much the same experience, if only a little less dramatic. White ladies, and there are several, have appeared to various people down the ages. To Keats she was manifest as *La Belle Dame sans Merci,* to Robert Graves she was the utterly ruthless White Goddess, while to Lucius Apuleius of second-century Rome she was none other than the great Egyptian goddess, Isis. All of this is made more intriguing when one considers that the Femina Alba, the White Lady, is actually a goddess of love. In this, the message of dire warning is a paradox that can be explained easily. The only other juncture in nature where love speaks harshly to the loved one is the parent-child relationship, of mother and child: the most important, most natural and sacred relationship in physical life. And so, we see that these visions share much in common. One wonders, do the sites, too, share commonality?

In this light, it is interesting that the specific locations of these visions should be at places of particular numinosity: temples, churches, and sacred sites in general. As we see, by virtue of being constructed in certain ways upon particular places, these places capture the subtle vibrating fields of the Earth and enhance them. These buildings are resonant places where acoustic patterns, including standing waves, can be set up. We know that standing waves of about 19 Hz resonate with the human eye, and I believe that, under those conditions, if there is a subtle form present, it may be seen. To clarify: this type of experience is not "all in the mind"; there is a presence that interacts with the mind, often being culturally clothed by it. Come the vison, cometh the church, but we should note that where there are churches there is always water.

There are many myths, stories, and actual experiences that relate to the power of water at sacred sites. Sacred waters are traditionally symbolic of the great Earth Mother, of her fertility, of the womb, and of life itself. A great majority of Marian visions have taken place at locations associated with water. Pilgrims still travel en masse to Lourdes to partake of its healing waters. The same is true of Knock in Ireland. There are many holy wells in Devon, Cornwall, and other British counties. I have personal experience of the languidness of the sacred wells scattered over Dartmoor, England. Holy wells are known for inducing sleepy states, a change of brain pattern. These waters are relatively high in natural radiation, as they seep forth from granite. Holy waters have been shown to be different from ordinary water—tap water for example—by the fact that they have different frequencies involved in their makeup. Using a technique called infrared spectroscopy, a beam of infrared light is shone through the water. In absorbing some of this light, the water can then be measured via its spectrum. Holy water absorbs light at different frequencies from ordinary water, and its properties were quite obviously known by the Ancients.

In early 1999, experts from Anglian Water were called to the site of Saint Withburga's well at East Dereham, Norfolk. The well is known over a wide area for its healing powers, notable in the treatment of skin ailments. The experts tested six liters of the water, and in their report they confirm that the water contains "naturally occurring minerals, such as potassium and calcium, which are known to be beneficial to health. The water is free of bacteria."[10] The presence of these minerals and the lack of foreign bodies is a testament to the extraordinary purity of sacred well water. Water from such sites has greater properties of spin than is usual. It is literally composed of tiny vortices that, in the process of spinning, give off an electric charge and hence an electromagnetic field; minute, but enough to transform it into something very different from ordinary water.[11] Incidentally, quite often, and usually for the use of kings and queens only, rock crystal bowls and chalices were carved to help maintain the purity of the well water.

The Ancients are also known to have deliberately increased the potency of water with the use of crystals. In 1974, Patrick Flanagan, an expert on the structure and properties of crystals, undertook research and found that "crystals of all kinds, such as quartz and precious gemstones, have a marked effect on water surface tension, a characteristic known to ancient Tibetan physicians who applied it to make crystal-affected water potions for their patients."[12] The water, when poured on food crops such as wheat, alfalfa, and mung beans, produced more vigorous growth and a substantial difference in taste. Flanigan believes that crystals are resonators of cosmic energy impulses and that the origins of these incoming energy fluctuations are supernovae and quasar activity.[13] Indeed, it is known that the rate of chemical reactions in water can be affected by sunspot activity (as discovered by G. Piccardi in the 1930s), solar eruptions, and incoming showers of cosmic rays. Water is incredibly sensitive to environmental stimuli and, in making up 80 percent of the human body, must therefore help make us sensitive, too.

Water also retains electromagnetic information and by the very act of drinking from a sacred spring, one is literally drinking in the information contained within it.[14] In 1988, Jacques Benveniste and his colleagues put themselves firmly in a position of heresy, for it was with some reservations that the science journal *Nature* published the results of a remarkable experiment. Benveniste found that "they could affect white blood cells in the blood with antibody solutions so diluted that not a single molecule of antibody could possibly be present in the water. They suggested that possibly water could act as a 'template' for the basic molecule of the antibody, and were convinced that 'transmission of the information depended on vigorous agitation, possibly inducing a sub-molecular organization of water.'"[15] In effect, homoeopathy.

Homeopathy uses incredibly diluted solutions of various substances, which are used to combat the presence of far larger amounts of the *same* solution within the body—fire to fight fire, in this case a match against an inferno. For various reasons, these principles are much derided by mainstream science, even though there is plenty of evidence to support them.* "The editorial of that issue of *Nature* made it clear why the results reached by Benveniste's team were unlikely to be accepted: they were "startling, not merely because they point to a novel phenomenon, but because they strike at the roots of two centuries of observation and rationalization of physical phenomena."[16]

Many others have since arrived at the same conclusions as Benveniste and his team, by the same process of adding a dilution and then halving and adding water, a process repeated, in some cases, countless hundreds of times: but with the same effect—the chemical message remaining in the water.

Further to these revelations, investigations are revealing that the energy field of water may, in some way, respond to thought. This is not as strange as it might seem. In our own physiology, water appears as a response to thoughts and feelings in times of sadness or joy: it flows from the eyes. The work of Masaru Emoto (a wonderfully appropriate surname) is a case in point: he took this observation to new heights in his work, which ably demonstrates the link between our expressed emotions and the patterns that appear in water when frozen.[17] What of other subtle interactions? Can weather patterns respond to thought?

---

*The Royal College of Veterinary Surgeons, for example, undertook a veterinary project in which a herd of cows was divided and one-half was subjected to a dilution, the other a placebo. Needless to say, the homeopathic cure was efficacious.

Any time-served shaman would say "Yes, most certainly." In fact, on November 28, 1999, to help end a drought, a group of rabbis boarded a plane and circled Israel, spiraling around it seven times while praying for rain. It rained!

Nowadays we recognize that this Earth is one organic, though globalized, whole. As it becomes apparent that we can interact with its component parts on the level of thought, we should seriously consider the possibility that the Earth is a conscious being.

The ancient traditions insist that the Earth—and indeed the heavens—will respond to prayer and affirmation. We must realize that these old sacred monuments were places designed and built specifically for this type of activity: interaction with the conscious being called Earth, and beyond that—not necessarily in distance, but in vibration— with the subtler universal consciousness.

The heavens descending in response might go toward explaining the mysterious thunderclap that was heard before the extraordinary vision of the Blessed Virgin at the Coptic Church of Zeitoun in 1968, as well as thunderclaps elsewhere. Gathered around the dome of the church was a strange mist of the sort that often seems to accumulate at sites of this nature. It is as if here we have an example of the descent of the higher—the godly—spiraling down into the frequencies of the lower, by somehow utilizing the site itself.

We are all aware, to some extent, that the dominating movement in the universe is the spiral. From galaxies to atmospheric conditions, where satellite photographs now show us spiral cloud formations in all their glory, the spiral is the natural way. In archaic myth, the Divine Feminine, the goddess, is often represented as a spiral, and this is a symbol to be found at a plethora of ancient sites—Newgrange, for instance. At particular sacred sites, it has been shown that "the vectors of electromagnetic force would have favoured spiral shapes twisting in one direction or the other."[18]

The spiral may also be seen in water as vortices spinning in clockwise or counterclockwise directions—yet again we have the property of spin, which gives off a magnetic field, adjusting to varying degrees the electromagnetic information retained within. Water is a component of

Fig. 2.1. Spiral vortex in water

blood; blood spins, and by consuming water from a pure source, we can adjust the varying frequencies and spin ratios of the blood, thus altering the information it contains,[19] a property announced by the medical world in 1998. It would seem, therefore, that the advanced chemical technology that is our body does, to a certain degree, create the proper circumstances for visions. Therefore, in tandem with the Earth and her electromagnetism, our bodies, particularly our brains and sensory faculties, have to be in a certain condition to receive information.[20]

Water, in having memory, is collecting and storing data wherever it is present upon the Earth, and when its frequency patterns meet other traces of water—in the human body, for example—under particular circumstances, sympathetic vibration can take place. We have seen how the builders of Newgrange seemed to encourage a damp atmosphere. They had a head start; the British Isles are world renowned for their damp climate; in fact, many seers and clairvoyants have said that these islands have so many visions and visionaries *because* of this. As already mentioned, spirals and vortices were also associated with water, which in every single ancient belief, without exception, was referred to as being female. In Hebrew, the sea is female, in European languages it is female; in French it is *la mer*, for example. Ships that float upon it are "she" and so on. Water is associated in myth with space, being referred to as "the primordial waters." This may be a metaphor perhaps relating the amniotic fluid of the womb to space as the womb of life;[21] but actually, water and the essential ingredients in its makeup are found in space. What I am indicating here is that we have all of this data stored in the feminine waters, molecule by molecule, and yet beyond mere data we have the haunting but interactive presence of the Virgin. The Virgin, though, is not the only interactive vision.

THE ABDUCTING ANGELS

All of humanity is interconnected by virtue of the common habitation of a single planet, Earth. But in evolving within her embrace we are even more connected than might seem to be the case. We are, by nature, mental, emotional, and physical bundles of water and carbon, vivified by healthy doses of magnetism and electricity, which are themselves forms of rhythmic vibration. We are like tuning forks that react to incoming vibrations; the music of the universe. At the same time we are trying to make sense of it all—and, by inquiry, we are aspiring. That which aspires is spiritual, it is of the spirit.

A series of experiments in the 1970s demonstrated subtle levels of connectivity in a surprising way. Russel Targ and Hal Puthoff of the Stanford Research Institute took pairs of people who were already fairly familiar with each other's habits and character traits—they knew each other well and they had an emotional affinity. They were separated and put in rooms at opposite ends of the building, thus ensuring complete isolation from one another.

At randomly selected intervals a rapidly flashing light was shone into the eyes of one of the pair. The effect was to reduce alpha activity, the brain wave pattern associated with relaxation. Meanwhile, the partner was supposed to say when the other was experiencing the flashing sensation. In the event, the responses failed to pinpoint when the effect was taking place, at higher levels than would normally be expected. This might confirm the general opinion that a relaxed condition is required for heightened ESP. However, although one was not able to say when the other was experiencing the flashing, both brains showed a reduced level of activity at precisely the same time. Remember, this was when only one was physically subjected to the flashing. A subtle link, not predicted, was showing itself. Could it be that we are linked to each other on many levels? If this is so, surely it is likely to be via the Earth and her subtle atmospheres. When it comes to weird phenomena, to coincidences, hauntings, and interactive visions, is it not the biosphere that we call Earth that is the key.

Research indicates that when the body's essential biorhythms are at a maximum, so is psychic activity. On a natural basis this peaking occurs at all three levels—physical, emotional, and intellectual—roughly every forty-five to forty-six years, but perhaps this state could be induced, almost religiously?[22] Religion, as we have come to know it, has been in decline in the West for years, with the telltale signs of a hardening of belief at the outer extremes—gross materialism and the New Age fantasy. Meanwhile, faith has to a large extent been replaced with technology. Science would seem to have won the day—or has it? We have already seen how an apparition will "fit" the society in which it is set, and, with the decline of faith over the years, the ability to see angels has been replaced by the ability to see aliens. What is of great importance is that if angels and the supposed aliens are the same, there appears to be little difference between the two—except that, in the newer guise of aliens, angels have become a lot more interactive, by some accounts even abductively so.

It is important to admit that the human mind is more incredible than we seem to realize; the UFO/visionary phenomenon has had an impact on regions of the brain about which we know so little. The fact that the form of visions seems to be dictated by one's own personal beliefs and cultural identity has great implications, and it is a shock to realize that what we perceive as being *out there* may in fact be *in here*. Jung had the notion that UFOs could be projections of the unconscious mind. This is not to reduce their importance as a phenomenon because, if life does exist elsewhere, our way of communicating with it will only come from a greater understanding of what is going on here, within ourselves and within our own planetary atmosphere.

The Cold War, at its height, induced a state of paranoia the world over. But of all the startling imagery of monolithic government, the black and white imagery of various military struggles, of good against evil, it is the science-fiction movies that had the greatest effect, an effect that remains ingrained in the consciousness of generations. We are still wary of the stranger from outside, the alien from the great unknown. Early science-fiction films include *Red Planet Mars, Invasion of the Body Snatchers,* and *The Day the Earth Stood Still.* By today's high technological standards, they seem primitive. But they worked.

Of course, the gist of it in the 1950s was that "they" were really the Communists. Many films from that era still have a McCarthy witch-hunt feel about them—the U.S. military were the good guys, the Soviets were the bad guys, and, actually at the time, a lot of the UFO phenomena was thought to emanate from behind the Iron Curtain. Many years later, when the Cold War had been consigned to history, the writer John Le Carre asked the Russians about their role in the paranoiac 1950s and '60s. They, too, had been convinced that they were the good guys. Furthermore, they were just as intrigued by the flying saucer phenomenon as the Americans.

UFOs quite often appear as lights in the sky and as silvery metallic objects. Generally dismissed as satellites in orbit about the Earth, or a trick of the light, these objects have been seen by too many people, and too often, to be dismissed out of hand. Recently they have come to be accepted by a wide body of the world's population. Disclosure seems to be a slow trickle, but it is seeping out. There has always remained a significant percentage that cannot be explained away. In fact, a national poll taken in 1998 by *Time* magazine suggested that 22 percent of the American people believe in visitors from another planet. More recent polls, twenty years later, are even more impressive. This same group shared the belief that beings from other planets have been in contact with humans; 17 percent said that intelligent life had abducted humans to experiment on them.

Since the 1970s there has been a huge wave of reported cases of alien abduction, literally hundreds of thousands of them. In the classic abduction case, the victim remembers seeing strange lights and that's about it: but then the nightmares begin and therapy is required. Often in an attempt to solve the problem, the patient is hypnotized and, during what is usually revealed to be a traumatic episode, can at last unlock a sealed compartment of the memory. However, for all its beneficial uses, hypnosis can reveal only what the subject believes might have happened.

Another aspect of abduction cases involves the implantation and later removal of fetuses for alien use: the bereft and confused victim being informed that they are helping to create a kind of racial improvement, a genetic modification upon the old model. Understandably, victims are left with painful emotions after these experiences—but also with a higher psychic sense and a greater sense of awareness. These apparitions or encounters are of an extreme psychological complexity, and all may not be as it seems.

The actual experience, as described by the many people who have had them, most certainly involves a change in awareness, where the individual is knocked out of this reality and into something that, under normal circumstances, would take considerable adjustment. Equally strange is the "healing" consequence of the abduction experience. Complaints that have been there for years simply vanish: arthritis, rheumatic disorders, even cancer can just disappear when the little "bug-eyed monster" comes out to play. But are abductees victims, or do such uncomfortable experiences have more to do with confused belief systems, uncomfortable readjustments of our internal programming, rather than an episode of interstellar intervention? At the Fatima vision of the Blessed Virgin in 1917, there were many instances of UFO sightings, all of which were attributed to the divine.[23] More than a hundred years later is it any different? The only thing that has changed is the nature of the belief. At the third sighting at Fatima, thousands of people heard a buzzing noise and heard an explosion, a sonic boom.

By the time that I had reached this stage of my research, I could not help wondering whether all of the divine visions and all of these UFOs are part of one and the same thing and related to earth resonances.

Through these pages, it will become clear that we can consciously interact with electromagnetic frequencies or energies with a view to transforming our awareness. But might these frequencies also be interacting with our subconscious, and if so, how would these interactions manifest themselves to our subconscious state? Furthermore, are these energies really separate from us? Perhaps by looking into space we are leading ourselves up the garden path. Surely the answer is much closer to home. If so, one would expect that religion and myth should have some, if not all, answers.

In February 1997 at a meeting of the American Association, a Vatican astronomer, Christopher Corbally, conceded that the discovery of alien life would necessitate a wider concept of God: "We need a proper sense of God, one derived in the dialogue between religion and science." This is stirring stuff but raises a point of contradiction. Given that the word *catholic* means "universal," what can possibly be meant by a wider concept of God? It seems to me that, in the light of recent scientific discoveries regarding the powers of the mind, the revelations of quantum physics, and so on, we should be looking for depth rather than width, experience rather than concepts. As mystics and saints have testified throughout the ages, depth of experience is gained by consciously looking within. Yet these words from the scientific arm of the Roman Catholic Church demonstrate that the church itself is looking anywhere else but there. Mainstream religious thought has for some time been given over to ideas of a God separate and "out there." It seems that officially, a change is not on the agenda. However, in the late 1990s the then Anglican Bishop of Oxford, the Right Reverend Richard Harries, brought some depth to the debate by describing angels as aliens. He could be uncannily close to the point.[24] During the Middle Ages, in a society that was deeply religious, any vision was very willingly interpreted in a biblical light. You already know my thoughts that what to us are aliens were called angels to medieval man.

Science and Western Christianity meet on two points:

1. Both are doctrinal, sometimes to the point of exclusion.
2. Both seem to view the universe and creation in general as separate from human experience. If human experience is informed by current scientific and religious dogma, then this more than implies a loss of contact, not only with ourselves but also with the planet and the greater reality.

We will certainly have to look more deeply to find our truths. But there is always help at hand. The great dictionaries are a treasure trove of insight. Here is an interesting fact that may yet reveal much: the word *alien* breaks into two very interesting and very ancient words: *el ayin. Ayin* is the sixteenth letter of both the Hebrew and Greek alphabets. It has the value of *o,* which stems from the ancient Egyptian hieroglyph for the third eye. The letter *ayin* or *o* represents and means "eye." *El ayin* means the "eye of god"* and refers to the third eye, or pineal gland. The third eye in ancient Egypt was the eye of Horus, the ultimate source of enlightenment. The symbol of the cobra, or uraeus, upon a pharaoh's brow is also representative of the third eye. As we shall see later, to the Egyptians angels and gods performed divine acts of healing, miracles were seen as being resonant with godly things, and the gods themselves were seen as resonances.

THE ONE WHO NAMES

Various theories have been put forward regarding the origins of shamanism. Some authorities believe that the first shamans came from the Altai Mountains in Siberia, others that they emanate from Tibet or other remote areas. However, it is now generally accepted that shamanism is a worldwide phenomenon whose beginnings are hidden in the impenetrable mists of deep antiquity. *Enuma Elish-la nabu shamanu* begins the great Mesopotamian story of creation, and in it there is a keyword, *shamanu.*[25] The whole line reads: "When on high, the heavens were not yet named."

The shaman is "the one who names," and this corresponds to native traditions worldwide, particularly those with a tradition that refers to the "first time," otherwise called the "dreamtime." This is an appropriate way to describe the "other" world, for here, at the finer levels of creation, where electromagnetic fields constantly interact with our senses, there is a blurring of identities. In this realm, the journeyman must keep a clear and focused mind, and like the hero, harrow hell to

---

*In addition, it may also refer to God's omnipresence.

enter the underworld, the region that underpins physical life and where spirits may be consulted. If the shaman is the traveler to other realms, then the sacred site is the gateway. In the previous chapter, we gained an insight into how sacred atmospheres can elevate sensitivities. They are a means of entrance to the nether realms.

In modern parlance, the shaman can be seen as a frequency modulator, tuning in to the wave band and traveling in consciousness, going where he or she chooses, even up into the sky. Consciousness, when unfettered by the body, can go where it wills. Certainly, many people have out-of-body experiences. I myself had a profound near-death experience in which I, my consciousness, left my body and yet I retained my identity. The realm of consciousness, though, is highly subjective and proving such experiences, in current scientific terms, is virtually impossible. Mainstream science will have to jettison its solely "objective" approach before serious study of the nature of consciousness can be undertaken. In the meantime, it is fortunate that a number of sacred sites support the claims of the shaman.

The most famous of the sites is the Nazca lines in South America. These vast geoglyphs were most certainly designed to be seen from the air—from the shamanic mind's eye. Unintelligible at ground level, except as a series of dusty pilgrim processional routes, the spectacular lines, from an aerial viewpoint, form spirals, vast animals, birds, and, most famously, a spider. They have been subject to many a hypothesis but are almost certainly *not* alien landing sites. The evidence on the ground at Nazca is safely situated high up in a desert region in southwestern Peru. It is an arid region with traces of ancient irrigation and thus early civilization. The markings, covering a vast expanse, lie on reasonably flat land that is perfect for sightseeing from above—aerial photography has long revealed the wonderful glyphs for what they are. The Nazca culture created the markings over a period of roughly one thousand years from 400 BCE until drought and the subsequent collapse of the culture brought all activity to a close circa 600 CE.

The Nazcans themselves lived not far away, on lower ground, closer to the sea. These locations were, and still are, water catchment areas, in complete contrast to the arid plateau some miles away. The water catchment areas have been called irrigation channels, which they undoubtedly are, but upon closer inspection they may also have served another purpose. By the indigenous peoples they are still called *Pukas,* a word that is extremely similar to the Gaelic *puca* (English *puck*), meaning "mischievous sprite." Many of these channels are fashioned as spiral walkways that descend gradually to the height of a well from wherein holy water was drawn. In these pages you will notice recurring instances of water, spirals, and folklore.

The local pots and shards that have been excavated picture some animals that are not local to the area: monkeys, killer whales, and certain species of bird. These animals are also depicted high up on the arid plain. Were these distant species first seen through shamanic flights? The most compelling piece of evidence, however, is the depiction upon pottery of a flying green man, spewing vomit and mucus from mouth and nostrils.

Sacred site and holy water alone did not induce the shaman to fly, for to travel to the gods, the local narcotic would be imbibed: the juice and flesh of the San Pedro cactus (named after Saint Peter—in his role of guardian of Heaven's Gate perhaps?). In flying to the realm of the gods, the shaman would negotiate various obstacles by taking the shape and aspect of many and various animals (very reminiscent of Mowgli in Kipling's *Jungle Book*). The shaman would use their various properties to fight demons and negotiate with the gods for various cures, for rainfall, and even a good harvest. Broadcast in 1997, a television program, *Flightpaths to the Gods,* presented by Anthony Spawforth, featured a prominent shaman and healer brought up within the Nazca locale. He stated, very significantly: "We need the vibration of the humming bird to help us tune into the universe, give us a better understanding of illness and the evil spirits that lay behind it. We need its beak to tap into the wisdom of the sun-god."[26]

The hummingbird form is one of the largest of the Nazca geoglyphs. The same shaman also told how they admired the condor because it spirals up into the sky: shamans say that they "themselves take off" in spiral fashion. The geoglyphs were processed or walked by the ordinary populace at large; music was played while walking the lines, and decorated pots would be smashed as if in sacrifice to the gods. This latter ritual also took place on a grand scale in dynastic and predynastic Egypt. The aim of these rituals was literally to massage the Earth, to placate her into providing water. These lines were the soul paths of the shaman, and lines like them are to be found the world over. A lot of them are pilgrim routes. But while the pilgrims walked the lines, the shamans saw them from above. It seems that shamans have ways of gaining knowledge of which we are not fully aware—invisible ways, sensitive ways.

And so it is that humankind *is* sensitive, and I would venture that the sensitivity was not necessarily drug dependent from the outset. Sophisticated ritual techniques and controlled dietary regimens seem to have been the favorite techniques within ancient societies. Archaeological evidence and analysis of sites and of individuals* have proved that fasting and dietary techniques were a hallowed part of ritual and of living.[27] Within this sphere, the spiritual "technology" appears to have been highly developed from the outset of civilization. If there is evidence of drug usage, it has a tendency to be late on the scene and reflects the general decline that befalls every civilization. As this happens, clarity is lost and so is the depth of knowledge. With this loss comes a facile interpretation of events, and even language takes on new, diminished, meaning. For example, the idea of the shaman and the hero harrowing hell can tell us a lot about this change of interpretation.

Earlier I referred to the underworld as the subtle substratum that underpins physical life. In ancient Greece, the underworld was called *hel* or *helle,* and here we see the multiple associated meanings that frequently appear in ancient languages: for these words also mean both "womb" and "sun"—in that womb and sun both underpin, or come before, physical life. A further extension of the words *hel* or *helle* means, equally simply but profoundly, "change." When you are going through hell, you are in fact changing. In its religious context, this is change associated with the element of fire—a baptism by fire, as one might say. To the Ancients, the sun not only upheld life, but it was also the great purifier, and to pass through hell was to pass through the fires of

---

*For example, the settlement of Qumran by the Dead Sea.

purification. Having been purged, their senses and systems purified, the "Pure Ones" were then able to commune with the gods, for they could now operate on a more refined level.

GODDESS AND SON

As we have already seen, the Greek name *Ge* or *Gaia* meaning "earth," was derived from the Egyptian *Geb*. It is echoed to this day in many familiar words beginning with *geo*. It is geometry that reveals the most about the Earth and her true connection to the sacred site. As revealed by Alexander Thom, the sacred site is not a haphazard siting upon an earth fault or spring, it is a precisely measured phenomena that adheres very closely to what Thom discovered to be a precise measure, the megalithic yard.[28] We shall not delve too closely into measure—at this point it is enough to realize that geometry goes beyond measure. It is the language of circles, spirals, and ellipses and therefore a language of movement, of heavenly bodies, a language that speaks cryptically of the true nature of the divine, particularly of the divine here on Earth. It speaks of the real and ancient history now expressed within the world's religions and mythologies, a language of pattern, rhythm, and number, obscured and unjustifiably dismissed.

See now how the mind can work through language. We have noted that the sixteenth letter of the Hebrew and Greek alphabets, el ayin, *o,* represents the third eye.[29] The Greek prefix *geo* thus reveals a hidden meaning: that as far as the Greeks were concerned, the earth, *ge* is only really perceived or measured via the gateway of the third eye—the letter *o,* which is a perceptual gateway between earth and the way in which we "have the measure" of it—ge-o-metry. The earth goddess is in this way the subject of contemplation, and that brings her measure to us, the measure of rhythms, vibration, and meter. To the human mind, such understandings were a guiding light in the darkness.

Further to Gaia, the Greeks had another name for Mother Earth, one that was older, one that linked her directly to heaven and the cosmos: the name is Maia, or Mara. However the name is spelled, this particular aspect of the goddess is the one that reveals her as the mother of the Redeemer, the one who will come, the hero. She is the Queen of Heaven, and, as noted previously, she has been witnessed in a vast profusion of visions.

Mary as the mother of the hero is famous and esteemed in the Roman Catholic Church as the Mother of God; indeed, she was proclaimed as the *Theotokos,* "God-bearer" at Ephesus in 413 CE by the newly powerful church. A significant point must be that Ephesus, where the Holy Virgin was assumed body and soul into heaven, was also the cult center of the goddess Artemis, another divinity often seen in visions. Given the way in which early Christianity absorbed and mimicked local pagan beliefs, the evidence would suggest that Mary was also an earth goddess, as was Artemis. To go even further, the Vatican Office of the Virgin described her as the primordial being, "created from the beginning and before the centuries." In the *Speculum beatae Mariae,* attributed to the thirteenth-century Saint Bonaventura, Mary is said to be "queen of heaven where she is enthroned in the midst of the angels, queen of the earth where she constantly manifests her power, and queen of hell where she has authority over the demons."[30]

The goddess is seen as a creatrix, a force, one of the two prime interacting aspects (male and female) of the Absolute. In native myths of the dreamtime, all life, all humanity, inhabited the womb of the goddess in an ever-present state of wonder and totality. The goddess was what science would call "causality"; from her came everything. Quite often within myths the goddess is seen as a huntress armed with a bow and arrows. In this guise she was called Artemis by the Greeks and Diana by the Romans. An essential aspect of the myth of the goddess was that each year she would give birth to a son, whose destiny it was to be cut down in his prime and, having harrowed hell, be reborn as the goddess's lover. She was the giver and the taker of life, and by the same token she could bestow immortality. Thus, in ancient societies, embedded deep within their mythos is the identity of the goddess as creatrix of all.

The Great Goddess has many names and forms, stretching far back into antiquity. As Cybele she makes frequent appearances in Greek mythology. She is known to have been the great mother goddess of the

Mediterranean. Her name is Anatolian* and is to be traced back as far as Çatalhüyük in southern Turkey. There, a shrine dating to 6500 BCE shows her giving birth to a bull's head with horns, the bull being symbolic of her son, the hero.

Cybele's name has a variety of meanings, each of which is valid, particularly as each goddess had a multitude of functions. One of her names is "She of the hair"; this is a specific reference to her portrayal as either having long hair or being covered in hair: just as the Magdalene and others are represented covered head to foot with hair. She (and probably a host of other goddesses) was the protectress of the chasms and cliffs of Gaia; it can be said that both she and Gaia are one with Mother Earth.[31] Within this remit she was also the guardian of the bears who hibernated in her caves during the long winter months—they, too, are hairy. Cybele had oracles at various sacred sites, the most famous of which was the sibyl at Delphi. *Sibyl* is another spelling of Cybele. The oracles were famous prophetesses of their day. *Cybele* splits into two syllables: *Cy* and *bele*. *Bele* means the same as its male counterpart, *Baal,* meaning "lord," but emphasizes beauty, the feminine aspect.† It is the first syllable that is of prime interest. One translation is "desert" or "barren place." I was surprised to learn, while reading one day through a book on the myths and legends of early Christianity that at least two Marys were associated with arid places: Saint Mary of Egypt and Saint Mary Magdalene, whose apparent desert experience was in the wilderness of southern Provence in the years after the crucifixion.[32]

Intriguingly, the festival of Cybele always fell on April 4 and was called the Megalensia. Given the similarities to the Magdalene, this seems to be a related name. It could be that someone, a Christian copyist perhaps, had sought to limit the similarities between Cybele's festival and the Magdalene by the simple but effective removal of the *d*. More intriguing is the fact that *Cy* also means, in Greek, "waves," hence cymatics—the study of wave forms. Cybele, in being the goddess, "lady

---

*After the goddess Anat. Egyptian Neit; Greek Athena.
†Note the similarity to *belle* and *bella*.

or beauty of the waves," seems to be making a direct reference to sound, a point borne out by the name of her son, who seems to have inherited something of her music. The son was called various names—Attis, Adonis, Anchises. In his original archaic Syrian homeland, he was Dumuzi—also known as Tammuz.

Tammuz is a shepherd god, one whose influence on later Christian literature has been noted by scholars. Called an early prototype of Jesus by Christian apologists, Tammuz is most frequently known as the "dying and rising god."[33] Shared similarities with Jesus include the fact that he was known as "the only begotten son," the savior, the healer, and the "heavenly shepherd." A month of the Jewish calendar is still named after Tammuz.

Tammuz, when divided into its component parts, reveals much: *Ta* and *muz*. The second syllable is more familiar to us as *Moses,* meaning literally "child of." *Ta* or *Te* is the seventh note of the musical scale. So, too, is *Si,* the little-known and infrequently used equivalent of *Te*.[34] Tammuz and his mother, the Sibyl or Cybele, share the same musical note: Tammuz is the child of the seventh (or the seventh son), and Cybele is the lady of the seventh. The number seven occurs countless times in religious literature. It was Jesus who, in one of the most famous episodes of his brief mission, cast out of Mary Magdalene seven devils. Are these seven related to musical notes? On the occasion of Tammuz's death, the Babylonians would ring out great ululations of despair, a tradition still held in Bedouin custom (under a different, Islamic, guise). The cry of the people, as described by Babylonian chroniclers, was "alalu"; in other words, "hallelujah." Tammuz had gone to intercede on their behalf. Was it not the *alalu* of Babylonians that went on to become the Te Deum, the supreme expression of rejoicing in the Roman Catholic and other churches?

Te Deum Laudamus, "We praise thee O God," hallelujah—is the true age of Christianity being revealed here?

Not surprisingly, there were many variations upon the theme of Christianity, some of which came to be seen as heretical. In the second century CE, a new heresy arose—Montanism. Its founder, Montanus, was a Phrygian of Anatolia, homeland of Cybele. Through Montanus, the Holy Spirit is said to have stated, "Behold, man is like a lyre, and I fly over it like a plectrum."[35]

By extension, it seems to me that the Holy Spirit is imparting an exceedingly precious morsel of knowledge: understand wave forms, frequencies, sound—the stuff of music—and not only do you have an understanding of the divine and its relationship to humanity but also a key to opening up that relationship. It seems that the Ancients used this knowledge fully in their architecture and in their wisdom. There is another clear demonstration of this wisdom, one that we can encounter now ourselves.

THE RISE OF THE GODLY

In the years preceding the First Crusade of 1095 CE, elements of a new and vital energy began to creep across the threshold of the newly emergent Europe. In the wake of the Dark Age terrors, when the power of Rome had become transformed into a nascent European cultural identity, the heaviness that was the Romanesque design of sacred buildings was about to be replaced by an abundance of lightness and measure.

The Romanesque seems to epitomize the darker age that preceded the medieval period. Light seems constrained within its walls, whose apertures barely deserved the description of windows. Within the Romanesque there is a limited sense of the soul and perhaps of the soul's purpose. There is a strange leaden feeling, a weightiness that plays upon it. Psychologically in the context of the period in which they were constructed, this is all too revealing. They are hallowed temples and deserve much good comment but are outdone by the Gothic—a framework within which light and sound do very profound things.

The Dark Ages had wrought much havoc upon the European landscape, but in their wake, much rose to wonder at. While the feudal system had become firmly established, transforming both culture and landscape into an organized and efficient system of management, the church itself, having undergone centuries of upheaval, was now subject to a more subtle revolution.

This was the era of the great Catholic hegemony, to whom even kings were answerable upon fear of excommunication and consequent removal. During this period (900 to 1500 CE) the Dark Ages were finally banished as a distant memory, as individual nation-states arose from the debris, guided by the principle of Christian law. Scholars often speak of the "death of the gods" amid the decline of the ancient world and of their subsequent resurrection during the early to middle years of the Renaissance. But surely there is an irony here? If the pagan gods had truly died out, then the ideas surrounding them survived: paganism was influential in all things Christian.

The Greeks had made the concept of the deity human; that is to say, they had gods with foibles and faults, they were fallible. According to this model, the concept of the gods and the language of myth is the history of life itself, it is the history of human impulse and the discovery of self and identity. Consequently, when the idea of one god came along to replace the idea of many gods, the many gods were easy to dismiss as mere human aspects of fallibility. But is divinity something that is purely superhuman?

According to many religious ideas, there is a part of God in each and every one of us. We are linked to the cosmos by the god who created it. God is in everything; therefore everything is related, everything is relative. Myth, with its many gods, is about relationship and is there to be used as a tool. It is about humankind reaching out and exceeding the state of humanity, to realize oneness with God. To clarify: the gods were more than a simple idea, a fad whose time had run out as Christianity came into fashion. In point of fact, their ideas *underlined* Christianity to the extent that the old gods were merely converted into fictional saints. It was recognition of a kind. How else can we explain the predominance of pagan imagery to be found within Christian sanctuaries? "How extraordinary were the adventures in space and time of these gods of the Middle Ages—these hybrid and phantom gods—and how much may be learned of the history of civilization from a study of their metamorphoses and reincarnations."[36]

By the same token, what is immediately apparent from a study of the history of the period is that quite a tangible change occurred in cultural, intellectual, and spiritual terms. Europe at last achieved self-identity. The rise of the Gothic is a starting point, for from its inception in the 1100s, it marked a transformation of conscious intent, the beginnings of a change in consciousness itself, in a way that is unparalleled in European history.

It was a full two decades before the First Crusade that elements of the Gothic arrived in Europe, at the Benedictine monastery of Monte Cassino, Italy, from 1066 to 1071. During this time Europe was still throwing off the physical presence of Islam, while adapting and absorbing into its own culture all that it had learned from the invaders. At its center in the Middle East, the Islamic culture was still far ahead in terms of science, philosophy, law, the arts, and a host of civilizing disciplines.

Exposure to this rich culture during the first of the crusades was to create an unprecedented explosion of creative thought and activity in Europe at the outset of the twelfth century. The Gothic cathedral was the foremost expression of this activity, and enthusiasm for the style was boundless. In France alone, between 1170 and 1270, as well as about five hundred Gothic-style churches, an astonishing eighty cathedrals were constructed. From the east to the west, medieval Europe was festooned in an abundance of gigantic palaces of light. These great towering masterpieces still display a richness as yet unsurpassed. They embodied profound ideas that serve to illumine the mystery of life.

Much that had been lost in the Dark Ages was being rediscovered, if not wholly reinvented, for a terrible price had been paid in knowledge lost during that time. In the words of the late William Anderson, the appearance of the Romanesque alone was an act that was nothing short of heroic; therefore the rampant march of the Gothic was, by the same standards, miraculous.

Contrary to much scholarly opinion, a view is arising, with evidence to support it, that the Romanesque appeared upon the stage only shortly before the Gothic.[37] William Anderson, whose work *The Rise of the Gothic* is a landmark, sees both styles born of

the explosion of new thoughts and new feelings that erupted in the twelfth century, with the birth of scholastic philosophy and the rediscovery of the eternal feminine that was manifested in various forms such as Mariolatry and the lyrics of the Troubadours. The sculptors of the Romanesque had also released the latent archetypal powers of the ancient gods and a new indigenous style was necessary to contain and reconcile these powers.[38]

Once the Romanesque style had released these powers, its job was effectively done and the stage was set for the total domination of a style devoted to letting in the light, for not only is the Gothic a considerable intellectual and aesthetic achievement, but greater than these is its sense of the spiritual. The observation that the Gothic was concurrent with the Romanesque leans toward the view that the latter arose out of historical consequence, while the former arose out of a school of learning where many strands of older beliefs and traditions were being drawn together into a unified whole. The sense that I got, particularly from that greatest of Gothic cathedrals, Chartres, with its plethora of unlikely imagery, was that fundamentally, this combined ancient knowledge was pagan, and from out of that Christianity had arisen. Quite possibly this is how it was meant to be read by the lowly peasantry of the day—that Christianity is the culmination of all that had gone before. But, as we have noted, initiating this activity was the influence of Islam.

In 1064, many thousands of prisoners were taken at the siege of Barbastro, a village in Aragon, Spain. Among these prisoners were many Muslim craftsmen. The Moorish Islamic style of architecture of the preceding centuries was very distinctive and made use of many features, foremost of which is the famous pointed arch. It was the sudden influx of these Muslim prisoners that set the scene for the transformation of European architecture over the next seventy years: their ideas would enfranchise and permeate the Guild of Masons. Again, in the words of William Anderson, "The architects who created the Gothic style were men of exceptional boldness and originality and they achieved something extraordinary. They were to form an international organisation that for four hundred years, from about 1140 to 1540, trained through succeeding generations artists of genius who could maintain the standards of their art and at the same time develop it."[39]

The continuity of Muslim tradition, allied to the new and exciting spiritual revolution of Christian Europe, began to yield swift results. A new style emerged, already with a surprising wholeness. At Monte Cassino, the Moorish tradition of Spain was linked harmoniously with the new knowledge then coming from the Holy Land. However, it was at the rebuilding of the cathedral of Saint Denis, north of Paris, that the Gothic really began to emerge, and in truly remarkable fashion. The rebuilding of Saint Denis was undertaken by the abbot Suger.

Born of a peasant family in 1081, Suger was a modest man who had become the confidant of kings. During the year 1147 he was even regent of France, while its king, Louis VII, was on crusade. By all accounts a good conversationalist and very good company, Suger withdrew from the royal court to concentrate on the rebuilding of his beloved Saint Denis. Intriguingly, his first action in undertaking such an immense task was to employ a group of monks to research the early Christian writings. He specifically enjoined them to look out for anything that commented or revealed any insight into the use of light and its magical and transcendent qualities. Abbot Suger adopted the motto *Ars sina scientia nihil est*—"Art without knowledge is nothing." But the abbot's passion stretched beyond knowledge and art for their own sake. He was reaching for something higher still. He was to write:

> When out of my delight and in the beauty of the House of God— the loveliness of the many-colored gems has called me away from external cares, and worthy meditation has induced me to reflect, transferring that which is material to that which is immaterial, on the diversity of the sacred virtues: then it seems to me that I see myself dwelling, as it were in some strange region of the universe which neither exists entirely in the slime of the Earth nor entirely in the purity of Heaven; and that by the grace of God, I can be transported from this inferior to that higher world in an anagogical [mystical] manner.[40]

Such intensity bore fruit at Saint Denis.

The new west front, begun in 1137, includes the first rose window.

The west front itself, the narthex, the choir, and the crypt are the only surviving elements of Suger's innovations, but these are enough to tell us that the Gothic had at last emerged. Here, instead of the old feeling of monumental architecture coupled with constraint and foreboding, we have a new climate, one of reconciliation and of consolation. Thus, the entrant into such an atmosphere feels reassured and at ease. The interior is unburdened by darkness, the columns in the choir are slim, the arches are high, and the space between them is broad. Furthermore, here is another element that is central to all Gothic architecture: the "axes" are aligned to the center of the choir so that colored light from the surrounding stained glass is able to penetrate, unhindered, into the deeper recesses. The simplicity of the design is startling, its approach quite breathtaking. At Saint Denis the new style is emergent; however, it is at the cathedral of Sens that it is born.

In 1128 the old cathedral of Sens was destroyed by fire and its archbishop, Henri de Sanglier, was confronted with a greater task than his compatriot, Abbot Suger at Saint Denis. At Sens, a phoenix arose from the ashes. Significantly, the new design incorporates the human skeleton as a major element. This is expressed through the use of ribs, arches, and piers (see figure 2.2). Here we see, for the first time, the

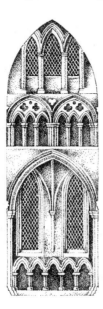

Fig. 2.2. Example of Gothic structure

interplay of forces that are in direct opposition to each other, to give a lightness of touch and of style. It is at Sens that we have the opportunity to appreciate its simple musical proportion. Its unity is expressed in the octave 1:2, an example being that the aisles are half the width of the nave. The structure rises upward in three stages, rather than the four of the Romanesque, and wherever one looks there is rhythm—vertically and horizontally. Also, there is a deliberate play on numbers. For example, with the three upward stages we have the Trinity and, with the use of the 1:2 ratio, duality. These buildings were becoming homes of true metaphysical thought. Within a short time, the new style developed and reached its apogee in one of the greatest masterpieces of historical architecture, Chartres.

A MUSIC FROZEN IN SPACE

It is a little-known fact that when Chartres Cathedral came to be built, monarchs from all over Europe gave financial support and manpower for its construction. Built over a period of twenty-six years, it is a visually stunning work of art, and yet, on approach, the first thing that assails the senses is something nonvisual. It is a great sense of the "power of place." So striking is this sense that, allied to the visual richness, one is considerably aware of a deep feeling of antiquity, an ancient sense of belief, of spirituality. There is a sense of "something other," a something far greater than that which is on display. The oldest and greatest of divinities is implicit in everything at Chartres. It is invisible—and yet it can be seen in the senses.

At Chartres, as at Reims and Amiens and a host of other European sites, the place is dripping with statues. Here, there is no uniformity, just thousands of individually carved faces; human, animal, and vegetable, figures real and mythical, with flora and fauna equally symbolic. Here, everything is crucial, everything has its role. There is a sense that all is vital and there to be read. But, stand back until all that has struck the eye as individual has merged into an overview, and then, only then does one gain a pure insight into the broad magnificence that is the Gothic. For rising out of the ground is a sense of something organic, something that has grown to towering proportions and yearns to grow further still. We are already aware of granite's crystalline properties, and we will look more deeply into these later. Parts of Chartres's interior are made of granite, particularly in the altar area. The exterior is limestone. The entire building is like a huge crystal, created to resonate to the sound of praise; praise to the Eternal, to Christ, and to Mary. True sacred sites are designed to be alchemical transmutation points of the soul. They are interactive, Chartres particularly so. Simply walking through the building can be an emotional experience. Gothic cathedrals have been called "books in stone." Chartres is more like a huge interactive library!

Here, as at Neolithic sites, sound is the interactive element, and the sciences of acoustics and geometry come together through the architecture to create what John D. Barrow sums up as "the unreproducible acoustic qualities of the Gothic."[41]

Ancient Newgrange gives its acoustic clues in simple carvings; Chartres's carvings are more sophisticated. Above the main door of Chartres, as well as a host of other cathedrals, can be found the tympanum. This takes its name from the Greek *tympanon,* meaning "kettledrum," and is derived from the verb *typtein,* "to strike." The tympanum displays Christ in all his glory, for there in its midst he is enthroned. Upon further reflection this design looks to all intents and purposes like

Fig. 2.3. Christ enthroned upon the tympanum at Chartres

a standing-wave pattern, with associated nodes and antinodes. There are yet more clues: gargoyles stare down, mouths open as if spewing forth sound. Above the central door of the royal portal is a musician-king who conceals under his robes a matrass—a long-necked chemical flask, symbolic of the alchemist. Also in the royal portal, Pythagoras can be seen in a pose of utter concentration, playing upon a tintinnabulum—a high-pitched bell. There follow more learned pre-Christian figures striking the appropriate themes: Euclid, Cicero, Aristotle, Ptolemy, Boethius, and Priscian. Interspersed among them are biblical prophets: Jeremiah, Moses, Isaiah, Elijah, and others.

Looking upward and away from the royal portal we see the magnificence of Chartres's flying buttresses, which serve to add to the impression of the whole as a heavenly crystal. The buttresses draw the eye farther up until, at the pinnacle of the tower, we see the full glory of the pointed arch. Within these pointed arches, illuminating the entire space of the interior, are an astonishing range of stained-glass windows. Gothic stained glass is very special and has peculiar properties. The secrets of its manufacture have since been lost to the world. In the light of this knowledge we have to reassess our view of the medieval craftsman. Was he aware of the fact that the stained-glass windows of Gothic cathedrals filter out the harmful ultraviolet of the sun's rays?

Glass is an insulator, a supercooled fluid that has turned solid upon cooling, without crystallization taking place. It is not a true solid, which is why it is often referred to as amorphous. The amazing qualities of Gothic stained glass are due to the fact that these windows are, literally, sandwiches of super-ground crystalline contents, though where this knowledge came from is an open question. The effect of this is vital upon the final quality of refraction: Gothic stained glass has a high refractive index. This gives the glass its exquisite jewel-like quality. It is widely believed that this is a purposefully created effect, one that serves to aid the state of mind by the application of only specific types of light, most of it colored. The work of Blanche Merz has shown that there is "neutralization in that zone which is directly related to the filtering effects of the stained-glass windows, creating a wavelength of light that can harmonize with the natural vibrations of human cellular tissue."[42]

Glassmaking of this caliber was a heavily guarded secret during the Middle Ages. It was the realm of the alchemist and of the secret arts. The results are strikingly impressive. This technology of colored glass sends the achievement of the Gothic up into rapturous heights. The rose windows, north, south, and west, are the glory of Chartres, each one a geometric masterpiece.

Chartres is a temple fine-tuned to an exceptional degree—and I use the word *temple* purposefully. As at Sens, the choir of Chartres Cathedral is a double square, "the ratio of 2 to 1 is exactly that of the Egyptian and Greek temples . . . likewise of Solomon's temple as far as the Holy of Holies."[43] This fine-tuning is also demonstrated in a phenomenon again reminiscent of Newgrange and other sites. Every year on June 21, the summer solstice—the day when the sun reaches its highest point in the sky—a ray of light shines through a spot in the stained-glass window dedicated to Saint Apollinaire. The beam descends to a specific place, a flagstone on the western side of the south transept. The stone, larger than those around it, has set within it a small metal tenon: it is this that is illuminated precisely at 12:45 p.m. local time. Because the height of the sun is now at its peak, this can only happen once a year, on this solstice day. Note that Saint Apollinaire is another version of Apollo, the Greek sun god.

Nothing was completed at Chartres without precise purpose and planning. So, too, the labyrinth. The labyrinth, as a symbol, is a curious phenomenon. It has its origins way back in the depths of prehistory. By the time its symbolic purpose was revived in the Middle Ages, its pagan roots had melted into the shadows. Of all the labyrinths at cathedral sites, the one at Chartres is the most famous. It was constructed as an object of mysterious ritual, said to be a penance path used by devotees as a substitute for an actual pilgrimage to Jerusalem. At Newgrange these enigmatic forms are to be seen everywhere. Evidence has also come to light at Nazca, suggesting that the labyrinth was a processional path. Indeed, the labyrinth at Chartres does have a ritual feel to it.

There was a similar labyrinth at Auxerre Cathedral. A well-authenticated story tells us of a ritual associated with it. On the afternoon of Easter Sunday, the dean and his chapter were to be found standing upon the Auxerre labyrinth, singing the hymn *Victimae Paschali laudes,* while the dean himself threw a golden ball to each member of the chapter as they danced the labyrinth. They danced in a triple rhythm to the hymn.[44] What I found of great interest was the probability that the ceremony dated back at least to the early Bronze Age.

The maze was an emblem of this age, although its origins may go back further still. In fact, there may be a link here to what is reputed to be the oldest hymn in the world, the Vedic Gayatri, known as the "triple song," or solar hymn. Here the golden ball of the sun is praised in a triple rhythm.* High above the labyrinth of Chartres, pillars rise up into the ceiling of nave, choir, and chapel, spreading outward like the branches of a great oak tree. Looking up at Chartres's vaulting I recalled that at Dodona, the oldest of the Greek oracle sanctuaries, the father god, Zeus, was worshipped as being immanent within the sacred oak; the rustling of its leaves in the wind was said to be his voice. Similarly, to me the great cathedrals are rather like a petrified forest echoing with the divine sound. Perhaps Goethe, the eighteenth-century philosopher, had this in mind when he likened architecture to frozen music. Maybe he knew that, during the course of construction, masons would quite often tap a pillar to hear its tone.[45] (I have done the same to Hatshepsut's fallen obelisk, which lies in the temple grounds at Karnak, Luxor, Egypt. Its tone is a low, resonant hum.)

That day at Chartres, I realized that the fan vaulting, when seen as a whole from end to end, resembled very closely an acoustic wave form.

SPIRIT-BREATH-WIND

If you think all of these acoustic connections are simply the result of my own imaginings, there are more, including evidence from experiments clearly linking church and cathedral ornamentation to the sci-

---

*In Sanskrit, the Vedic language believed by many to be the root of the Indo-European family of languages, *Gaya* is "song," and *Gia* is "earth" (Sir Monier Monier-Williams, *Sanskrit-English Dictionary*).

ence of sound. The experiments also answer the question I had often raised about some of the curious forms to be found carved into many an ecclesiastical interior. Generally, these forms are organic—plantlike. In fact, they bear a reasonable resemblance to ferns found in a forest. My question was: "What do they really represent?" For everything in a Gothic cathedral serves a deeper purpose.

The original experiment was performed by Father Andrew Glazewski, a Polish physicist based in Britain. Father Glazewski was also a Jesuit priest, whose main interest was crystallography and plant growth—and their relationships to sound. All organic matter is made up of crystals and compounds. What Glazewski set out to demonstrate was that plants, composed of specific crystals, followed the same harmonics as the crystals themselves, extending offshoots from their stems proportionate to the same musical phases and tonalities.[46] His experiments showed that the flutings on the columns of Gothic cathedrals also follow the harmonics. He demonstrated this by use of a smoke-filled chamber placed near to a music speaker. The frequencies emitted from the speaker passed through the thin transparent walls of the chamber and affected the smoke. Various photographs show the response to a well-built violin, a badly bowed violin, a flute, and so on. When the harmonic is played, the smoke looks just like our fern. "There is strong evidence that they reflect the same ratios in the swelling and narrowing of growing plant stems, as well as the flutings on the columns of Gothic Cathedrals such as Chartres."[47]

This experiment reflects those undertaken by the Princeton team at Newgrange and other Neolithic sites. The conclusion is the same: the builders had an in-depth knowledge of the science of acoustics—and they left their clues in the ornamentation. Sacred buildings are created to bring the faithful closer to God. Souls enter to be unburdened and uplifted by refined atmospheres. It is very clear that these beautiful Gothic cathedrals were dedicated to creating such atmospheres and that to do this the buildings were specifically designed with frequencies in mind: frequencies of light and sound. There is no shortage of evidence that sound was the interactive level.

One summer, while visiting Corpus Christi College, Cambridge, I was given further evidence. In the library there is a text written in 1230, with information dating back to 1140, the era when cathedral building was at its height. In *De Organis,* Wulfstan of Winchester exudes praise at the building of a powerful organ. He goes on:

> You have built here such an organ as is nowhere else seen, built with a double foundation. Twelve bellows are combined above in order, and below lie fourteen. With alternative blasts they *produce the greatest of resonances.*
>
> Seventy strong men operate them, pulling their arms and being drenched with much sweat; eagerly they each exhort their companions so that with all their strength they may drive the blast upwards so that the blast bag may roar with full throat.[48] (My italics)

An organ of such giant proportions is fit only for a cathedral; what is more, it would be capable of very low frequencies, which would create powerful resonances within the building.

> And the seven distinctions of sound strike up the "jubilus" with the chant of the lyric semitone mixed in. And in the manner of thunder an iron voice struck the ear, so that besides this the ears take in no sound. The sound roared to such an extent, reverberating here and there, that everyone would close his ears with his hands, in this way being able to tolerate the roar when drawing near which the various sounds produce in their noise.[49]

One expects that the power of such a resonance would have been stronger than the deep Hindu and Buddhist "Aum," which, when practiced correctly, produces refined atmospheres and clarity of mind. Wulfstan goes on to intimate that the organ did indeed have a cathedral purpose. The original document was composed in Latin, and the Latin word for resonance is a very descriptive one, *spiraculum,* which, of course, is related to the word *spirit*. We have looked at *spirit* before and we shall look again. However, it is instructive to investigate early understandings of the word. Much of what has come down to us from the archaic past suggests that for *spirit* we should perhaps read "breath," as in holy breath, and "movement," as in vibrations.

The lack of a clearly defined distinction was ably summed up by the late Owen Barfield.

> When we translate Latin "spiritus" we have to render it either as "spirit" or as "breath of" as in "wind" . . . But early users of language would not have made any such distinction between these meanings. To them a word like "spiritus" meant something like "spirit-breath-wind." When the wind blew it was not merely like someone breathing: it was the breath of a god. And when an early speaker talked about his soul as "spiritus" he did not merely mean that it was "like a breath": it was to him just that, the breath of life.[50]

To extrapolate: spirit is the breath of life, it moves, and as the breath of life, it is within the individual. This sounds like the qualities of frequency or vibration to me, for we know that without this movement, there is nothing; where there is no vibration, there is no life. This vibratory movement of life is present on all levels, from the orbit of an electron around the nucleus, to the movement of great fields of energy radiating out from the sun or the Earth, or even out from the beginnings of time and space. It is also the movement of localized yet potent fields, for instance, emanating from a type of stone or the human voice. Everywhere this energy is moving, everywhere it manifests differently, but its attributes, so aptly described by those early scientists as spirit-breath-wind, are always the same: life giving, life sustaining, and life transforming. Now, as then, the wise give praise at the wonder of life and thanks for the mercy of sustenance, but they also hope and pray for the joy of transformation.

The aspirations across the ages are the same. The difference is this: the masons and their ancient predecessors knew what they were doing—whereas we are only just beginning to discover their secrets. One of the most important things that we now know in this regard is the fact that our own alpha states can bring us into alignment with Mother Earth's own resonant atmosphere. Such is the effect that, as far as frequency is concerned, humankind and our environment can effectively become one and the same thing. Pressures and stresses fall away, and the mind enters a state of transcendent calm, while at the same time remaining alert. Traditionally this is the condition wherein the third eye, el ayin, begins to open up, and with dedicated practice, the subtle realms can be entered. "The brain and mind are not a monolith: they have multiple structural levels, and the highest of those levels creates instruments that permit the observation of the other levels."[51]

When we look upon the carved spirals of Newgrange and the fern-like designs from the Gothic, we need no longer ask, "How did these people know what to carve?" They would have no need for the assistance of smoke chambers and the like, for mystics and clairvoyants throughout the ages have told us that when the subtle realms are entered, the finer energies can be seen. By extension, one who had developed such a gift could witness the relationship between sun rays, earth energies, and specific local emanations. By introducing humankind's own interactive tool—the human voice—best design, best vicinity, and best practice for further transformation could be ascertained. It is quite possible that a few historical individuals may well have attained this level of knowledge in the past. They would have known beyond a doubt that form is a function of frequency. But as initiates, they would swear secrecy, and this powerful knowledge would remain hidden.

This ancient, sometime pagan, sometime Christian, knowledge was not devil worship, it was a divine science. The people of spirit-breath-wind knew the power of the word, for their terms also describe the movement of great fields of energy radiating out from the beginning of time and space, as in: "In the beginning was the Word, and the Word was with God."[52]

THE WORD OF GOD

Knowledge of the origins and meaning of words is fundamental to a quest of this nature. Unfortunately, we are all so close to language that we take meanings for granted. Therefore, without constant inquiry and a shift in mind-set, the obvious passes us by, unrecognized.

Thus as I contemplated the above quotation from the opening of Saint John's gospel the obvious appeared. Some key words hide the truth so simply: the word *universe* means literally "one word." Who or what, therefore, is God? Here was the very motive of my quest.

The word *god* has an interesting past. Its etymology reveals an extraordinary origin. The letter *d* in many old languages of Middle Eastern or Indo-European origin is interchangeable with a *t*. Each is a version of the other—soft and hard sounds. Hence the English word *god* becoming German *gott*. This derives from the Old Norse *goth,* and this in turn gives us the familiar term *Gothic*. However, there is a much more intriguing and ancient origin to this word. It can be traced back to the much older Old High Norse *guth* . . . and *guth* literally means "voice." This is saying that the Gothic is the godly, it is vocal, it is resonant, it is the Word of God.

Thinking back to Goethe's quote about architecture being frozen music, where then is the *un*frozen music? Music, as in the "music of the spheres" is all around us. The universe is this music, and the Ancients were saying that it is God's sound, from God's voice. They were saying, *are* saying, that God is voice, God is manifest in resonance.

Have we now revealed that the true meaning of the word *God* is "voice"? If so, Jesus, it would seem, is the son of the Voice. Could that be the sound itself? We know that *son* is the feminine of *sound* in French, and that "sound" in Scandinavia is *sund*. These are quick thoughts, but they bring to mind the hero's link to the sun and reaffirm the connection to the goddess and to sound. However, one must be more circumspect. *Voice* in itself might mean "sound," then the son of sound might be a particular harmony, a frequency, or conscious effect—or something more abstract. Certainly, there is more to be unraveled.

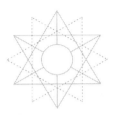

3

The Power and the Place

The nature of God is a circle of which the center is everywhere and the circumference is nowhere.
ATTRIBUTED TO EMPEDOCLES, 493–433 BCE

It has served us well, this myth of Christ.
ATTRIBUTED TO POPE LEO X, 1475–1521

THE SANCTITY OF PLACE

In April 1980 the Vatican, perhaps in recognition of the rise of the new environmental awareness hitting the West, adopted Saint Francis of Assisi as the patron saint of ecology. The significance of this gesture is considerable, and it is reasonable to suggest that, in choosing Saint Francis, the Vatican was tacitly giving its sign of approval to the idea of the Earth as a living organism. After all, Saint Francis is renowned for his canticle of "Brother Sun, Sister Moon," wherein these heavenly bodies are addressed as beings. This is not only supportive of Sir James Lovelock's Gaia hypothesis, which postulates that the Earth is a single organism and also introduces the idea of divinity into our relationship with the planet. Gaia, of course, is the Greek name for the goddess of the earth.

Of all the saints in Christendom, it is generally agreed that Francis, the merchant's son from Assisi, was the nearest, both in spirit and personality, to Jesus. Francis renounced earthly wealth, but to the rich in

96

spirit, material poverty is no great hardship and it can open the eyes to what is real and essential: the greatest treasures of Mother Nature herself. In the medieval period it was Saint Francis who most successfully espoused the sheer beauty of the planet. "Brother Sun, Sister Moon" speaks of the planet, the cosmos, and life in general as a miracle beyond knowing, and in a biographical way we get a great sense of Francis's awe at the wonder of creation.

> *Oh, Most High, Almighty, Good Lord God,*
> *to Thee belong praise,*
> *glory, honor, and all blessing.*
>
> *Praised be my Lord God, with all his creatures*
> *and especially our brother the Sun,*
> *who brings us the day*
> *and who brings us the light: fair is he,*
> *and he shines with a very great splendor.*
> *Oh Lord, he signifies us to Thee!*
>
> *Praised be my Lord for our sister the Moon,*
> *and for the stars,*
> *which he has set clearly and lovely*
> *in the Heaven . . .*[1]

The medieval period had opened up as one of foreboding for the inhabitants of Europe. There had been a sense of doom in the air, a premillennial fear that at any moment the four horsemen of the apocalypse would spring forth from the gates of hell, bringing with them the last judgment of God. The air was abuzz with rumor: and the freedoms that had emerged from the ruins of the old Roman Empire were threatened by dark clouds of doom. Not only were the Muslim hordes knocking at the door of the Christian West, so, too, was the Antichrist. This sense of despair and the occasional accompanying outbreak of hysteria can quite firmly be put down to the medieval sense of millennium, the feeling that a time of tribulation was near.

In retrospect it can be seen that millennial tensions only served to focus the mind upon the intangible. For if doom was in the air, it was in competition with a spate of miracles. The blood of long-dead saints miraculously turned to liquid upon their feast days: the ampoule of blood belonging to Saint Januarius, patron saint of Naples, still does this every year on September 19. Visions of the Holy Virgin occurred and some even claimed to have seen whole armies of angels, led by none other than Saint Michael himself.

When we look at these disturbed times, we may see a mass of superstition, rumor, and belief giving rise to illusion. However, we also see the numinous in the period—and how the ultimate mystery of Christian revolution had, in the majority, evoked a sympathy of the soul to ritual that, quite simply, is unintelligible to us today.

The year 1000 came and went, and gradually, very gradually, the millennial tensions ebbed away. Indeed, the lifetime of Saint Francis (1181–1226) can be taken as a marker for the emergence of a new era, one in which ritual was given accentuated potency through power of place. The cathedral-building program was beginning to electrify whole peoples in a new act of mass awareness.

Now, gloom was giving way to divine possibilities, to the transcendent experience of becoming at one with the sense of God. In this way, identity could almost be left behind as a result of divine possession. Even today, one can sense this when visiting one of the great cathedrals or listening to inspired music from the period. It was possibly within this context that, toward the end of the short life of Saint Francis, there began to be manifested upon him one of the more inexplicable of the great miracles, the stigmata.

Stigmata are marks that appear upon the body, and they resemble the wounds that Christ suffered upon the cross. However, this enigma is not restricted to the pious: the marks have also appeared upon the bodies of those whom we might call "less than holy." Could these extraordinary manifestations be caused not so much by faith as by place? *The Oxford Dictionary of the Christian Church* describes how the phenomenon is often accompanied by other, seemingly paranormal, events such as levitation, bilocation (the ability to be in two places simultaneously), and telepathic faculties. Furthermore, stigmata disappear as quickly as they appear, without leaving the slightest sign of their presence. Neither do the wounds become septic, and when treated, they do not respond.

The fact that the great majority of these phenomena actually occur inside churches, cathedrals, and temples has, to a large extent, been taken for granted. While I agree that all of the symptoms bear the hallmarks of the numinous and a sense of the divine, I would suggest an emphasis on the divine in situ. I have myself seen blood seeping forth from wood in a lych-gate within a particularly powerful church site. I am convinced that the key to the phenomenon lies within altered states of consciousness brought about by exposure to these sites.

We have already seen how the divine is intimately linked with the sacred site; in human terms such intimacy is an acutely powerful thing, so powerful that the human electromagnetic field changes dramatically in its presence. In fact, changes in field structure occur *before* any change in brain wave frequency.[2] The implication is that it is the field that gives the body's primary response to any outside interaction. Interestingly, the measurement of this field by telemetry also reveals a person's response to his or her environment. When near the mountains or by the sea this field has a tendency to expand, whereas by contrast, a person's individual response to rushing winds makes the energy field contract.

The difference between these conditions can be attributed to positive and negative ions. In a field of negative ions, the body thrives. It is for this reason that ionizers have been recommended for use in certain environments, as a means of recharging the air with negative ions. Furthermore, research and experimentation performed over the past fifty years, beginning in the 1970s, have shown that the human electrical field is, in certain situations, more than capable of responding to the wider atmospheric field; that is, with the Earth itself. In induced circumstances of electromagnetic deprivation, the body's response was dramatic—interaction and thinking were at best confused. However, when normal electromagnetic levels were restored, the body responded positively, thinking became a lot more lucid. Going further, when electromagnetic fields were increased beyond normality, subjects reported an increased awareness and heightened consciousness. This demonstrates that, amazingly, the whole process of bodily movement and coordination is closely linked to the interaction with the electromagnetic pulse of the environment.

It has been tacitly accepted by a group of scientists in Britain and the United States[3] that the Earth and the human body have meridian points. External electromagnetic energy penetrates these points in order to flow throughout the entirety of both. In the human body, such energy penetrates at the acupuncture points and flows through the connective tissues. Connective tissue is to be found in the body, most notably in the form of tendons and cartilage, as well as in the nonmuscular structures of arteries, veins, and bone. Valerie Hunt, professor emeritus of physiological science at UCLA, writes, "We concluded that the electromagnetic environment is a milieu in which life and physiological happenings occur. Apparently, for all systems to be 'go,' a rich electromagnetic field must be present. . . . If further studies verify these findings, the implications are staggering."[4]

Acupuncture uses a knowledge of electromagnetic fields, the origins of which are lost in the deep recesses of history. Perhaps the knowledge goes back as far as Megalithic and Neolithic times. At any rate, it can now be shown that the importance of the electromagnetic environment was not lost on the sacred site builders of the past. We shall see that they were aware of the power of radiation and sought to use it to their advantage.

To recap: everything in the universe resonates in some form or fashion; resonance can be described quite simply as the radiation and absorption of energy by vibration. Our planet sings to the rhythm of the cosmos by being struck (or some would say, played) by cosmic waves, much like a bell being struck by a hammer. In this way the Earth gives off her own vibratory rhythm, a rhythm to which, as testified by the brain's alpha waves, we may become attuned. The prescribed place of attunement for the Ancients was the sacred site.

These are places of power; the sites are, in their own way, speaking to us and so are the builders—from across time and space. This has enormous implications.

THE BUILT ENVIRONMENT

Sacred sites are effectively meeting places of time, space, and experience. They are totally oriented to the pragmatic, and it is with this in mind that we must regard not only the site but also the mythos associated with it. Not only are thousands of these sites dotted across the landscape of Britain and Europe, they are to be found across the entire face of the planet. Almost all of them are affected by high levels of natural radiation arising from the decay of uranium both in the ground and in the rocks used to build them.[5] Significantly, on the site of the world's greatest megalithic complex at Carnac in Brittany, France, there is a high level of granite, so, too, uranium.

Granite is a radioactive rock—it has a higher than normal emission of radioactivity. It is to be found at a high proportion of sacred sites across the world, from Stonehenge to the pyramids of Egypt. Granite's use by the Ancients is a testament to its properties and to their sensitivity and innovation. How interesting that so many sacred buildings should be constructed of granite—and should also be located in close proximity to uranium deposits. Atmospheric tests inside these monuments register radon levels that are often two or three times higher than normal and in some cases, as with the Sekhemkhet pyramid, Saqqara, Egypt, more than thirty times higher.[6]

These constructions manipulate the electromagnetism of their environment in a very subtle and beneficial way. However, the effects sometimes give results that are confusing to the uninitiated; for those who are stressed, heightened states of awareness can quickly give way to symptoms of fatigue. Basically, when individuals attain a high level of attunement with the sacred environment, the body becomes subject to forces that have, until recently, been little understood by science. Given their effects, it is hardly surprising that these places should be associated with spirits and divine presences. Even today, the enhanced state of environmental energies, aided and abetted by ritual, can create the perfect example of the sanctity of place and demonstrate the ability of certain locations to change our perceptions. One need not be mystically inclined to bear witness to this power. In fact, it is most often the no-nonsense, non-mystically oriented who are most vociferous in their praise of these sites.

Granite itself gives clues to its powers in a very down-to-earth, visual way. The rock, because of its radioactivity, produces fertile, higher than average growth in plants. Bulbs, by being planted near granite, can be induced to flower early.[7] Is it not likely that rapid growth in the vicinity of granite had been noted by early civilizations? Surely, then, the effect would be associated with divine beneficence and its potential further explored.

Evidence of the use of natural radiation in the sacred landscape is not confined to Britain and Europe: it is to be found at sites throughout the Americas, Australia, Africa, and Asia, where, for example, some of the greatest of the Tibetan monasteries are located in places of higher than usual radioactivity. That being the case, perhaps it is not surprising that certain mountains, held in great esteem by Buddhists and Hindus, have been found to be rich in uranium deposits.

A mysterious and seeming deliberate use of granite's properties of radiation was related to me in a remarkable story that, coming from any other man I would have dismissed. A. J. Scott-Morley told me of a phenomenon to be found at the temple of Karnak in Egypt. The story reminded me of the fabulous monolith in Arthur C. Clark's *2001: A Space Odyssey.*

Tony was invited to go to Egypt by a group of friends who were investigating various scientific aspects of the Karnak temple. On the last day of the visit, he was asked to take some magnetic readings of one of the altar stones. To his immense surprise, when it came to applying the magnetometer to the stone, Tony could only get a reading off one side of the stone, a seeming impossible occurrence. When he related the story to me, he was at great pains to point out that he had checked and rechecked his equipment—it was in full working order.

In Clark's novel, and to great effect in Kubrick's film, the enigmatic monolith does precisely the same thing. Prescience, or has

Clark been to Karnak and seen the stone? I have yet to find out.*

If the Ancients were clever enough to make a stone behave in this way, or sensitive enough to find one that did this naturally, then they knew a thing or two that we do not. Certainly, their ingenuity is underlined by the incredible lengths taken to get materials of exactly the right resonance and also by the construction techniques used to optimize the atmosphere at sites. Materials used in the construction of a majority of these sites came from far and wide. The example of Stonehenge is among the most renowned, with blue Preseli sarsen stones being hauled form the extreme end of southwest Wales, across the Severn estuary and overland to Salisbury Plain. Even much of the stone used to build the Giza pyramids is not indigenous to the Giza Plateau. It was hauled hundreds of miles up the Nile.

As we have seen, the materials were silicon-quartz rocks, and granite was especially prized. Silicon is a semiconductor, and electrical circuits can be created within the material itself; thus, they are sensitive to electromagnetic fields. Also, they resonate with the Earth.

The distances traveled show how important it was to get the right stone. However, having the right materials is only part of the story. What is done with them in terms of construction is also important. For instance, prehistoric burial chambers, usually sited at power spots, are constructed of layers of organic and inorganic matter. It is believed that these "clays, built sandwiched between an inner stone layer and the outer grass covering, act to shield the inside of the chamber from geo-magnetic fields." Serena Roney-Dougal gained this insight from her days at university, where she undertook research inside a Faraday cage. Faraday chambers, in being made of a fine copper mesh, screen out

---

*Granite has other properties. One of these is to be found within its magnetic record. Analysis of this record has uncovered evidence of past changes in the polarity of the Earth's magnetic field, via a process of "fossil" magnetism. As molten magma hardens upon exposure to the air, liquefied iron minerals solidify. In hardening they become magnetized at a certain critical temperature—the Curie point. It is this fossil magnetism that gives us a record of past ages. Regarding enhanced sensitivity to magnetic fields at sacred sites, it should be borne in mind that many types of rock contain iron-bearing minerals such as magnetite, also known as lodestone, and thus magnetic impulses can be measured from them. Among these rock types is granite.[8]

unwanted electromagnetic waves such as those from radio and television.[9] An atmosphere "cleaned" of all interference would be all the better for the performance of sympathetic resonance—acoustic and ritual chants. The cleaner the atmosphere, the greater the change of state, the more ecstatic the feeling of "oneness," or attunement with the Earth.

The mound at Knowth, Ireland, and the covered pyramid at Silbury Hill in Wiltshire in England are classic examples of this layering technique. Perhaps the greatest example of all is the Great Pyramid of Giza, Egypt. It, too, has layers, the inner layer being granite and the outer limestone, which is the main material used in the construction. At this stage, let us witness a more simplistic view, one that to my mind reduces the achievement of the Ancient down to a matter of aesthetics: "The mound at Knowth was decorated not only with quartz and granodiorite but also with blue and white striped siltstone cobbles. These colourful cobbles were collected from near Dundalk and were found near the entrance to the Eastern and Western tombs. It is not clear exactly how they were used but evidently they were chosen for their attractive colour."[10]

The problem with such a comment is that this is about as far as it goes. Did megalithic man really go shopping for attractive colors, or is there more to it? Wilhelm Reich, who earned notoriety in America for his unfashionable and "unscientific" ideas, died in a U.S. penitentiary because of a little thing called orgone energy. Reich maintained that the orgone was the life force, and he constructed orgone accumulators to prove it. Interestingly, these featured layering similar to that found in mounds and pyramids. Whether he truly was mad is a matter of conjecture; what cannot be doubted is the strange properties of his little box. When looking into an orgone accumulator, what can be seen quite clearly is a strange spiral of greenish light. Such a light has never been adequately explained.[11]

Just like Reich's box, pyramids, mounds, cathedrals, churches, and long barrows all have their hidden dark-as-night chambers. Very often witnesses speak of a greenish light—this includes reports from within the Great Pyramid. It was from this interior that Napoleon Bonaparte is known to have emerged visibly shaken. He had expressly commanded that he be left alone there for some time, in imitation of Alexander the Great. Reich's work has not been taken up by mainstream science, but another aspect of ancient knowledge is at the leading edge of scientific application: that of quartz and silicon technology.

At a number of sites, quartz is used alongside granite. At Newgrange the easterly walls are entirely faced in white quartz. Archaeological evidence suggests that in its heyday it may have been covered everywhere with this material. To various of the world's indigenous peoples—for example, the Australian Aborigines—quartz was actually "solidified light," while to some Native North Americans it was "holy ice." Quartz was held to induce or enhance states of consciousness. Rock crystal was particularly useful for this purpose. It was believed that spiritual and emotional forces could, by the use of rock crystal, be transformed. In recalling that when a charge passes through quartz, pressure arises and it begins to vibrate at higher frequencies,* one wonders, did ancient traditions have knowledge that we are only just beginning to recover?

Within the world technologies there is a rapidly growing awareness of the use of crystals to transmute and transform electromagnetic energy. It is estimated that the silicone-based technologies of computer systems will be replaced by the use of a new species of laboratory-produced diamond of great purity (they have no faults) that can easily outdo silicon.[12] If, as the work of the Ancients seems to be telling us, there is a connection between natural forms of radiation, consciousness, and the structure of buildings, the effects of incredible breakthroughs in technology may also serve to transform our ideas of consciousness and, by extension, our understanding of sacred sites and the technology of the spirit. For example, it was a ruby crystal that was fundamental in the breakthrough of the first laser technology. This led to the production of the first holograms, which are essentially energy interference patterns, and holograms have now been adopted by scientists in developing and explaining such theories as memory storage in the brain and holographic models of the universe.

Looking at the evolution of the scientific development given above,

---

*An electrical charge arises within it. This is piezoelectricity.

one can imagine a scenario for the early development of sacred sites and their rituals. The evidence tells us that our ancestors were aware of the sites and, with the limited building technology of the time, they did their very best to enhance them. This awareness may have been derived initially by a particularly sensitive woman or man standing at one of the sites and imbibing the heightened atmosphere, which then led to construction programs aimed at "capturing" this feeling.

As the Ancients became more adept at building, they would be able to enhance the feeling with precision, using, for example, the power of quartz-bearing rock. A continuing quest for finer attunement to Mother Earth and the cosmos would in turn lead to acoustic-based practices that would "bind back the soul to its source" (*religio*).

If this seems improbable, we only have to look at the resonances involved at the sites to see that they are very specific indeed. Newgrange is not the only example of purposely attuned resonance. Incredibly, no matter where in the world we go, sites fall within bands of earth resonance. From Newgrange, Knowth, and Dowth in Ireland to Stonehenge, West Kennett Long Barrow, and Chun Quoit in England and the pyramids of Egypt, Mexico, and elsewhere the resonant frequencies are found.

Also, worldwide, there is an underlying mutuality of form. Where structures are not circular, they are cruciform and, very often, pyramidal in some way. Many of these buildings are oriented toward the midwinter solstice. At Maes Howe in the Orkney Isles, as at Newgrange, the solstice sun sends its golden rays through the entrance passage to illuminate the deep, dark interior of the inner sanctum. Surely, here is a message of power and enlightenment that underpins our own festival of the midwinter season—the celebration of the birth of the light-giving hero, Christ.

PAGAN CHRISTIANITY

Whatever the religion, whatever the mythology, each had its concern with the power of place, and early Christianity shared these concerns, making use, on site, of whatever preceeded it there—and for much the same purpose. It is necessary at this point to generalize, for Christianity has rarely spoken with one voice. Attitudes to Megalithic and Neolithic monuments have varied considerably from use to abuse over the years, uttered by the superseding religion. Robert Graves made the point that "Demons and bogeys are invariably the reduced gods of priests of a superseded religion."[13]

Where there are signs of that from which it has emerged, myth is a good starting point for any historical detective work. So, too, the study of linguistics and the origins of various national terms and phrases, for myth and language are very intimately linked. To use one of my favorite mixed metaphors, it is a virgin field, pregnant with possibility.

Take the cairns, or *si-uns,* of the British Isles, many of which have interiors cruciform in plan. The term *cairn* is *carn* in Gaelic. It means simply "rock pile." The name of the county of Cornwall in England tells us it is a place of rock piles. In Cornish the name is even more explicit: Kernow. What interested me most about the Celtic si-un is that, apart from looking very much like the Hebrew *sion*—with much the same meaning, it brings us, yet again, back to the musical scale, through *si,* the seventh note. In the north of England si-uns were known as laws. Through "God's laws" we are brought back to the idea of Zion and the covenant with the Divine. It's as if the idea of God is implicit within the place and that one had to go to these places to be granted a covenant. Psalm 84:4–7 is supportive of this: "And they pass on from outer wall to inner, and the God of gods shows himself in Zion."

Moses went to a "high place" and returned with the Ten Commandments. Abraham's covenant was signed and sealed by his remarkable composure in the attempted sacrifice of Isaac—at a "high place." Perhaps it is not surprising that in one of the earliest references to God in the Old Testament he is called El Elyon, "God Most High"— "Most High" being both metaphorical and literal, as myth seems to imply.

In Islam, the rock of Sion, over which is built one of the most famous pieces of ecclesiastical architecture, the famous Dome of the Rock, is actually called *es Sakra,* meaning, literally, "the rock," this being one of the few recognized holy rocks in Islam. "Rock" in this sense of being holy means "sacred." Now, Jesus's words to Peter take on new significance: "Upon this rock I will build my church" (Matthew 16:18).

As we have seen, the internal structure of a majority of cairns, or si-uns, is cruciform. At Newgrange the internal structure is not only cruciform but also reflects both physically and acoustically the layout of the greatest of the European cathedrals. For centuries, mainstream religion has looked down upon the pagan. The term *pagan* means "country dweller" and does not at first glance signify anything overly hidden. All it seems to imply perhaps is that *civilization* was mostly an urban affair. *Civilization* comes from the same root as *city*—Latin *civitas,* thus showing that cities and civilization go together. In contrast the country is not civilized, in the sense of *citified*. However, *pagan* means more than mere country dweller: it can also translate as "rural district," "the country," or, significantly, "a landmark fixed in the earth."

Nowadays these landmarks, the ancient sacred sites, are still associated with pagans, and pagans themselves are generally taken to be those who predated Christianity and indulged in nature worship and superstition. The sacred place was a point of contact to the divine no different from Christianity—or, should we say, that in this respect Christianity was no different from the pagan.

The early church fathers knew that these sites were potent and held in veneration by locals. Even though the core of the ancient "science" had already been lost, a tradition remained. Therefore, the church literally built upon this tradition. Churches and cathedrals were erected on sites of power. Megaliths are quite often to be found inside churches; it is as if what had been there before had then been updated—the site, in effect, converted to Christianity. It is for this reason that the rise of the parish church in Britain and Europe was so overwhelming and such an obvious way for Christianity to spread.

Wherever churches were built, the local deities were absorbed only to reemerge later as Christian saints, Saint Brigid being a case in point. Originally a Celtic goddess, her curriculum vitae was altered to make her an Irish nun who founded a convent: all the best girls went to convents. Her name was changed to the more chaste sounding Saint Bride. Moreover, the same Bridget or Bride was, according to legend, midwife to our Lady, Jesus's mother, the Blessed Virgin Mary.

Wherever the saints were, in fact, the superseded gods of an older belief, so, too, were the secrets of the site consumed, to be forgotten. The irony is that these ancient secrets were part and parcel of Jesus's original knowledge—part of the curriculum, the three Rs as one might say: radiation, resonance, and ritual.

Regarding the original knowledge, we have seen how the internal plan of many sites is cruciform in nature. What is not immediately apparent is the true Christian association with the cross. As A. N. Wilson points out, the crucifix did not exist as a Christian symbol until the high Middle Ages, *but the cross did.*

However, it did not represent the death of Jesus—it was a symbol of the triumph of God over death,[14] a symbolism that could be attributed to almost any of the older religions it superseded, for all religions have used the cross. Crucially—and I use this word guardedly for it comes from the Latin *crucis,* "cross," hence the crux of the situation—the Latin term *predates the advent of Christianity* and yet contains the same meaning of God over death.

In an article in a British newspaper in late 1998, the bishop of Lincoln, the Rt. Rev. Robert Hardy, spoke out against the rise of paganism in the midst of Anglican society. The bishop claimed that other churches had "virtually abandoned the countryside." In the same breath he went on: "Large areas are unrecoverable for Methodism. It is a case of the Church of England or paganism."[15]

We look down on paganism. It is a general term for the religions that were superseded by Christianity, and we have learned to associate it with devil worship. The devil, in this context, is widespread throughout not just Britain but the entire globe. Is it really possible that he could spread himself so thinly? The answer lies in words that reflect those of Robert Graves: "The God of one religion becomes the Devil of that which replaces it."[16]

The ancient sites have even had their component parts buried in the attempt to be rid of "Old Nick." At Avebury and other places, former standing stones are still being excavated, many years after they were buried in the ground by the Christian faithful. Perhaps the Avebury site was too vast for a conversion: no church or even cathedral could be built that would dominate the site. However, other sites served the Christian religion well.

In the year 601, Pope Gregory the Great wrote a letter to Saint Augustine of Canterbury. In it he explained:

> We have been giving careful thought to the affairs of the English, and have come to the conclusion that the temples of the idols among the people should on no account be destroyed. The idols are to be destroyed, but the temples themselves are to be aspersed with holy water, altars set up in them and relics deposited here. For if these temples are well built, they must be purified from the worship of demons and dedicated to the service of the true God. In this way, we hope that the people, seeing that the temples are not destroyed, may abandon their error and, flocking more readily to their accustomed resorts, may come to know and adore the true God. And since they have a custom of sacrificing many oxen to demons, let some other solemnity be substituted in its place, such as a day of Dedication or the Festivals of the holy martyrs whose relics are enshrined there. . . . They are no longer to sacrifice beasts to the devil, but they may kill them for food to the praise of God, and give thanks to the giver of all gifts for the plenty they enjoy.[17]

This letter is an astute example of politics in its purest form. The Christianization of megaliths and megalithic places became de rigueur. Saint Gregory, as he was soon to be known, had instituted a carefully crafted policy of conversion, convincing in its authority to this day. For example, every year on the last Sunday in July (called Garland Sunday) up to six thousand people make the ascent of Croagh Patrick in County Mayo, Ireland. Here, one of the Western world's great pilgrimages is made along a path first beaten by Saint Patrick in 441 CE. The pilgrims take with them all manner of things to absorb and also to exalt the holiness of the place.

Quartz crystals, called cure stones, would be carried to the summit, perhaps to be recharged by contact with the sacredness of the site or by the unseen presence of "the man himself." Others would cut strips of cloth and tie them to ash saplings for luck. Woe betide those who did not. At the first station, pilgrims recited seven Our Fathers and seven Hail Marys and then recited the Creed, while walking around the stones seven times, in an ever-decreasing spiral. Crystals, pieces of tree, spirals, and seven again! What is more, the peak of Croagh Patrick is topped by a chapel where before there was a si-un. The entire area is a testament to humanity's megalithic and mystical past. Nearby is a long-stone and, close to that, the old rock of Boheh, upon which is carved a plethora of concentric ring markings. Farther along the way is a site known as Music Hill, where it is said that the fairies dance at night. However, perhaps the most relevant testament to the staying power of older pagan beliefs lies in the story of Sean na Sagart, or Sean of the priests.[18] Sean was a priest killer, responsible for the death of dozens of priests in the eighteenth century, all of whom were tracked down and murdered for the £5 bounty on their heads (issued by the British government).* Eventually, Sean himself was killed by a peddler avenging the death of his brother, who had been one of Sean's victims.

It is Sean's burial that is of interest, as it reveals beliefs that are far older than the eighteenth century. Sean was interred beneath a split ash tree in the graveyard at Ballintubber. He lies facing north so that his soul will never again see the light of the rising sun. Elements of these older beliefs are present in the choice and construction of church sites.

Any suggestion that the men who constructed the old churches and cathedrals could not build straight walls or construct true right angles is manifestly absurd. Why, then, did they deviate from conventional rules, and for what reasons did they decide whether a chancel should be twisted to the North or to the South, and to what degree?

. . . For no obvious reasons many country churchyards are raised several feet above the surrounding land. Indeed, the very position of many churches is inexplicable. Though there appear to be more convenient sites, some are half a mile or more from the villages they serve, and some are built on ground which is quite unsuitable for heavy structures.[19]

---

*Paid out by the incumbent Protestant authorities.

The last point is amply illustrated by the construction of the Norman cathedral (pre-Gothic) at Old Sarum, near Salisbury in Wiltshire, England. Old Sarum is an earthwork of deep antiquity. Until the Middle Ages it was important as an administrative, religious, and military center. It now lies totally ruined and abandoned, a curious site not far from the cathedral city of Salisbury. In fact, it was to Salisbury that most of the inhabitants decamped after a debacle unparalleled in the annals of cathedral building.

In 1070, William the Conqueror held a victory parade at Old Sarum. Having recognized the importance of the site, the Bishop's See was moved from Sherborne and a new cathedral was constructed. Five days after its completion and consecration, the cathedral was blown down in a severe storm. This warning of the inappropriateness of the site passed unheeded: the need to build upon the old earthwork must have been great indeed. Work was begun upon the construction of a new cathedral under the next bishop, Roger the Norman. The cathedral was again completed, but the weather at the particular spot was so severe that, even in the building's deepest recesses, one could not get away from the noise of the wind, which hurled itself against the cathedral walls without abating. Very soon, the same walls began to show signs of severe erosion. What is more, Old Sarum then found itself with a shortage of water. In the early thirteenth century the ecclesiastical authorities gave up: a new Gothic cathedral was built on lower ground at Salisbury, and Old Sarum went into terminal decline. Why the Normans should have persisted with the Old Sarum site is something that had intrigued me for a long time: Why not go to a less exposed site?

The answer is in, and on, the ground. Old Sarum is built upon a fault line, or fissure in the rock deep beneath the earth. Furthermore, it is also aligned to other sacred sites. Even the later site of Salisbury Cathedral was built into this pattern. "From Stonehenge, the spire of Salisbury Cathedral, the successor to Old Sarum, can just be seen in the distance. Precisely in line between the two places is the central citadel of Old Sarum, and when one stands at the center of Old Sarum and looks south beyond Salisbury's spire, one sees the line continue to a corner of the wooded Iron Age earthworks of Clearbury ring."[20]

What we have here is a sacred alignment and, as John Michell makes clear in *Sacred England,* the legend that the building of Salisbury Cathedral was given as an instruction to the bishop in a dream is, in all likelihood, an allegory of a system of divination by which these sites were traditionally chosen.

In the year 1200, Lambert, the parish priest of Andres, put quill to parchment, having borne witness to the work of a master. What follows is of great interest.

> Many oftentimes came together to see these great Earthworks; for such poor people as were not hired labourers forgot their penury in the joy of beholding this work; while the rich, both knights and burgesses and oftentimes priests or monks, came not daily only, but again and again every day, to refresh their bodies and see so marvellous a sight. For who but a man stupefied or deadened by age or cares could have failed to rejoice in the sight of *that master Simon the Dyker, so learned in geometrical work, packing with rod in hand and with all master's dignity, and setting out hither and thither, not so much with that actual rod as with the spiritual rod of his mind, the work which in imagination he had already conceived?*[21]

ONWARD, CHRISTIAN DOWSERS

The above quotation is a clear reference to dowsing, which has for a long time been dismissed as unscientific. However, in the light of a greater understanding of the forces beneath our feet, it is at last being taken seriously as a responsive tool. The tenet of this research shows that water diviners, or dowsers as they are better known, are sensitive to electromagnetism.[22] A little less than two hundred years later and the Rev. Lambert's observation of dowsing in practice would be dismissed as witchcraft and the practice of heresy. In other words, paganism had outlived its usefulness—but how useful it had been.

Guy Underwood, in a posthumously published book, *The Pattern of the Past,* was among the first modern commentators to observe that those who built sacred monuments, be they henges or churches, knew more about geodesy—earth measure—than has been generally assumed. Before the construction of a monument could go ahead, water had to be located. The most natural siting of any place of power was near a spring or well. Water, of course, makes the land fertile. It was also seen as having curative powers in the treatment of illness, and the water at these sites is indeed more potent, both minerally and electromagnetically. The water shortage at Old Sarum must have played a larger role than has been recognized previously. As Underwood points out, the building of a church could only go ahead once a source of water or aquifer had been located. An aquifer is a fault line in the earth's crust, a rock formation in which water has become trapped and is, to a certain extent, recoverable. The water gives off a charge, being a carrier of electromagnetism, and this charge is sensed by the dowser as he walks over the site with the tools of his trade: usually a hazel rod or rods, sometimes copper rods or ordinary steel wire. Underwood sensed that beneath each altar spot he came across there was an "altar spiral." Altar spirals are effects in dowsing caused by terminal blind springs; terminal in that they came to a halt. Blind springs are the esoteric centers of the old religion—and, to some extent, the new.

In the ancient period, the blind spring seems to have formed what I shall call the "G-spot." *G* in this case is for God on the earth—and thus around this spot a monument would be erected. Before churches came along, sites of this nature would be marked by the erection of a phallic stone, demarcating the spot's particularly resonant and fertile nature. The stones were known as betyls, which is derived from the Semitic *beth-el,* or "house of God." The church or kirk—both words derived from the early Greek *kuklos,* meaning "circle"—would be marked out and distinguished as holy, consecrated ground. Gods and heroes are the stuff of such places—a brief consultation of any book on Greek mythology will very quickly reveal these sites as places where the oracular hero was consulted, oracular in the sense that he would speak via an oracle. In myth, the connection to water is often strengthened when the hero himself consults, or is even married to, water nymphs. Perhaps the most famous example is the legend of the god Pan, in which Pan encounters Syrinx, a water nymph. (The legend is wonderfully expressed by

Debussy's music of the same name.) Pan encounters Syrinx one day, bathing in a woodland stream. He is at once smitten. Shocked by his half-man, half-faunlike appearance, Syrinx reacts in panic (note the word derivation) and flees. Pan pursues her all the way from Mount Lycaeum to the river Ladon, whereupon she invokes the gods to aid her escape. The gods, with more than their usual irony, turn her into a water reed. Pan is distraught, for he cannot determine one reed from the next, and, clutching at straws, he cuts a whole sheaf of them. These he transforms into the now famous panpipes. Now, it should be reiterated here that, aside from his appearance, there was another aspect of Pan that made people react in panic: his voice, which was heard as a high-pitched squeal or a low resonant boom, and at the most unexpected of times. In the myths of Pan, he is described as sneaking up on people unawares and terrifying them. Also related to this is the fact that Pan's voice was quite transformed when it was diverted musically via the panpipes; what is of interest is the fact that the panpipes are made of seven component reeds. Seven again!

Pan was later demonized by the Christians. In some cases, he became the devil incarnate. As already mentioned, the name of Tammuz, another pagan god demonized by the Christians, means "child of the seventh." It is no secret that the seven relates to frequency of light, as in seven colors, and more particularly regarding ritual application, the frequency of sound. Music and chanting, we know, were used to heighten the effect of sacred sites at particular times. These times, as any dowser will relate, are connected to the cycles of the moon and sun. This implies a changing charge of electromagnetism and, indeed, there seems to be an alternating charge at various points on any site. At the Rollright stones, these rings of charge spread out from the center of the circle in a series of concentric circles of alternating charge. In fact, as Tom Graves points out, "all areas can also be polarised in relation to others, mainly at ground level."[23] This is a very widely recorded phenomenon that only adds to the complexity of the sacred site. Mineral veins and the movement of water through faults and fractures cause small perturbations on a geophysical level, and these in turn can affect the strength of the magnetic field. It is thought that what makes dowsers more than able to pick up signs of these perturbations is a higher than average sensitivity within the pineal gland.[24]

It is fascinating to see dowsers at work, to see their rods moving inward and outward, as if by magic, as they slowly pace up and down, tracking lines of charge. When we consider the subject of dowsing, it is apparent that Western science has been very reluctant to even consider such degrees of sensitivity within humans. Life, however, came into being and developed within the Earth's geomagnetic field and naturally has a sensitivity to it. The incredible sensitivity of life on Earth to even extremely low levels of magnetism should therefore come as no surprise. Fortunately, due to the early groundbreaking work of scientists like Duane Chadwick, this area of study is now being taken more seriously: "In 1970, Duane Chadwick of Utah State University conducted a series of experiments that fully matched the critics' requirements. He tested 150 people and he carried out control experiments. The results were startling; the dowsers' rod movements showed an apparent link with tiny changes in the intensity of the Earth's magnetic field."[25]

The charges picked up by dowsers at various sites indicate an uncanny sense of knowing on behalf of the early Christian authorities, for whereas the charges at stone circles such as Rollright fluctuate on an hourly basis, "churches and Christian sites in general are different: the altars of those that I've tested are almost invariably positive and stay that way."[26]

Oddly enough, in the same text, Tom Graves goes on to relate an exception to this rule: the many Lady Chapel altars in cathedrals are fixed at negative. Therefore, in cathedrals we have the same alternating positive-negative charge. As Guy Underwood was to note, this displays all the functions of a pattern, radiating outward like a halo in steps of negative-positive charge. As Tom Graves notes, "[church buttresses], particularly at the East ends, for some reason, are more like standing stones, as their charges wander somewhat; and other points within churches that tend to be strongly polarised are fonts and piscinas."[27]

Buttresses . . . more like standing stones? Do we have here a possible further reason why, across Europe, many churches have included standing stones into the fabric of their construction? At the church of

Chapelle des Sept-Saints, Planaret, France, this is precisely the case, as it is at Maplescombe Church in Kent, at Saint Mabyn's church in Cornwall, the Church of Buhel at Latsch, Germany, at Arrichinaga in Spain and a host of other sites.[28] Quite plausibly, the belief may have been, at the inception of these buildings, that the incorporated megaliths would somehow recharge the environment on an occasional basis, for this is indeed what appears to happen. Following the lunar cycle, energy moves up and down the stones in a system of ever-increasing energy release. The lowest points of the cycle occur on the sixth day after the new and full moons. Underwood noted, in *Pattern of the Past,* that this coincides very accurately with the actual structure of the Celtic calendar. According to a bronze tablet of the first century, found at Coligny in France in the nineteenth century, the months started on the sixth day after a new moon.[29]

These incorporated stones would, of course, originally have been used by the Ancients, not superstitiously but scientifically, for evidence is growing of their scientific interests. Some rather simple rock carvings at the Neolithic site at Knowth have been shown to be a representation of the moon. Canadian scientist Phillip Stooke carefully placed a copy of the carvings over an image of the full moon and came up with an exact match. The carving is approximately five thousand years old. Interestingly, the carving seems preoccupied with the major "seas" of the moon rather than an overall outline of the whole body. Dr Stooke observed that "the people who carved this moon map were the first scientists, they knew a great deal about the motion of the moon. They were not primitive at all."[30]

Perhaps one of the most famous of incorporated megaliths still resides at the place to which it gives its name. The Latin verb *vaticanate* means "to prophesy" or "to utter prophetic inclination." In the year 204 BCE a stone was brought via the good offices of King Attalus of Pergammon to Rome. There it was led in triumphal procession to its new home, high on a hill—the Vatican hill, the hill of prophecy. The stone was the holy aniconic—that is to say, "representing without image"—stone of the Great Mother herself, Cybele. Her new temple was exactly where Saint Peter's Basilica stands today. The goddess who this stone represents gave birth to a son, who by tradition is born yearly every December 25. Because of the winter solstice on December 21, the days begin to get longer around this time, light increases, darkness is in retreat—it is a time of renewal. The son, the god Attis, otherwise known as Tammuz, was unsurprisingly worshipped by the Romans as the redeemer.

Cybele's stone has since disappeared, and where it might be to this day nobody knows for sure. It may still be where it was originally placed—though nobody knows where.

THE JOURNEY OF THE HERO

Unfortunately, the many generations who have looked upon the ancient stones and their markings have assumed that our ancestors were ignorant savages, particularly if the markings were simple line drawings such as spirals—which are common at labyrinths—and in caves.

Caverns are wonderful places and were held in the highest esteem in ancient times. How Gothic these places look. They were thought of as entrances to the womb of the Mother herself. If you ever see a river flowing through a cavern you cannot but help notice the spinning vortices the water creates in its journey through darkness. This water was seen by the Ancients as the lifeblood flowing through Mother Earth's veins, in much the same way as blood flows through our own. But those spirals of the dark waters also imitate the passage of energies flowing through the cosmos. It is easy to understand why troglodytes were drawn to these places and left their marks and their paintings. These caverns seem to be telling us, as are the spiral markings of the people who once frequented them, that the forces of nature—from galaxies to the flow of the blood in our veins—move in spirals, as does sound, the voice of the divine.* That these things were understood by our ancestors seems clear enough. Besides the very many spiraling and labyrinthine markings on petroglyphs, the turning of the circle and the spin of the spiral are age-old fundamentals to ritual dance—and also, we have the

---

*Sir James Jeans calls this movement "little whirlwinds" (*Science and Music,* 126).

spirals of pilgrimage. At Nazca, not only were the famous lines walked and processed but so were the many spirals, and for thousands of years pilgrims to the sacred site of Mount Meru in Tibet have approached it in the path of an ever-decreasing spiral.

More recently, since the advent of Islam, pilgrims to Mecca have approached the Ka'aba in the same way, as do pilgrims on the path to Croagh Patrick—seven times around in a slowly decreasing orbit. The ultimate intention is that such should be the focus—that the mind, the body, and the spirit become as one with God. It was for this reason that the labyrinth was an important part of the cathedral interior. Labyrinths were often processed barefoot, as a sort of ritual dance. This is borne out by the testimony of the tradition at Auxerre. However, the most famous labyrinth is at Chartres. Forty feet across, it is to be seen set in the floor of the nave. Apparently, it used to have a bronze plaque at its center with a depiction of Theseus and the Minotaur— Theseus being the hero of the most famous labyrinth myth of all. "For a Christian justification of the use of this ancient Cretan legend we have to go back to Lucca, where a similar labyrinth is incised on a pier of the west front with an inscription speaking of the labyrinth of sin which is so easy to enter."[31]

This is a little difficult to swallow in light of the phenomenal use of pagan imagery at these places, for we have noted other non-Christian images, Pythagoras, Aristotle, and the like, among the traditional Christian figures at Chartres. It seems to me that what is really being expressed is a unity of belief and therefore unity of purpose; in the case of Theseus, belief in the hero and the purpose of the hero.

The Ancients appreciated that human beings are, first and foremost, individuals and that the journey of the hero is an individual act where success is based upon the calling up of great inner resources. However, in contrast, today in the West we live a cosseted and increasingly uniform life. Mass production and mass marketing have changed our lifestyles and attitudes, and not necessarily for the better. Modern medicine works to generalized measures of human tolerance, mass inoculations, and summary dispensation of antibiotics to which we are becoming increasingly resistant. Little room for heroics here, even with the common cold. In this and other spheres, the much-vaunted rights and strengths of the individual appear to have been overridden by scientific uniformity. It is almost as if, in failing to conquer himself, his own weaknesses and feelings, man (and I use the masculine purposefully, for here we are discussing male-dominated actions) opted instead to conquer the environment, with the disastrous consequences of pollution and global warming that are now upon us.

This viewpoint that man is somehow separate from nature and can therefore conquer with impunity still dominates uniform scientific and corporate practice, where, until lately, scientists have been obsessed with the reproduction of key findings *in the laboratory*. Such attitudes have prevented science in general from recognizing the extraordinary power of the sacred site and its effect upon people as individuals. For when it comes down to the power of place, it is the entire planet that is the laboratory, and man can no longer be a detached observer. Thankfully, the development of acutely sensitive instrumentation, and the work of a relatively few, dedicated mold-breakers, is waking us up to our connectivity to everything. We have been caught in a labyrinth of our own making, a labyrinth of ignorance, and at last, we are becoming alert to it, we are recognizing our responsibilities. And by this very act of recognition, beginning to absolve ourselves of some of the more criminal excesses performed in the name of science—and almost always, for great, though short-term, profit.

The labyrinth at Chartres takes us from the outer to the center, not in a direct spiral but instead in a complex series of arcs that quite quickly bring us tantalizingly close to the center, but then take us away, back toward the perimeter—the path seems to reflect the "two steps forward, one step back" of life. Moving away from the center without instant fulfillment, one learns to let go of the goal. Indeed, as we let go, the procession becomes all important and in the here and now, the atmosphere of the sacred site is allowed to do its work.

Whenever I have walked the labyrinth at Chartres,* I have always felt, toward the end, in a particularly passive, open, state of mind.

---

*Unfortunately, this is not so easy of late, as seating now usually covers the labyrinth.

From a spiritual point of view this serves as an excellent way to focus the whole sense of being upon the question of God. The placement of Theseus and the Minotaur—the famous monster that was half bull and half man—at the center indicates that the pilgrim was undertaking the journey of the hero, in this case Theseus, to the center of the labyrinth, overcoming the monster inside. Theseus's ingenuity overcame this problem by the use of a thread supplied by Ariadne. This gave him the ability to retrace his steps; he retained his composure, and the Minotaur was easily overcome. Furthermore, Theseus was a hero; his name means "he who lays down." I was struck by the similarity of the name Theseus to that of Jesus, and the comparison is strengthened by its meaning. Jesus is most famous for having "laid down" his life.

My experiences have led me to wonder, is this really to do with the loss of self as one gains perfect peace at the still center of being and reality? Is this where we realize God? Meanings within meanings. The enigma of Chartres is a revelation. The labyrinth of Chartres was called the Jerusalem. The heavenly Jerusalem is called the City of Revelation, and it seems that each of the European cathedrals has its own particular "Jerusalem." At Aachen Cathedral in Germany, the Jerusalem is to be found in the form of an orb suspended from the center of the octagonal dome, which itself surmounts an octagon. At noon every June 21, the summer solstice, the sun's rays enter the dome and strike the orb (which hangs on a chain above the Barbarossa chandelier). In contrast, the rays of the midwinter sun alight at noon upon a mosaic, situated above the north window. The mosaic depicts the Chi-Rho symbol of Christ situated between the letters alpha and omega—the beginning and the end. However, the mysteries of Aachen cathedral go deeper.

Fig. 3.1. The Chi-Rho

About twenty years ago, Hermann Weisweiler discovered a reflection of Britain hidden within the construction of Aachen Cathedral. He noted that both Aachen and Stonehenge were very near the line of 51° north, Stonehenge slightly above and Aachen slightly below. Looking for further connections between the two, he soon discovered that the inner trilithon horseshoe pattern of the sarsen stones at Stonehenge was of the same dimensions as Aachen's Palace Chapel ground plan, in fact an extraordinarily close fit. This correlation is further underlined by additional comparisons of the environment of the cathedral to that of Stonehenge. There is an alignment that runs northeast at Aachen, running through some very significant sites, all of which are places of power: wells, springs, churches, and various Megalithic tumuli. To me, these things again infer something of the utmost importance: Are we being told by the masons that Christianity is much older than is generally believed?

But there is more, in this case a phenomenon that manifests physically, one normally only related to Egyptian sites—that of mummification.

PYRAMID POWER

The idea of mummification extends over a far wider area than is generally assumed. Mummification zones are to be found at various sites worldwide, including parish churches and cathedrals in Britain and Europe. Mummification occurs in organic material that, instead of decomposing, desiccates. Desiccation is preservation by drying. In *Earth Currents: Causative Factor of Cancer and Other Diseases,* Gustav Freiherr von Pohl quotes Edward Bush from the *British Society of Dowsers Journal* 1994–1995: "Bodies buried above underground streams of Geopathic Zones do not rot in the normal fashion but become mummified."[32]

Geopathic zones are areas of natural electromagnetic vibration distorted by the action of subterranean running water. All of the sacred sites so far mentioned have an association with subterranean running streams, the pyramids of Giza included. The bodies found at a number of sites in a mummified condition are exceedingly well preserved and have a very leathery look about them. These mummification zones can be partially explained by looking at the construction elements of the monuments. These same elements are to be found as the fundament of almost all the significant sacred sites of the world, and, where they are not found, there are variations that move toward the same effect.

The phenomenon occurs at Neolithic sites such as the long barrow at West Kennett, near Avebury, and Winterbourne Stoke, near Stonehenge, as well as at the sepulcher of Kerlescan, the tumulus of Mustair, and the Dolmen of Mane Groch, all at Carnac, Brittany.[33] The mummification zones in the Brittany mounds are the size of the whole mound area. These zones are no accident or coincidence of construction. They are there by design in Christian churches and cathedrals, too. At the ruins of Glastonbury Abbey, this zone may be witnessed as the burial ground situated between the Lady Chapel and the Abbot's house. Another such zone exists at Westminster Abbey, burial place of the English kings. It stretches over the King Henry VIII chapel and Saint Edward's chapel. Significantly, all the royal tombs are within the zone.

When workmen of the Caliph Abdullah al Mamun (ca. 820 CE) broke into the Great Pyramid they were to discover nothing but empty chambers constructed in a supremely skilled yet mysterious fashion. Furthermore, these walls were described by subsequent explorers "to have been coated with a layer of salt sometimes as much as half an inch thick."[34] Salt, as we know, aids the process of desiccation, but it is not only to be found in such dramatic form at the Great Pyramid: it occurs, in the same form, inside the crypts of many of Europe's churches and cathedrals.

The site upon which the three pyramids of Giza stand is perhaps the most powerful geodetic center on Earth. It is a site of great potency, underscored by the existence of watercourses that radiate from the Nile. This, along with local seismic activity, makes Giza exceptionally active with electromagnetic emissions. This activity has then been enhanced by the construction of the most fabulous monuments on Earth. The overall effect is one of energy enhancement, both inside and outside the buildings. Indeed, the Pyramid Texts tell us that the Nile, aside from feeding crops, was being used to enhance the spirit and the health of humankind.

In the 1950s a Frenchman, Monsieur Bovis, visited Giza and while in the Great Pyramid could not help but notice that the more organic rubbish, left there by untidy tourists, had a tendency to dry out rather than rot. He also noticed the dead bodies of bats, rats, and cats, all of which were mummified. Returning to France, Bovis constructed a small wooden pyramid to the scale of Khufu's pyramid in Egypt: he aligned it north–south, took his final ingredient—one dead cat—and then positioned it in the pyramid, approximately one-third of the way up. He left the body there and went back three days later. What he discovered has continued to intrigue scientists the world over. Monsieur Bovis's dead cat was mummified. Bovis went on to try a whole catalog of organic comestibles from sheep's brains to other inglorious items; the effect was always the same.

Chasing quickly upon the heels of this new idea, a Czech radio engineer, Karl Drbal, took out a patent, no. 91304, and proceeded to manufacture cardboard Cheops Razor Blade Sharpeners.* These devices utilize the same proportions as those found in the Great Pyramid—the ratio of base-to-sides, 15.7:14.94 cm—and, famously, they work. "My record so far with Wilkinson Sword blades is four months. I have a feeling that the manufacturers are not going to like this idea," writes Lyall Watson. He points out in the same text: "The pyramid shape itself is very much like that of a crystal of magnetite, so perhaps it builds up a magnetic field."[35] Indeed, the electromagnetic field in and around the pyramids of Giza is strong enough to promote crystal growth, hence the quantities of salt inside the monuments. It is crystal growth that sharpens razor blades, with the help of Mr. Drbal's device. As regards the effects of pyramidal geometry, it has been explained that "structures having these geometrical proportions have been described as 'generators of energy.' Any electromagnetic rationale behind such propositions must derive from similarities with the geometries of acoustic and electrical resonators."[36]

---

*Cheops is the Greek for "Khufu." Both are alternative titles for the Great Pyramid.

Interestingly, the polarized light that comes from the moon has an undesirable effect on razorblade sharpeners. This is because polarized light vibrates in one direction only. For crystal growth to take effect it has to vibrate in a variety of directions.

> Crystals are found in only seven symmetrical forms. In decreasing symmetry, these are called cubic, tetragonal, hexagonal, rhombohedral, orthorhombic, monoclinic, and triclinic. There are no more than seven crystalline forms because crystals consist of units of atoms. These units, all of the same shape and size, must pack together without leaving gaps between them. Five-sided units and all other shapes of six or more sides cannot be interlocked without leaving gaps that are unsatisfactory according to the laws of crystal structure.[37]

There is no polarized light deep inside a pyramid, but there is plenty of electromagnetism, and so the crystal growth is made to come "alive" because it is able to resonate within a given shape into a particular form: salt. (In Britain many secret sites are aligned upon what were, until recently, called salt lines.) The various shapes—spheres, semispheres, pyramids, and squares—act differently as resonators of the various energies around us. In human terms, the sphere and the pyramid forms are particularly good. Architects around the world are realizing this. "Architects in Saskatchewan, Canada, have created trapezoidal rooms and unusual corridors in a hospital for schizophrenics and found that the new environment was beneficial for them."[38]

Further to this, in late 1999, I heard an article on a BBC radio program that described the activities of a Russian man who is building a fiberglass pyramid for use over quite a wide area. His stated aim is to change the energies of the Earth. A crackpot? Maybe, but with the support of the Russian army behind him I would not dismiss the story too lightly, for the army believes that he is doing precisely what he claims.

Up until quite recently, the popular conception was that pyramids were confined to Central America and Egypt. However, during the 1990s this limited view was transformed. It is now known that pyramids or pyramid remains are to be found worldwide. There are some fantastic examples in China, not far from Xian—125 in all. In New Zealand, pyramid remains of impressive size have been uncovered. The local Maoris, principally the Ngapuhi tribe, refused to go near them. They have declared them to be *tapu*—that is to say, sacred—and thus forbidden. These pyramids are like the Central American step pyramids, though a little smaller and with a large flattened top. Remains of ancient pyramidal structures have also been found in Peru, Brazil, and southern Africa, as well as off the West African coast, on the island of Tenerife, where the great anthropologist Thor Heyerdahl has been responsible for the restoration of six ancient step pyramids. The remains of similar monuments have also been located in Italy and Turkey.

Elsewhere, ancient mound builders have used a similar approach to proportion and construction. A British example is Merlin's mound, in the grounds of Marlborough School, Wiltshire. It is said to be the gateway to King Arthur's tomb. Again, in Wiltshire, a much larger mound, Silbury Hill, has a huge circular terraced structure beneath its earth covering. A very interesting article in the British antiquarian magazine *3rd Stone* in the summer of 1999 pointed out some interesting similarities between it and various stupas of the Indian subcontinent.

Silbury is much like some of the Buddhist stupas. R. W. Morrell points out that, although Buddhism is much too young for any influence to have been extended Silbury's way (ca. 600 BCE for Buddhism, as compared with ca. 2750 BCE for Silbury) the cult of stupa mound building is inherited from a much older belief system. He further points out that at Silbury a large piece of wood was excavated and came to be known as the "Druid Oak." Likewise, at the stupa mounds at Lauriya-Nandangarh stumps of wood were found, pillars that bore all the signs of having been placed in these positions.[39]

In Bhutan there are stupas with eyes painted or carved on the apex, in a symbolism that echoes the Masonic "eyed pyramid" to be seen famously on any U.S. one dollar bill. Furthermore, in association with local belief, there is an annual display of the secret curtain that covers the holy of holies, a rite that echoes the story of the youthful Virgin Mary in the nativity story. An involvement in it is apparently an excellent way of cleansing away sin. Pyramidal shapes and their proportions are worldwide phenomena, and perhaps unsurprisingly they are also to be found within the tradition of church and cathedral building in Europe. They are proportions that resonate favorably with us. This goes beyond artistic taste and into electromagnetic interaction. Furthermore, as I have established, the earth below as well as the building materials give off favorable vibrations. We are invited to resonate with them, for sound, in the form of chant and hymnal, is prescribed at sites the world over as an important part of ritual. I should add that what is not generally known is that the Egyptian pyramids had their own choirs.[40]

The King's Chamber in the Great Pyramid is known for its incredible acoustics and, needless to say, the builders have made sure that the radon count is high. "Radiation counts taken of the air . . . were in fact higher than on the surface of the granite walls, presumably as a consequence of radon being emitted from all sides."[41]

The effect of standing inside the King's Chamber is a good example of the power of this particular material. Much of the emotional and spiritual heightening felt by many there is likely to be in response to the high radon content. As the poet and painter William Blake (1757–1827) once said: "Every rock is deluged with deity."[42]

I was beginning to understand entirely what he meant.

THE POWER TO UPLIFT

Returning to the architecture of the medieval masons, churches and, particularly, Gothic cathedrals have an aspect not usually shared by the more ancient buildings—that of space and, with it, light. This is why I personally believe that sacred architecture reached the height of power and of glory in the form of the Gothic cathedrals: the space, and what those master masons did with it!

The power of the sun streams into the space through stained glass; the atmosphere is optimized—this time by radiations of light.

There is one more particular aspect of churches and cathedrals to which we should attend. The masons had set up the atmosphere through the choice of site and materials, and by filling the space with light.

However, that space was created to harness the acoustic science of the Ancients. We have seen how it worked in those far-off days, but those older structures did not usually have much space. In fact, acoustic technology can be used to great effect in small areas, as we shall see later when we return to the King's Chamber; the medieval buildings, however, were made for larger groups of people. Not only are these buildings wide, they are also tall. Added to that, though, are the soaring structures to which we pay so little regard in the modern world: bell towers.

The standard explanation is that bells in towers are essential to call the populace to prayer. I wondered: Why build such mighty towers, and mighty expensive towers at that, at great cost to life and limb, when an external, simple structure, perhaps even made out of wood, would have sufficed? A quick survey of an ecclesiastical interior confirms that the calling to prayer was only a part of the answer. A bell tower helps the fabric of the building to become a giant sound box. You will remember that this is how Newgrange was designed, only on the horizontal. Looking around at the décor and seeing how many mysterious patterns and shapes carved by the medieval masons resembled sound waves, I felt that the entire cathedral-building program had been a giant exercise in the practice of cymatics—the study of wave forms—and that this motive could arguably be extended outward to the fabric of sacred architecture worldwide. All of the sites were directed to the glory of, in one form or another, God, the Great Mother, and their son, the hero.

The faithful came and would need little encouragement to chant and sing: for practiced aright, the effect would be most uplifting. Congregations still have words to sing in praise of God, but there is something amiss, something has changed: the sounds themselves were very important and part of an underlying science, which has been long forgotten. In fact, surveying the evidence now emerging regarding sacred sites, we seem to have forgotten a great deal.

MUSIC IN SUMMARY

At all of these sites so far discussed, people quite often see strange lights and hear otherworldly music. The viewer has a tendency to see not so much a form but a large globe of golden light, such as the one at Knock in Ireland in 1879, which appeared to the accompaniment of what was described as "a heavenly choir." These appear to be a common feature and tell us of induced brain wave states.

Significantly, in mythic and other traditions, the otherworldly is associated with music. Pan played his pipes; Orpheus, the harp; there were sirens of Greek myth; fairies who use music as a means to charm; and the Pied Piper, who uses musical magic to lead children into a new world within the rock. There are so many tales bearing the hallmark of music, the gift of the Muse. Music is frequency in harmonic form.

The eight notes of our musical scale are familiar to all. Each sound is different because of pitch, the number of vibrations per second, and these in turn are related to each other via a series of fractions or intervals. One might think that the interval between notes is spaced perfectly, but it isn't. The fractions are one and an eighth, one and two-eighths, and so on. It is these fractions that are important in that the music that these people have heard at the sites seems to be an expression of these fractions.

Gerald Hawkins, author of *Stonehenge Decoded* and a number of other groundbreaking works, is a respected astronomer and understands the importance of Euclidean geometry both in terms of astronomy and other disciplines. He realized that fractions of the musical diatonic scale were appearing in the makeup of these visions.[43] Hawkins also realized that we are dealing with something intelligent. Furthermore, new evidence has come to light that we are dealing with a phenomenon of infinite harmonics, of vibrations within vibrations, of frequencies within frequencies. The noise found at these sites is, in fact, rhythmic vibration, which to all intents and purposes is birdsong speeded up. What is extraordinary is that birdsong is symphonic.

The work of David Hindley, a Cambridge musician, has shown that the songs of birds relate precisely to the rules and principles of human musical composition. Birdsong is a highly compressed sequence of notes. A forty-eight-second song of a skylark ran to nearly thirteen minutes when transcribed to human speed by Hindley. Also, it became apparent that the composing skills of Mozart and Beethoven had found a sparring partner.

Following on from the earlier revelations of an ancient acoustic science, this brief journey into the strange world of visions at ancient and sacred sites should help us to understand that sound, electrical, and magnetic forces can be applied together with intelligence to give rise to form—not only in the mind, where the tendency is to mentally clothe the subtle images, but also physically. Here we approach the ancient dictum that, in truth, name (sound) and form is the realm of the hero.

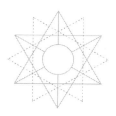

4

Enigma at the Heart of Mystery

God and I both knew what it meant once: now God alone knows.

FRIEDRICH KLOPSTOCK, AS QUOTED IN
THE MAN OF GENIUS BY CESARE LOMBROSO

When words lose their meaning, people lose their freedom.
CONFUCIUS 551–479 BCE

Heroism is a much-used term in Western cultural iconography. The idea of someone who can, often by virtue of necessity and in almost impossible circumstances, exceed their own humanity to become a superman or superwoman is a familiar one, although the concept has come to be seen in largely cinematic terms, it is frequently debased and overused in the media.

In Eastern cultures, the message is deeper and clearer. The hero is more than metaphor, he is more than just a cultural icon, he is hope, he is justice, he is the rescuer of humanity from the ravages of time, decay, and inevitable defeat. He is a poetic image that appeals directly to the spirit as a model of initiation, transition, and inspiration. There may seem little difference between East and West in these aspirations, but there is one subtle difference that stands out, and it is a modern one: in the West we seem to have allowed ourselves only one universally recognized superhero—Jesus. By contrast, in the East, there are many such heroes and this has been the case for many thousands of years. This is a difference that can be overcome in the light of the hero's true identity.

Sir Isaac Newton held the view that all nations originally subscribed to one universal religion. Because of what we have seen that links the hero to secret sites on a worldwide scale, this may not be far from the truth. Furthermore, there is a great sense in every one of the world's religions and mythologies that the hero is born of the earth, and yet, while tied to it, the hero's journey is one of the breaking of bonds in the name of freedom. The great mythologist Joseph Campbell believed that human beings are the consciousness of the earth, and it is compassion for the earth that will call forth from the depths of our shared humanity a new "mythology of this unified earth as of one harmonious being."

It is in this way that the heroic cycle plays its role within the human condition—by aspiring to go beyond ourselves we are linking into something profound, gaining a momentary glimpse into the nature of everything, for the hero is both mortal and divine. Though doomed to die a terrible death, in death the hero will set the legions of hell atrembling and afeared, and in doing so will inspire humankind beyond the deathly state and upon the road to immortality and paradise. More often than not he is the son of a goddess and a not-so-divine father; although in Western mythologies there is the tendency to make him a son of God, bestowed upon a virgin mother.

The tone of the myth is one of culmination; in any body of religious tradition the creation is described and then various series of events unfold beyond that. These events describe more or less the affairs of the gods or the angels and the wars of mankind. Very common to all of these traditions is the intervention of the divine. It is with the story of the hero that such intervention comes to a head and an air of both inevitability and anticipation sets in; it is precisely this sense of anticipation that defines the hero myth and marks it out as different from anything else.

This is a distinction that I have noticed across the wide spectrum of the world's mythological traditions, but further, too, into the heartland of Western culture itself, in its histories, its songs, its paintings, even in the way that sports journalists comment about football matches with anticipation for the game quite often centering around the deeds of one particular player. If a certain star player succeeds in scoring a goal, great emphasis is placed upon words such as *glory, heaven, hero, heroism, heroic,* as if the player concerned is more than human. These are terms that in the common interpretation go beyond mere metaphor and take on a sheen of brighter, almost divine, splendor.

However, the glory of the modern hero is an ephemeral one: the moment is gone, so, too, is the hero. For the mythic hero, though, that glorious moment is ever present in that it transforms all of him and he experiences apotheosis: he becomes one with God. In the modern world that mythic moment is associated with yesterday, it has become almost a sentiment, and yet many still wish for such heroes, hoping that some great soul may fall, like a bright bolt, earthbound from the heavens. Myth encourages this. For example, at the end of the Arthurian cycle, the mortally wounded king is taken away to the Isle of Avalon by three mysterious queens, to be healed in time to come again, "the once and future king." And so the anticipation is kindled.

Throughout the latter half of the twentieth century, as the millennium became paramount in Western consciousness, some religious groups became more and more fervent in their anticipation of Jesus's second coming, and they remain forever hopeful. The first, and perhaps the last, resort of these believers is prophecy. However, prophecy is rarely a straightforward matter. In early Greek history, prophecy was divine advice from an oracle. Often the oracles were prophetesses who, when consulted, would speak as the particular divinity concerned. The fundamental problem with such prophecy was that it was not always as clear as it might have seemed; for the oracles often spoke in riddles, so it was all a simple matter of decoding the riddle in a fit and proper fashion in order to get at the kernel of wisdom that it might contain.

A good example is to be found in the legend of Croesus, the last king of Lydia, in what is now modern Turkey (ca. 560 to 546 BCE). Croesus's empire was prosperous and strong, and he had ambitions to expand his territory farther. With these ideas in mind he put in place preparations to campaign against the might of Persia. As a note of caution, he consulted the Delphic oracle, who famously replied that "if Croesus attacked the Persians, he would destroy a great empire."[1]

The oracle went on to advise, in all fairness, that the Lydian king should ally himself with the most powerful of the Greek states. This should have brought an air of foreboding into Croesus's mind, but, in the best traditions of political spin-doctoring, he was relayed the advice in optimistic tones and was thus delighted with it. Needless to say, the oracle proved right. Croesus's vain expedition destroyed his own great empire. Correct interpretation is all, and without it, we are on a road to ignorance.

The story of the hero is fascinating and has much to tell us. It begins as all quests begin: a need arises. It is never stated but is implicit within the story. In the Gospel of Matthew, the coming of Jesus is anticipated by lengthy genealogies, whereas in Mark, Luke, and John it is through the mouthpiece of John the Baptist that we learn of the needed hero's imminent arrival. In other texts, ranging from the Greek and Egyptian myths to the Sanskrit Puranas, the land is ruled by a tyrant, a wicked king whose ruthlessness knows no bounds; the air is thick with oppression, and darkness reigns—in weather and in mood. The stage is set for miracles and wonders, including, of course, the birth of the miraculous child.

NATIVITY

The father god, Vishnu in the Hindu myths, Zeus in the Greek legends, Jupiter in the Roman, and Yahweh in the early Christian tradition, miraculously inseminates the prospective mother, a virgin, in a process famously known within Catholicism as the Immaculate Conception. In some versions of the myth, the mother is a goddess, and, after delivery of the child, she renews her virginity by bathing in a secret pool or river. In the myth of Perseus, Zeus appears to Danae in a shower of gold. In a legend of Buddha, his mother, Maya, conceived him after having had a dream in which she witnesses Buddha descending from heaven and entering her womb in the form of a white elephant.

The imminent birth of the hero is proclaimed by the heralds—a title derived from the word *hero*—much to the displeasure of the reigning king, who can only see it as a threat to his kingship. Little does the king realize, however, how close to home the threat lies. It is usually a daughter or son who has given rise to the birth, and there is a flavor of what is to come in the announcement of the forthcoming child as "a king in waiting." The earthly, nondivine father is soon disposed of as the forces of darkness gather, and the mother of the child is forced into exile. The exile is brought about because of a decree banning the rearing of male children. This is a motif familiar to the stories of Sargon, King of Akkad (ca. 2370 BCE), Alexander the Great, and Augustus, Emperor of Rome.[2] In the myth of Perseus, Danae and her babe-in-arms are cast adrift in a box, whereas in the Old Testament story of Moses it is the child alone who is cast adrift. As we can see, there are variations upon a theme.

Amid great wonders a bright star shines forth, and angels herald the miraculous child. Three kings come to visit and pay homage to the child, and, in some versions, three shepherds are told by an angel of the wondrous birth, as time stands still. In the mysteries of Adonis and of Osiris, a star "of salvation" dawns in the east.[3] In many traditions this star is actually the planet Venus, known in ancient Egypt as a representation of the goddess Isis. The stories, of course, are full of metaphor and symbolism.

In the early Christian period, there was already a well-entrenched tradition of teaching the secret mysteries through the use of drama. This is a very ancient practice, one that was common to the Orphic mysteries and the Eleusinian mystery rites, as well as to Christianity. Within them we get a good idea as to the use of symbolism, particularly that used in association with drama.

> *The third scene, for instance, opens in the Bethlehem*
> *stable on a darkened stage.*
> *The Cock (crowing): Christ is born!*
> *The Bull (lowing): Where?*
> *The Ass (braying): In Bethlehem.*[4]

To an untrained, unfamiliar eye, two thousand years later, the above scene looks straightforward enough, if a little simplistic. The three animals concerned are the same three animals to be found near the manger of the holy child. But they are sacred animals, and it is what they represent that is important. The cock announces the dawn, the coming of light after the darkness. It is sacred to Hermes, conductor of souls. This is interesting, for the hours before and during dawn are the hours when the frail and the elderly are most likely to pass away.* The cock crowing three times is an omen of death. The bull is a symbol of the coming holy child, who is the bull, or *Baal,* of his father. In Canaanite-Phoenician mythology the son of the father god, El, is called Baal and is symbolized by a bull, hence the word.

The ass or onager represents all that is wild and disorderly and is symbolic of the Egyptian god Set, mistakenly called the god of evil or the representation of evil by many writers. Set personified ignorance and all that was uncultured. He was man in need of a makeover. In some versions of the hero myth, the ass is ridden by the hero on the road to his coronation, a sign that he has tamed his wild self, that it is under control. Jesus rode an ass into Jerusalem on Palm Sunday; Lucius Apuleius is transformed into one in *The Golden Ass.* All of these animals were used as a means of relaying to initiates the trials and tribulations to come. This is the reason that they appear in the story of the nativity and in nativity plays the world over.

The hero is brought forth in a grotto. In the Gospels or in early Christian nativity traditions, Jesus is either born in a manger or a cave. The time of the year is important, as the hero represents light emerging from darkness. In various traditions the dates are close: ranging from December 21 to January 7. Significantly, the day celebrated as the birth of Osiris-Dionysus, January 6, is also the likely date, though heavily disputed in early Christian circles, of Jesus's birth. The dispute centered on whether the birthday was December 25 or January 6. December 25 was

---

*According to Richard Knight in *A Discourse on the Worship of Priapus* there is, hidden away in the Vatican, a bronze image of a cock with the head of a penis. Appended to this Graeco-Roman statue are the original words: "the Saviour of the World."

also the birthday of the god Mithras, who also represented light, truth, and justice. An epithet of Mithras was Sol Invictus, the invincible son.

In Britain in the year 1752, people did not take kindly to the introduction of the Gregorian calendar and stubbornly clung to the old date of Christmas—January 6. This is now the feast day of the Epiphany, when Jesus manifested himself to the three wise men of the east.

The birth of the child hails the dawn of a new age. With the advent of Christianity, the sun began to rise at the spring equinox in the sign of Pisces the fish, whereas some two thousand years before, it began to rise in the constellation of Taurus the bull. That this is beyond supposition is borne out by the reliance upon myth, during these respective eras, of the signs of both fish and bull.* The symbol of the early Christian community was that of the fish, whose geometry of two interlocking circles is also known as the *vesica piscis*: ✕◯.

In Greek, the word for fish is *icthus* and was used as an acronym of Jesus Christ, Son of God, Savior. Icthus was also the Greek name for Adonis.[5] Interestingly, as the sign of Pisces formed the background to the rising vernal sun, its opposite sign was the sign of the virgin—Virgo. According to many commentators, ancient and modern, this provides the reason why many in the pagan world expected the coming savior to be born of a virgin. There is, however, more to the role of the constellation of the Virgo that will arise further on.

It is at the point of the hero's birth that the old king, the tyrant, has a portent whereupon he issues his decree: both mother and child go into exile. In the story of Isis and Osiris, Osiris has at this point been killed and Isis, in order to protect her newborn son, Horus, from the ravages of the jealous Set, decamps to the Nile delta, where she hides in the papyrus reeds, away from the prying eyes of Set and his spies.

Again, there are variations upon a theme. In the Gospels, the family of Jesus cannot return to Palestine until Herod is dead. In other texts, the family must wait for the child to grow older, stronger, and vengeful. To facilitate the child's waxing strength and his mission of revenge, he

---

*During the age of Aries, preceding that of Taurus, ca. 3000–2000 BCE, it was the ram that was the favored symbol.

is educated either by shepherds, whereupon he becomes a great hunter with a bow and arrow, as in the cases of Orion and Apollo; by centaurs, with much the same result, as in the case of Achilles; or by wise and learned men, as in the case of Alexander the Great, who was tutored by Aristotle. Similarly, it seems from some of the noncanonical Gospels that, as is hinted in the Gospel of Luke, Jesus spends a part of his time being tutored by the doctors of the law.

The divine child grows and matures to full adulthood. There is surprisingly little within the mythologies that covers the intervening gap between childhood and adulthood. Not surprisingly, the Gospels are no exception to this rule—and a considerable industry has built up surrounding Jesus and the "missing years." Some say he was in Egypt for the duration, others that he was in India, and yet more that he journeyed to Britain. With the passage of time the hour of the hero comes, for the old king must pass away and a new king must take his place.

At the due moment, almost without exception, from out of foreign quarters, the hero returns to his homeland, the land of his birth, as a stranger in a strange land. The old king dies or is killed, and his place is taken by the very child he has tried so desperately to be rid of. The hero undergoes the rite of anointing, and this necessitates, at some stage in the proceedings, a full immersion in water or sometimes fire. In the myth of Achilles, he is rendered invulnerable except in the heel, the place where his mother, Thetis, held him—the place untouched by the immortal fire.

Baptism is more familiar to us as a water-based activity. As Joseph Campbell observes: "The rite of baptism was an ancient rite coming down from the old Sumerian temple city Eridu, of the water god Ea, 'God of the House of Water.' In the Hellenistic period, Ea was Oannes, which is in Greek Ioannes, Latin Johannes, Hebrew Yohanan, English John."[6]

The Buddha bathes in the waters of a stream before going on to sit under the tree of initiation. Like fire, water plays the role of purifier. It is in the waters that the goddess bathes to renew her virginity. In the legend of Actaeon, as so brilliantly portrayed by the artist Titian, the unfortunate hero spies the goddess Artemis bathing in a stream—and stays to watch. What follows is poetic justice. So that he should not be able to boast of his adventure, Artemis transforms him into a stag, whereupon he is torn to shreds by his own hunting dogs.

At the Eleusinian mysteries, initiates ritually purified themselves in the sea.[7] The narrator of *The Golden Ass*, Lucius Apuleius, also underwent a purifying bath and was anointed by sprinkling.[8] It is after the baptism, wherein the holy child is an anointed savior and becomes "God made man," that the most miraculous period is entered into.

MIRACLES

As with Mithras, Adonis, Osiris, Dionysus, and a host of other heroes, Jesus gathers about him twelve disciples. These are generally assumed to be symbolic of the twelve tribes of Israel, which, as Sir William Drummond points out in *Oedipus Judaicus,* are in turn symbolic of the twelve signs of the zodiac. As Timothy Freke and Peter Gandy observe: "The zodiac was an extremely important symbol in the pagan world as Osiris-Dionysus is symbolically represented as the still spiritual centre of the turning wheel of change represented by the twelve signs. As Mithras, Dionysus, Aion, and Helios he is often depicted at the center of the circling zodiac. During the initiation ceremony in the Mysteries of Mithras, twelve disciples surrounded the godman, just as the twelve disciples surrounded Jesus."[9] Quetzalcoatl, the Central and South American savior god, had twelve disciples or companions, too.

Surrounded by his disciples, the hero begins to manifest the superhuman aspect of his character. Being a savior "god-man," it is only reasonable that the most dramatic of the hero's powers should be his aptitude for healing. Miracles of healing were attributed to Apollo, Dionysus, and Hermes as well as to other, non-European-based hero-gods, from the South and Central American saviors Viracocha and Quetzalcoatl to Krishna in the East. Perhaps the most famous non-Christian healer was the first-century Apollonius of Tyana. Apollonius was a contemporary of Jesus, his biographer Philostratus the Elder placed him during the years 3 BCE to 97 CE. Apollonius was a neo-Pythagorean philosopher, and both Pythagoras and Apollonius were associated with the performance of miracles. So impressive are his feats of miracle working, that he has been described as a plausible "pagan Christ."[10] Apollonius shares other, similar traits. "There is, however, no reason for us to suspect an invention, save as regards the details of the biography recast by Philostratus in the third century. It is likely enough that he was a devout Pythagorean, a student of medicine and astrology, a universalist in his creed, and a believer in immortality. He may conceivably have travelled in India, though there are no details available."[11]

Apollonius was a miraculous seer and is mentioned as such by the early Christian writer Origen. In an even closer echo of the gospel narrative, Apollonius is said to have brought back to life the daughter of a Roman consul. Beyond this he is also credited with the casting out of evil spirits and the feeding of a multitude in much the same way that Jesus, from a few loaves and fishes, fed the five thousand.[12] Another shared miracle is the feat of turning water into wine. This was performed by Dionysus at his marriage to Ariadne. Dionysus was the Greek god of wine, called Bacchus in Rome, and celebrated for the famous Bacchanalia. The feat of raising people from the dead is another common aspect of the hero myth. Isis raises Osiris from the dead after he is torn to pieces by his wild brother, Set. In Greek mythology, it is Aesclepius who performs the same deed, as well as other acts of healing. Another of the famous healers was Pythagoras. In his stories of Pythagoras, Iamblichus tells us that he was born of a virgin, had a golden thigh, called upon the birds—who obeyed him—and that he could calm stormy waters. In some versions of the stories, Pythagoras walked upon them.

However, the tales of Aesclepius's feats of healing go further. He is made to pay by an angry Hades, the king of the underworld, for stealing away his prospective denizens, for, in bringing them back from the dead, Aesclepius was taking away from Hades that which he considered his own. Aesclepius was a son of Apollo. His punishment was to be killed by a thunderbolt from Zeus, and, in revenge, Apollo killed the Cyclops, the one-eyed giant. When it came to writing down this story, I could not help but notice a connection between the legend and the letter of the Greek and Hebrew alphabet, el ayin. In the symbol of this letter, as we have seen, is *o* and is derived from the hieroglyph of the eye. This is a reference to the third eye, the pineal gland, and I wondered whether the legend was relating that with the death of Aesclepius the healer; the sun, represented by Apollo, was on its way to the summer solstice—June 21—and thus, fullness. With the death of the hero, the sun's (and the son's) rise to full glory is assured, for the hero and the father, the sun, are become one.

This is where we come to the most poignant aspect of the hero myth: his death and resurrection.

DEATH AND RESURRECTION

Quite often in mythology the hero is accompanied in some way by a dog. In certain parts of southern Europe and the Middle East, dogs are associated with death to such a degree that death is known as "the power of the dog."[13] In other areas of the world similar associations are to be found. In myth, dogs are carrion eaters, they dispose of dead things. In nature, this is precisely the role of the jackal and other canine species. It is the dog Cerberus, meaning "Spirit of the pit," who guards the gates of hell in the Greek mythology. In Egypt, the dog-jackal was Anubis. In Celtic myth, *Dormoth,* "Death's door," and so on. In the story of the hero, the fact that he has a companion who is a dog is a metaphor for the hero's journey in the underworld; for as part of the process, all savior hero-gods must pass through hell.

Many of the hero myths are unspecific about the reasons for the hero's death. Remember, though, that we are in a fantastic domain and not in our comfortable world of reason. However, there are motives beyond the later interpretations of classical pagan and Christian commentators. The myth is the story of life and its journey through eternity, and, therefore, the earliest forms of these myths were guides to the potential of the soul in life and in death. Hidden within this format was the arcane element that speaks of astronomy, mathematics, and geometry as a link between man and God. It is in this light that we must look upon them and, ultimately, upon their profound significance.

Our hero has been through many trials and tribulations, but in death he comes to the greatest trial of all. The hero at the end of his mortal life becomes, in effect, a scapegoat. Upon his shoulders he bears the burden of the people: their sins, their crimes, their guilt. As part of this process, he is expelled by the people; he becomes the messiah rejected by his nation. "They will make fun of him and spit at him and whip him and kill him" (Mark 15:17–20).

The hero willingly goes to his terrible death—crucifixion, either by being nailed to a cross or a tree or, as some motifs describe it, being hung on a tree. There is a remarkable similarity in world mythology about the mode of death. It begins with a feast, one that is familiar to every Christian as the Eucharist. The Eucharist is officially the Sacrament of the Lord's Supper: "He who will not eat of my body and drink of my blood, so that he will be made one with me and I with him . . . shall not know salvation."[14] This is how a Persian Mithraic text runs; it is almost indistinguishable from the words of Jesus in the Gospel of John 6:53–58. This aspect of the eucharistic meal, wherein the bread and wine represent the body and blood of the savior god, is a common motif to be found in the myth of the hero's death. It figures within the major mythologies of the Middle East, southern and northern Europe, Central and South America, and Africa. The Dogon of Mali, famous for their extraordinary rituals revolving around the star Sirius, worship the figure of a crucified man, a resurrected savior, of whom different clans "partake" in a symbolic but unmistakable Eucharist.[15] Incidentally, the idea of the condemned man's last meal comes from this body of tradition.

This sacred meal was, in all actuality, a holy communion with the god concerned; it was, to a certain extent, the act of taking upon oneself the powers and presence of that god. It was for this reason that priests and priestesses at the many oracle centers would eat, drink, or imbibe intoxicating substances before "communing with the god." We know the gruesome fate of the Christian hero following the Eucharist—crucifixion. There are further gruesome trials in world mythology.

In the legend of Dionysus, the eponymous hero is torn to shreds and boiled in a cauldron, while pomegranate trees, symbols of the underworld, sprout from the soil where his blood has fallen.[16] There are echoes here of an agricultural purpose. The Eucharist and death of the hero-god are as inextricably linked as humankind is to the soil. In a less painful episode in China, the emperor, after having plowed the first soil of the year, would turn to the east and to the west, in precisely the same manner as the genuflecting priest in the Catholic mass. This "ceremony of the Guiding Light" took place at the same time as Easter in the West and was much the same as its Catholic counterpart—even down to the niceties of the dance that is the Holy Sacrifice of the Mass.

Furthermore, the musical accompaniment is similar to Gregorian chant, and in light of the effect upon the brain that Gregorian chant has, we can see how powerful the ceremony would have been. At its culmination, three grains of rice were placed upon the emperor's tongue and a cup of rice wine was drunk, thus completing the Eucharistic ritual,[17] a ritual that paid homage to the earth. This is a ritual that has echoes all around the world. In South America the blood of sacrificial victims was given to the earth as an act of fertility, and sometimes the body and blood of the victim were consumed eucharistically. Cannibalism has largely been eradicated, with the odd exception, but this was often the thinking behind it. The word *cannibal* is derived from *caribe,* meaning "brave and daring," the properties of the hero—hence the *Caribbean,* the area from which the etymology emerged. By the same token these occasions were, and still are, celebrated within religious communities with carnivals, from Latin *carn* and Italian *carne,* meaning "flesh"; *carnival* meaning literally "the cessation of flesh eating."

The most ancient record of the Eucharist followed by the death of the god is to be found in the Egyptian Book of the Dead. The initiate "eats" the gods so as to take upon himself some of their powers. The main objective of this, particularly where the deceased pharaoh is concerned, is to become one with Osiris. In the Egyptian legend, Osiris, too, has been torn to pieces by his wild brother, Set. In the version related by Plutarch,[18] Set tricks Osiris into a sarcophagus that is duly sealed and cast adrift. Landing at Byblos, it sprouts leaves and becomes a tree. In some mythologies, it is a tree that the hero is hung from or nailed to. In others, heroes are crucified. Dionysus is crucified, as is the Gaelic Esus. The god Attis is known as "He on the tree,"* while images of Adonis were often hung on

---

*Attis is sometimes held to have died under a tree, to have castrated and killed himself—this is a euphemism for crucifixion.

trees. In Scandinavian mythology, Woden goes through a bizarre process of self-crucifixion as an act of renewal. In other legends, the image of the tree or cross is very much in evidence, but the hero does not die upon it; instead, death is usually delivered in the form of the hero falling upon a sword,[19] or being stung in the heel by a scorpion, as in some versions of the story of Osiris. It is Achilles's heel that is his downfall. He is "stung" by a poisoned arrow,[20] so, too, Krishna, when mistaken for a deer by a hunter. Attis, son of Cybele, is gored to death by a wild boar. In some versions of the crucifixion story, it is not the hero who suffers crucifixion—fate has something else in store for him. Instead, it is his brother who is crucified. Yet another version of the legend of Osiris has his wicked brother Set crucified.[21] Set, by virtue of being wild and uncultured, would appear to be symbolic of man's lower self, and so the act of crucifixion represents an act of transcendence, of rising above bodily desires. As stated in chapter 1, a motive for the death of the hero is atonement, literally at-one-ment— the ultimate act of communion with God.

Atonement for some great sin brings the hero to this state. For example, in some legends the hero must pay the price, or atone, for the great sin of, albeit unknowing, committing incest with his sister. The most famous case of incest in the history of the West passes almost without notice because we are inclined to think of the protagonists as mere man and woman. Yet, in catechisms and the writings of Aquinas, we are told that, in the eyes of God, not only are they man and wife, but they are also brother and sister, and the result of their actions is not only the human race but *the* original sin as well. These sources tell us that Adam and Eve were cast out of Paradise for this very crime.

Jesus was called "the Second Adam" so that he would atone for his predecessor's sin. This is also what happens in the Arthurian mythos, where King Arthur has acted in the same way and must rectify the deed, laying down his own life in the process.[22] In the Finnish legend of Kullervo Kalevanpoika, the hero unwittingly sleeps with his sister and later falls on his sword.[23]

In the Gospels, Jesus utters his harrowing last words: "My God, my God, why hast thou forsaken me?" In the original Greek of the Gospel of Mark (15:34) these words are *Eloi, Eloi, lema Sabachthani?* This echoes *"Euoi, Euoi,"* the last cry of Dionysus, the savior god who was also sacrificed. The Roman version of Dionysus is Bacchus. He, too, is sacrificed: "After Bakchos [*sic*], who cried 'euoi' is struck, blood and fire and dust will mix."[24] These words echo the earthly nature of the sacrifice, that it is from the earth that we come and back to the earth we shall go. In a peculiar but somewhat significant way, the last words of Jesus are echoed in the Mayan tongue. According to a Guatemalan monk, Antonio Batres Jaurequi, the equivalent in Mayan is: *Hele, Hele, lamah sabac ta ni,* meaning "I faint, I faint and my face is hid in darkness."[25] Needless to say, this is a much-disputed record.

The hero harrows hell, and meanwhile on Earth his body is placed in a sepulcher. In the legend of Orpheus, the hero walks the infernal paths to rescue his beloved Eurydice, as does Dionysus, who plunges into the dark depths of Hades to rescue his mother, Semele. On the third day after his death, the hero reappears in the greatest miracle of all, resurrection. "When night had fallen, the sorrow of the worshippers was turned to joy. For suddenly a light shone in the darkness: the tomb was opened; the god had risen from the dead; and as the priest touched the lips of the weeping mourners with balm, he softly whispered in their ears the glad tidings of salvation."[26] In ancient Rome, the carnival celebrating the rise of the god from the dead was called the Hilaria. The date was March 25 (Lady Day in the Christian church) and the particular hero-god concerned was Attis/Tammuz.

In this brief overview of the myth of the hero, I think that we have now run through sufficient examples to see that it is an extraordinary and worldwide phenomenon. As has transpired with others before me, the more I thought about it, the more I uncovered many correspondences that were beyond mere coincidence. Unfortunately, because of the similarities and the reliance upon particular dates, such as the solstices and equinoxes, the various heroes of world myth have been reduced in status to mere vegetation deities, in the view of Sir James Frazer, or psychological archetypes, in the opinions of C. G. Jung and his followers. I do, to a certain point, agree with these estimations, but there was for me, from the outset, a growing doubt that these really were answers in any complete sense.

THE ETERNAL HERO

The world has always needed its heroes, in whatever shape or form, but more than this, the hero is relevant to all societies and all cultures throughout history. History is a progression of humanity and its ideas, from out of the mythic into the real. And yet, wherever we look for real meaning, the hero stares back at us poignantly, and with a smile, transcends time and defies linear "progress." Even among the least spiritual, the power of the hero's story is grasped.

Communism in the form of the Soviet state was repressive, monolithic, and totalitarian; it was also officially atheistic. How ironic then that the architect of the regime should have been interred in a mausoleum that has all of the hallmarks of a pyramid. As if to confirm these suspicions of Lenin as some kind of religious focal point, a book published in 1938 asserts that like a mythic hero, Comrade Lenin, too, had a golden portion—his shoulder—and that he was born of the moon and the sun, and so on.[27] This, of course, is a perverse exploitation, and yet it serves to show that everywhere we look there is the hero staring out at us, beckoning us on. Needless to say, despite the Communist aspect, the hero was not simply some primitive political ploy leftover from the archaic past.

Earlier we encountered Sir James Fraser and the idea of heroes as vegetation deities. While there is some substance to this thinking, the view of the hero, and of religious belief, as mere metaphors of agriculture and the turn of the seasons has been aptly summed up by many writers: "The religious experience of primitive peasant societies was conditioned by their close contact with the mysteries of life and growth and by their dependence upon the rhythm of the seasons and the fertility of the soil. Whereas pastoral and hunting peoples thought of the divine powers as formidable and incalculable, to be propitiated or obeyed."[28]

That this is the case cannot be denied, but the real solution to the enigma of the hero lies very deep within this answer. We have seen that at certain places, due to radon emissions, plants thrive and that this, having been noticed, could have led into the development of a science of sacred sites. The stuff of the hero is the very life force, which naturally both includes and transcends the idea of fertility within the soil. The very presence of a science of acoustics in sacred architecture is a testament to this transcendence and gives the lie to claims that agriculture is as far as the hero goes.

It seemed to me that there was a source for the myth and that it was something actual, not merely symbolic or metaphorical. It was in this light and with this attitude in mind that I approached some of the supposedly wilder claims of the early Christian fathers, though not before gaining a little context. "When, as frequently happened, pastoral peoples overran agricultural communities, the old vegetation gods were considered as having been overcome by the sky gods [worshipped by the incomers], just as the peoples were subjected to their new warrior overlords. From the fusion of cultures which resulted, new syncretistic pantheons were developed in which the old gods tended to be relegated to the underworld, where they ruled over the dead."[29]

There are two points to be reminded of here. First, as society changes, so, too, does our view of our ancestors. The idea of the Ancients being anything other than primitive, ignorant savages was, until recently, widely dismissed as romantic. Observations about their monuments have, to a considerable degree, changed all that. It seems that the more we discover about ourselves the more that there is an echo of it in the past.

Point number two is the observation that the underworld, the substratum, is exactly where the hero belongs—he went there of his own volition. The later chapters of this book will show just how important this point is. For now, it suffices to say that it can be more or less ascertained that sky god and earth god were seen as intimately and necessarily connected so that they could be transcended. This is what the hero myth is all about, and this also is the very point of Christian belief. When one takes a look, as we have, at ancient sacred technologies that relate to consciousness—the experience of altered states and the manipulation of the experience to higher levels—this shows us that the real primitivism lies in our "progressive" view of history. We are now rediscovering much, and this indicates that civilization and the knowledge that goes with it are cyclical: "What goes around, comes around." This phrase is an apt illustration of the cycle of birth, life, death, and rebirth that the hero seeks to rise above.

I was never more aware of this cyclical nature than after I had read the works of Saint Augustine of Hippo (354 to 430 CE), one of the great forces of the early church: "That very thing which is now designated the Christian religion was in existence among the ancients, nor was it absent even from the commencement of the human race up to the time when Christ entered into the flesh, after which true religion, which already existed, began to be called Christian."[30]

These words made me more than a little suspicious about the claims of syncretism. Syncretism is the reconciliation, or the attempt to reconcile, different systems of belief. As cultures developed and met other cultures, syncretism took place. The similarities and differences of each individual cult became merged. The reasons could be innocence or political purpose. This latter was the modus operandi of the Romans. They did it to a certain extent to the Greek gods: Zeus became Jupiter, Aries became Mars, Aphrodite was Venus, and so on. *Syncretism* is a modern definition of what Saint Augustine of Hippo realized was an underlying and extremely archaic theme. These individual points of similarity seemed to begin at the point of civilization, at the very point of humanity becoming self-aware. In short, they are a record of the past, a record of an ancient knowledge, a science even—though not in the way we might understand it today.

Augustine's words have a ring of truth about them, but needless to say, in the early centuries, there was much propaganda that exasperated the pagan sense of reason. Celsus, the second-century pagan philosopher, is the author of the oldest literary attack on Christianity of which details have survived. Among other comments that were perhaps too severe, the following is clearly justified: "Good Lord! Is it not a silly sort of argument to reckon by the same works that one man is a god while his rivals are mere sorcerers?"[31]

Certain of the Christian rites, beliefs, and gospel stories were already very familiar to the first Christian critics. In fact, many of the early Christian commentators are insistent that their rites are exceedingly ancient, inferring that Jesus was there all the time. Justin Martyr

(100–165 CE) writes: "In saying that the Word was born for us without sexual union as Jesus Christ our teacher, we introduce nothing beyond what is said of those called the Sons of Zeus."[32]

These admissions, while not negating that there is a uniqueness to what we call the Christian experience, indicate one that echoes outward from deep antiquity. This was a fact very well concealed and known by few, even within the church. What Saint Paul had started out with, Constantine had completed, and the rise of the church became inexorable; but in the light of politics and the aims of empire building, the archaic past of Jesus in myth was to remain hidden until, at last, the whole question of the mystery could be approached more dispassionately, as now. It is Jesus himself who raises the question in the Gospel of Mark (8:27–30): "Jesus and his disciples set out for the villages of Caesarea Philippi and on the way he asked his disciples, 'Who do people say I am?' They answered, 'Some say John the Baptist, others Elijah.' 'And you,' he asked, 'who do you say I am?' Peter replied, 'You are the Messiah.' Then he gave them strict orders not to tell anyone about him."

It is indeed odd that the Jesus of the Gospels is so overt and yet so secretive. The whole mystery is nonetheless blown wide open by an explosive comment from the one-time editor of the Dead Sea Scrolls team. Professor John Strugnall had been working on the enigma of the scrolls since 1952. In 1990, however, he was sacked in a somewhat dramatic move that cited failing health and emotional distress as the reason for his removal. There may have been another motive, an arcane but obvious one.

The Dead Sea Scrolls are a very sensitive matter, given their position within the volatile cauldron of Middle Eastern politics. In 1990, a few weeks before his dismissal, Professor Strugnall gave an interview to an Israeli newspaper in which, with a grave air of insouciance, he made some remarks that, to Jews at least, would have been extremely offensive: "Judaism is a horrible religion . . . based on folklore . . . It is a Christian heresy."[33]

Whether or not Professor Strugnall was under a certain amount of distress, it is hardly believable that a scholar in his position should forget what most of us take as a simple fact of history—that Christianity postdates Judaism. Or does it?

Clearly, from the researches that I have conducted, a doubt had arisen about the true age of Christianity, a significant doubt. The reasons noted so far center on the power of place and the consequent Christian adoption of pagan sites and either the conversion or demonizing of the old gods. But what of the focal Christian personality, the Son of God? It was time to take a closer look at the myth of Jesus in relation to the hero.

The figure of Jesus seemed to me to be at least half as old again as the Christian era within which we live. Thus, if Jesus is far older than the dates given, there must be a variety of reasons. The notion of syncretism was a point in question. It has been postulated in a remarkable thesis[34] that, via this process, Egyptian deities became transformed into Greek deities and these in turn influenced the identities and practices of the Roman mythos and furthermore, that the Egyptian influence spread out over a very wide area indeed. I was beginning to think that the idea of religious pluralism had indeed stemmed from a single definable source and that even this source was derived from something exceedingly archaic.[35] Jesus's words "No one can come to the Father but *through* me" (John 14:6) began to take on a new hue. The words are intriguing. Jesus was no idolator, so what was it that spoke through him? The answer was to come as a huge surprise, though not without some digging.

A. N. Wilson in his book *Jesus* makes the observation that "theology has concerned itself almost exclusively, from the beginning, with the risen Christ."[36] Certainly, the early church was at pains to distance itself from the mythic Christ. "When Eusebius recorded his memorable boast that he had virtually made 'all square' for the Christians, it was an ominous announcement of what had been done to keep out of sight the mythical and mystical rootage of historic Christianity."[37] Moreover, Eusebius "and his co-conspirators did their worst in destroying documents and effacing the telltale records of the past, to prevent the future from learning what the bygone ages could have said directly for themselves."[38]

Despite borrowing arguments from Greek mythology to bolster the concept of the Immaculate Conception, to Justin Martyr the similarities between the pagan mystery religions and the church were "the results of 'diabolical mimicry' which blind the foolish to the essential differences between Christianity and paganism."[39] There is more than a hint of motive when we view the extraordinary claim of Saint Ambrose, Bishop of Milan, ca. 370 CE. In an attempt to refute pagan claims of plagiarism, he announced that Christianity had not borrowed from Plato, rather that Plato had borrowed from Moses,[40] whose assistant at Sinai was a figure called *Joshua*—the same name as Jesus. (One is a direct English translation of *Yeshua*—*Jesus* in its original form—the other is English, from the Latin *Iesus*.) So, while Professor Strugnall thinks that Judaism is a Christian heresy, Saint Ambrose says that Plato, representing the pagans, borrowed from Moses.

By now we know the similarities of the hero stories worldwide. Surely, everything is all too interchangeable. Many scholars, from the first to the twentieth centuries, have commented upon the similarities of the world's many deities—and it is well accepted that the gods represent attributes such as love and wisdom and also aspects of nature such as wind and fire. However, this explanation does not solve the riddle of the heroes—Jesus and those like him.

THE NAME OF THE HERO

During the first century, by a shift of focus, Christianity transformed itself from a religion based upon the heroic cycle, the wheel of birth, life, death, and rebirth, into a religion based upon the power of love. Unfortunately, there was a downside, for this removal from the traditional hero cycle marks the beginning of a loss of knowledge—that God is within everyone, that everyone has the potential to be a true hero and undertake the journey back to divinity. Nevertheless, the love ideal proved most potent. What some of the early martyrs suffered upon the basis of this remarkable philosophical principle was extraordinary. That they should have endured, suffered, and died, often in the most appalling agony, is a great testimony to the power of this focus on belief.

This is not the first manifestation of such a focus. The religion of love bears many striking similarities to the Orphic philosophy, which was one of high moral purity. The ethic of noninjury and of purity is there, as is the theology of redemption and the doctrine of original sin. The focus on love is manifested in the form of the Greek god Eros. The Orphics believed that at the creation Eros was the first god to emerge from the womb of darkness, in which a silver egg had been laid by Rhea. Eros, once born, set the universe in motion.[41]

Significantly, Eros was, according to Plato, not only the oldest of deities but also the one who inspired strength enough within the soul to ascend the heavenly heights after death. In Greek the word *eros* denotes a hero or demigod. Eros was the beginning and the end, the alpha and the omega—and the original hero. What the Orphic myth makes quite clear is that the hero goes hand in hand with the concept of love: you cannot have one without the other.

It is clear that the word *Jesus* is from the same family of names as *Eros* and *hero*. The link is made clearer with the help of another rendition of the Jesus story: this time with the name *Esus*. Lucan, a first-century authority, makes mention of the Druidic god Esus as "uncouth Esus of the barbarous altars." He is also a crucified god; in some accounts he is even depicted as a woodcutter.[42] One can grasp the Jesus-Esus-Eros hero link, both through similarity of story and name. But much more is concealed.

At this juncture it is worth remembering that many of the Ancients—the Hebrews and the Egyptians, for instance—did not generally write down vowels. Consonants only were written. When consonants are uttered, the vowels emerge gently and naturally, depending on the precise shape of the mouth and lips as the consonants are sounded. We all know how regional accents make the same words sound different: some accents are nasal, some guttural, others rounded, and so on. In this way, for example, one's *bill* is another's *bell*. Among the many reasons why names change from region to region and era to era, this is one of the most important.

To return to the connectivity of names, Esus is known as Hu-Esus— Hesus or Hu for short. Hu was an exceedingly ancient god and of prime importance in the mythologies of the ancient Britons and Gauls. The link with Eros is that "hu" is a primal vowel sound in the European languages of the period. It indicates darkness, in terms of the womb of the Earth. The form of the name *Hesus* means literally "healer."

Hesus·is also associated with the bringing of music, in Celtic myth. The indication here is that music is light, for *Hu* also means "light." This is a complex theology, but it is also science in the sense that science means knowledge.[43] The feast day of Hu still bears his name, Hogmanay. It is of course the Scottish New Year. In the past it was celebrated as a "kind of Saturnalia."[44] But where does the name Hu take us?

Hu also means "lord"—"the Lord." It is to be found in the long Hebrew form of Jesus's name Jeho*shu*a, meaning "the Lord is Salvation." The variant *Yahu* means "Lord God"—*Ya* is "God," *Hu* is "Lord." In this way the heroes are inextricably linked to the identity, the very sense of being that we call God. *Hero* is also one of the oldest words still extant within the English language. Its source is shared by the name of one of the greatest heroes of classical myth, Hercules. *Hercules* is the Latin version of the original Greek *Heracles,* meaning "glory of Hera." There is, however, an older derivation. I believe that the Greek *Hera* is a later feminization of a god name that is derived directly from ancient Egypt. It is a god name that was used to describe the very identity and nature of the pharaoh himself. For Heracles, perhaps we should read *Heru-Akhty*—the "living image of Horus," known more familiarly today as Horakhty,* whose image is famous to the world as the Great Sphinx. When linked to Ra, a composite name is formed, Ra-Horakhty, which is illustrated by the hieroglyph of a hawk-headed male divinity. Thus, the word *hawk* also contains the element of "the Lord." *Horus* is the Greek spelling of the Egyptian Heru—Horus is the oldest known form of the hero.

In the Greek myth, Heracles ascends to heaven in a blaze of glory, as does Jesus in the Gospel of Mark. In the legend, Heracles is reborn

---

*The god Horus, the lord whose name is Hu, meaning "the," became a higher form—he became a stellar deity, a cosmic god, a higher form of "the" as it is used on earth, an all-encompassing "The," in the divine sense of ultimate realization, "I am." He was the former Horus now become Osiris, a "Gesu," the man who had become god, giving hope to humanity, "Gesu" being a composite of Horus the earthly, and Osiris/Asar, the father who art in the heavens.

into heaven to sit at the table of the gods. In Egyptian myth, Horus is the original holy child, the son of Osiris and Isis. He was the embodiment of the Egyptian pharaonic kingship. A pharaoh was "the living image of the Lord," of Horus, and when a pharaoh died, he ascended to the heavens as Osiris—the father of Horus.

Taking our investigation just a little bit further, in the Old Testament book of Genesis, Abram, by virtue of his loyalty to God, changes his name to *Abraham:* the insertion of the *ha* element is significant, for in Hebrew *ha* means "the," it is the definite article. *Hu, ha ho,* and *he,* as in Heru and all the other variants, at root, mean precisely the same thing: The Lord. For "the Lord" read "the The."[45] This is reminiscent of the Vedic definition of the divine: "I am That, Thou art That, all This is That." There is no doubt that each pharaoh was a hero whose goal was to move beyond the realm of "the" and into a higher realm, the realm of "The." The *ha* is a reference to higher things.

The hero goes from the small self to the universal self. Jesus, in being a hero, was partaking of this same mystery. His story relates more than can be read literally. In understanding the meaning of the term *hero,* we can understand more about the nature of the mythic Jesus and some of the extraordinary correlations with archaic Egyptian myth.

Some of the early iconography of the church is strikingly similar to that which is found in Egyptian myth. As an example, the Madonna and child could easily be Isis and the infant Horus sitting on her lap. Jesus is quite often to be seen with his arms abreast in a gesture of pure welcome; a pose that we see in pharaonic Egypt particularly from the tomb of King Tutankhamun. Jesus and Horus are both called "sons of the widow." These are but a few examples.[46]

That Jesus is a hero in the vein of the Ancients is borne out by the mythic quality of his story. A summation of the telltale elements demonstrates just how obvious this really is, even down to the finer points:

- Jesus is the anticipated Messiah, the Son of God, God made flesh, the savior.
- He is born in the grotto amid wonders, of a virgin, on either December 25 or January 6.

- He is attended by three kings.
- He goes into exile, pursued closely by agents of the jealous king.
- He is baptized by a wild man, after being tempted by the devil.
- He has twelve followers or apostles.
- He performs the miracle of turning water into wine at a marriage ceremony.
- He raises a man from the dead and performs other miracles, such as walking on water.
- He rides into town on an ass and is triumphantly acclaimed as king.
- He is put to death upon the cross at Easter.
- His lifeless body is placed inside a rock tomb.
- He harrows hell and releases the spirits of the dead.
- He rises from death three days later.
- He is found by the woman who mourned him, having escaped miraculously from the rock tomb. These events are then celebrated with the feast of the Eucharist.

We have looked at some of these aspects across myth in a general manner. As to the Jesus story—this time using the gospel stories as our springboard—Jesus, by being brought forth from a grotto, in symbolic terms represents light from darkness. This was the nature of the original Nazarite myth. Samson is born in a cave whose entrance is so bright that it cannot be seen; Samson is the sun. The myth is also denoting that Jesus is a son of the earth, in Greek, Gaia or Ge, in ancient Egypt, Geb, the father of Osiris. By the same token Matthew 27:45–53 describes the events of the crucifixion in equally earthly terms.

> From midday a darkness fell over the land, which lasted until three in the afternoon . . . and at about three Jesus cried aloud, "Eli, Eli, lema sabachthani?" which means, "My God, my God, why hast thou forsaken me?" Hearing this, some of the bystanders said "He is calling Elijah." One of them ran at once and fetched a sponge, which he soaked in sour wine and held to his lips on the end of the stick. But the others said, "Let us see if Elijah will come and save him."

Jesus again cried aloud and breathed his last. At that moment the curtain of the temple was torn in two from top to bottom. The earth shook, rocks split and graves opened; many of God's saints were raised from sleep and coming out of their graves after his resurrection entered the Holy City.

The last sentence is interesting, given that Christian saints were only numbered from the inception of Christianity and there are few in the canon who predate Jesus.

Jesus, before ascending to heaven, must cleave at the crack of doom, at the very doors of hell itself. *Hell* is an interesting word and does not mean what it implies, certainly in the dogmatic sense. We have touched upon this before, but I now shall reiterate from a slightly different angle. In Greek, and in certain of the old European languages, it is *helle*. Greece is, to this day, the land of the Hellenes. The Hellespont is the natural land bridge that conjoins southern and mainland Greece. Helle herself was a maiden and a sister of Phrixus. It was while fleeing abroad with the fabulous Golden Fleece that Helle fell overboard and was celebrated in the name of the Hellespont.

This is significant, for it was Jason, another form of the name Jesus, who went after the Golden Fleece with the twelve Argonauts. As we have seen, the meaning of *helle* is "womb," in the sense of "womb of regeneration," and Hel is also associated with rebirth, usually through purification, often by fire, hence *helios,* Greek for the "sun." Hel or Helle was a goddess. The Jewish version of Hell, Sheol, is also feminine, as are other versions. There are times when we "go through hell"—going "through hell" is about change and renewal: it is about transformation.

Heracles ascended Mount Olympus in glory. Buddha ascended the mountain Pandava, in Sri Lanka, whereupon heaven opened and he was bathed in light. He shone as the brightness of the sun and moon. Wherever we look, we see comparisons. The temptation in the wilderness? Look no further than the personal account of the devil tempting Zarathustra on the Mountain (ca. sixth century BCE). Walking on water? We have mentioned Pythagoras, but again in the Greek mythos, Poseidon does the same.

Besides the similarities in storyline, there are also the comparisons to be made between Jesus's words and the words of early heroes.

JESUS:

> You know the commandments: You shall not murder; you shall not commit adultery; you shall not steal; you shall not bear false witness; you shall not defraud; honor your father and mother.

-Mark 10:19

BUDDHA:

> Abstain from killing and from taking what is not given. Abstain from unchastity and from speaking falsely. Do not accept gold and silver.

-Khuddakapatha 2

JESUS:

> Do to others as you would have them do to you.

-Luke 6:31

BUDDHA:

> Consider others as yourself.

-Dhammapada 10:1

JESUS:

> This is my commandment, that you love one another as I have loved you. No one has greater love than this, to lay downs one's life for one's friends.

-John 15:12-13

BUDDHA:

> Just as a mother would protect her only child at the risk of her own life, even so, cultivate a boundless heart toward all beings. Let your thoughts of boundless love pervade the whole world.

-Sutta Nipata 149-50

When speaking of himself, Jesus says, "I am the Alpha and the Omega." Buddha says, "I am the letter A." In another example, Krishna says, "I am the light, I am the life, I am the sacrifice," words echoed by Jesus. We are in heroic territory and a comparison of the facts of the myth demonstrates that Jesus is a hero in the ancient sense. No matter what the early church fathers may have said, or how they said it, their refutations appear to fall flat.

Some of the above comparison of quotes comes from the work of Gerald Massey. Massey was an interesting man, unique in fact. He was primarily a poet and came from a poverty-stricken background. He was born in 1828 to illiterate parents who are described as having lived in a hovel, barely able to feed and clothe their children. Massey's education was a few months at school, where he learned to read and write, but at the age of eight he was put to work in a local silk mill for twelve to thirteen hours a day. In the ensuing years, Massey's main secondary occupation was reading. When he was fifteen years old, he moved to London, where though always struggling against the hardships of poverty, he thrived intellectually, producing some very important insights into the origins of historical Christianity: "According to the unquestioned tradition of the Christian fathers, which has always been accepted by the Church, the primary nucleus of our canonical gospels was not a life of Jesus at all, but a collection of the Logia, oracles or sayings, the logia Kuriaka, which were written down in Hebrew or Aramaic, by one Matthew, as the scribe of the Lord."[47]

Unremarkable stuff, you might think, until we come to Massey's identification of Saint Matthew.

> The logia or sayings are *mythoi* in Greek. . . . The sayings were the oral wisdom and, as the name implies, that wisdom was uttered by word of mouth alone. They existed before writing, and were not allowed to be written afterwards. . . . In Egypt the sayings were assigned to various divinities, that is, mythical characters.
>
> Among the sayings of Jesus, or Logia of the Lord, is the saying that "the very hairs of your head are numbered"; in the [Egyptian] Ritual, every hair is weighed; also, the night of the judgment day is designated that of "weighing a hair."

Matthew alone of the evangelists represents this drama of the Egyptian ritual.

Massey goes on to relate the copyist of the Egyptian version.

These sayings, or Logia of the Lord, were written by Hermes or Tat [Thoth], the scribe of the Gods.

This is the part assigned to Matthew, the called one, the Evangelist and Scribe, who first wrote down the sayings of the Lord. *Now, the special name or title of Hermes in the particular character of the recorder is Matthew in Egyptian—that is, Mathiu.*"[48] (My italics)

This is remarkable. In the same paragraph, Massey goes on to relate that the original sayings of the Lord are none other than the original sayings of Horus, whose very name means "the Lord." Were the apostle's names merely symbolic titles of ancient Egyptian deities? That this is a distinct possibility I will demonstrate further on.

The savior called Adonis, whose name also means "the Lord," comes from Syria, and his name is a Greek variant upon the magical name of the Hebrew god Adonai. *Adonai* means literally "Lord God" and is a variant of the name of God, Yahweh, sometimes known as Jehovah. The myth of Adonis is exceedingly ancient and not surprisingly there are many familiar hallmarks. He was born in Bethlehem, the same Bethlehem as Jesus. Earlier I mentioned Hogmanay. The celebration is never complete without the oatcakes, sometimes wheatcakes, that are symbolic of Hu. Oats and other cereal crops figure quite prominently in the story of the Lord. Bethlehem means "house of bread." Adonis was a god of the corn[49] and was symbolized by loaves of bread. He was born of a virgin, Myrrha, sometimes Mari, and shortly after his nativity he was hailed by cries of "The Star of Salvation has dawned in the East."[50] The date of his nativity is January 6 or December 25.*

The legend is particularly beautiful in its description of an event

---

*Due to changes in the Caesarean calendar, ten days were taken off the year during the reign of Pope Gregory in 1582, giving a confusion of dates. Thus old Christmas—January 6—and new Christmas—December 25—all the same.

known as the Adonia. Adonis died at Eastertime; he was gored to death by a wild boar, and where his blood dropped, there sprouted anemones, the flowers of the underworld. Adonis harrowed hell and rose again from the dead. During the days that he harrowed hell, the sea was literally blood red. In the ritual of Adonia the blood element was supplied by the spring rains loosening particles of hematite—bloodstone—in the mountain soil, which would then be washed downstream and into the sea. Maidens would sail out to sea at this time and there sprinkled the "blood" with anemones, which grow wild in the Syrian hills.

Significantly, Adonis's sepulcher is exactly the same place where Mary nursed Jesus, the same cave that that was Jesus's sepulcher. Adonis was particularly revered by the Orphics, who composed hymns to him, and, like Jesus, Adonis and his variants, Attis and Tammuz, are all "sons of the widow." Interestingly, this is also a term common to Freemasonry.

It is now, as we come again to look at the myth of Attis and his mother, Cybele, that we arrive closer to home and get a sense of scale as we approach the very heartland of the mystery. The cult of Attis had a considerable influence upon early Christianity, and for a good reason: its headquarters were what is now known as the Vatican.

The goddess Cybele, mother of Attis, was brought to Rome in 204 BCE from Phrygia (now Turkey). The earthly incarnation of Cybele was Nana. By eating an almond, Nana conceived Attis, the savior. In the old rites Attis was eucharistically eaten as bread, which was washed down with his blood, wine.[51] Attis's father was never known, hence he was called "the Virgin's son." After his death, he was resurrected as "the Most High God, who holds the universe together."[52] His epiphany, his ascent into heaven, is greeted with the words, "Hail Bridegroom, Hail, new Light."[53]

Significantly, Attis was crucified on a pine tree, whence his blood, shed upon the earth, redeemed humankind and the fertility of the soil. The date of his passion is March 25, Lady Day in the Christian calendar—the feast of the Annunciation of the Blessed Virgin Mary. The church father Lactantius puts the date of the crucifixion of Jesus as March 23 and the 25th as his resurrection. As if to hammer home the point, the day of Attis's death is a Friday, called Black Friday, the Day of Blood. The cry of Attis upon the cross is recorded in a hymn of the Naasenes of the third century CE: "Euhai, Euhan, he is Pan, he is Bacchus."[54]

Attis's cult remained at its Vatican hillside for at least two hundred years into the Christian era. The Vatican is described by a Roman church man called Gaius, who lived in the time of Pope Zephyrinus (198 to 217 CE), as the place where you will find the trophies of those who have founded the church—rather an interesting statement considering the importance of Attis to this site. According to Pliny the Elder (23 to 79 CE) the Vatican had long been an unhealthy, poor, and somewhat squalid area outside the hills of Rome. Pliny complained that it was full of snakes, and, according to Martial (41–104 CE), it was renowned for its disgusting wine.

However, in the era of Caligula and Nero it soon found itself fashionable again. It was during the persecutions of Nero that it is believed that Saint Peter suffered crucifixion. In the 1940s, during excavation under the high altar of Saint Peter's Basilica, organic material was discovered, including fragments of a skeleton and the remains of various animals. Margherita Guarducci, an archaeologist and epigraphist who worked on the material in the 1950s, believed that she had found the remains of Saint Peter himself. Given her devotion to her faith, her enthusiasm and professionalism were very highly commendable. However, the remains were those of a mouse, a cock, an ox, a sheep, and, interestingly, a pig. All of these animals were sacred to Attis, particularly the pig or boar—the very animal that had done Adonis so much injury. (The pig or boar is one of the disguises of the Egyptian Set, brother of Osiris.)

The explanation of these remains was that they were pilgrim offerings—pilgrims come to adore the Lord, perhaps? But which lord? Attis? Adonis? Tammuz? Bacchus? Dionysus? Mithras? Hercules? Heracles? Horus? Hu-Esus? Esau or Jesus? Rome was a very cosmopolitan place—and all roads led there. Excepting the latter of course, none of these names could be tolerated by the early church fathers, even if comparisons with Christianity was admitted.

IN THE NAME OF GOD

So many ancient holy names, linked through common sounds and ancient tales, will now draw us into the relationship between spoken and written words. Mythology and the alphabet, as I was soon to discover, are inextricably bound together in a way that is deeply profound. Many commentators have made this observation over the years and it is nicely summed up by Leonard Shlain*: "One explanation for the Israelites' intense hostility towards images is that, having discovered the immense utility of alphabetic writing, they considered iconic information to be a threat to their new-found skill. Learning to think without resorting to images is indispensable to alphabetic literacy. 'Make no images' is a ban on right-brain pattern recognition."[55]

The Judaic, Christian, and Islamic religions all banned the worship of images and became more and more dependent on the written word. Inevitably, the oral tradition suffered. In contrast to the written word, the oral tradition's riches lies in the personal transmission of poetry, metaphor, and symbolism. Such ingredients entrain both left and right hemispheres of the brain and encourage mental imagery and emotional response. Thus, the spoken word can make the intangible seem real and give rise to greater overall awareness. This is a gift of spoken mythology, but more influential still can be a single word of appropriate vibratory power, used knowledgeably and purposefully. This is the true role of words in religious ritual. We are reminded of, and inspired in, our purpose through correct understanding of myth, this understanding and strength of purpose then helps empower the words of ritual. Myth tells us of our origins and the ultimate source of all—the divine. The practices of religion bind us back to that source.

There is, for instance, an element of this in a very interesting correlation of names of power that I came across when looking again at the

---

*Shlain submits that Christianity never could have gained a foothold in Egypt if the Gospels had been written down in hieroglyphics (*Alphabet Versus the Goddess,* 259). This is, I feel, contentious, but possibly, if only partially true, as Christianity could not have existed in the first place without hieroglyphs; many of these glyphs entered Christian myth in a much misunderstood form.

Celtic god Esus. Esus, in being a Celtic god, is also a son of God. In the Celtic creation myth, God pronounces his name with "the Word"; this is remarkably reminiscent of the prologue to Saint John's gospel. The letters of the holy name are called the three columns of truth. These three columns /|\ spell out God's secret name as *A W E*.[56] How close this is to Yahweh—all that is really missing is the *y* sound as the mouth opens to enunciate the *A*. Yahweh (sometimes Jehovah) is still recognized as a name of power today in Judaism. Can this similarity of Esus and AWE—Jesus and Yahweh—be put down to mere coincidence? When we look further, the coincidences begin to look more than a little familiar.

The greatest of the Irish gods, the Dagda, is called the Good God, not good in a moral sense, but "good at everything" (*RuadRo-Fhessa,* or "Lord of Perfect Knowledge").[57] However, the Dagda has an alternative name meaning, "Father of all"—*Eochaidh Ollathair.* The pronunciation of *Eochaidh* begins by compressing the first two vowels together in order to pronounce them. The last two consonants are relatively silent, giving us a soft *d* sound verging on a *th*. The result of all this is to give us a name remarkably close to that of Joseph, who in the Gospels is also called a good man.

As if in a confirmation of these correlations, in the Irish Christian tradition it is Saint Brigid who is the midwife at the nativity. On that night she was miraculously transported to Bethlehem, where she assisted beautifully at the birth. In Irish myth it is Brigid who is the Dagda Eochaidh's consort. At the coming of Christianity, Irish writers refused to demote their powerful goddess to the rank of a mere saint, insisting that she was the Queen of Heaven herself, Mary. She was called "Mother of my Sovereign, Queen of the South," "Prophetess of Christ," and significantly, "Mother of Jesus."[58]

Just across the water from Ireland, in Wales, the mythologies, though very much censored by the ravages of evangelism, have been preserved in a body known as the Mabinogion. This comprises tales from the White Book of Rhydderch and the Red Book of Hergest. The Mabinogion[59] has been dismissed as a meaningless term related to "juvenile romances"; however, it has been suggested by Gwynn Jones that, rather than being meaningless, *Mabinogion* means "tales of the son of a virgin mother"[60] and that the son concerned is the child always born at the winter solstice. While we are on the subject, the Welsh term *Dodman* means "the man who came" and is the same in Arabic. The name Dod is given in the Old Testament as one of King David's forebears. It is the same as *David*—in Arabic, *Daud*. Thus, we have in Welsh a "man of David," a Dodman. Dodman is also a system of measure;[61] in other words, the spirit of Dod is upon one when using it.

All of this is intriguing, but where does it bring us, beyond mere similarity? Myth is trying to tell us something of exceptional importance. That these names, so familiar to us, are to be found over a wide area is fascinating, but their very presence denotes an importance of meaning. Names in myth are not given in a random manner, and it seems that, when the first scribes set to work, they were most concerned to record the terms correctly.

> Now that documents of the earliest ages of writing are available, one is struck with a wholly unexpected feature. Those first predecessors of ours, instead of indulging their whims with childlike freedom, behave like worried and doubting commentators. . . . They move among technical terms whose meaning is half lost to them, they deal with words which appear on this earliest horizon already "tottering with age" . . . experts have noted the uncertainty prevailing in the successors of old texts, the attempts in them to establish correct names and their significance.[62]

It was at this point that I was struck by a great sense of the obvious, as if everyday things are so shrouded in mist that we are rarely able to see their significance. I had been poring over a book about churches in Rome[63] when I noticed the church that is the headquarters of the Society of Jesus, the Jesuits. It is called the Gesu. Something about the name struck me as being plural, though I could not quite put my finger on it, something that tacitly implied many Jesuses, not one. In the sense of the order being a missionary one, the many Jesuses would be spread out worldwide, all of them being in essence the same figure, but again with regional variations. I was fascinated that the idea of an archaic worldwide Jesus hero figure was again being reflected here.

I was further intrigued when I discovered that inside the head-quarters resides one of the great masterpieces of the Renaissance. It is a mural by Bacciccia.* If the word *god* means "voice," then the title of this masterpiece spoke volumes; it is titled "The Triumph of the Name of Jesus." In the mural, where the name should be, there is a glorious flash of light and, barely discernible in its midst, a cross.

Again, this reflected my thoughts about the power of names and of acoustic phenomena. Inspired, I strode forward in my researches with a renewed raison d'être.

THE GISA CORRELATION

The name Jesus is the Greek form of the Hebrew Joshua, sometimes spelled Yeshua. This in itself is a foreshortening of the longer Yehoshua. *Jesus* means savior, "the Lord is Salvation." Sometimes, in very old writings of the Jews, called Toldoth, there are variations such as Jeshu or even Jesha, but they are the same name regardless. My eyes wandering through the Old Testament, I noticed that there seem to be variants of this name throughout, but one that came to my notice most readily was the name of Esau, the unfortunate brother of Jacob who, in one of the grubbiest episodes in biblical mythology, has his birthright stolen through a cheap trick played by his younger brother upon their father, Isaac.

In phonetic Arabic script, the names of Jesus and of Esau (who gave his name to the land of Edom, from whence Herod came) reduce down to Gi-sa and Gi-su, respectively.[64] Now, if all of this seems unnecessarily complex, there is a method to my madness in that this transcription raises an interesting anomaly. The name of Esau, to which the name of Jesus is clearly linked historically as well as linguistically, has no Hebrew etymology or source. In other words, both names precede Hebrew history and are therefore very, very ancient. This age indicator is significant and is supported by another venerated source.

---

*Bacciccia was born Giovanni Battista-Gauli and executed his masterpiece between the years 1672 and 1685 in the Baroque decorative style.

In the Koran, one might expect that the name used for Jesus would be derived from the Hebrew, Yeshua. Quite clearly, it is not. Yeshua in Arabic would be transliterated as "Jeshu," but the Koran calls Jesus *Issa*. This is reflected even in early Christian-Arab translation of the name: instead of *Yasu* we have *Isa*. The Jesus of the Koran is different from the Jesus of the Gospels: he is older by far, for the Nasara, the Nasarenes of the Koran, are a sister sect of Judaism, *not an offshoot of it*.

A god called Issa was most certainly recognized during the first century period right up until the coming of Islam,[65] and, going back further, Isa (pronounced "Eesha") is the name of a Upanishad of the Indian Vedic tradition. The Upanishad scriptures were recorded between 800 and 200 BCE.

The Isa Upanishad focuses on Self or "being." Throughout scripture the favored definition of being is "I am." These are the words uttered at the outset of many a creation myth worldwide, and this act is described in the same myths as the act of creative purpose. On an individual level, the statement "I am" means that the speaker is at one with the divine. This was a favorite term of Jesus. "I am the light" would therefore mean "I am at one with the light that is divine"—light, of course, is vibration at a high level. "I am the way" would be understood as "the attainment of heaven is only available through the divine."

The attainment of the divine was the goal of the hero and from that place of power, great things could be accomplished. As if to confirm the ancient understanding of at-one-ment and of the hero, in some languages we can see a very clear link between hero names and the term *I am*. For example, the French *Je suis* is close to the name of *Jesus*.

There are some further interesting correlations that lead us on to the most controversial aspect of the thesis. The Greek mythological horse, *Pegasus* means "source of water." In the legend, Pegasus flies to Mount Helicon and strikes a rock with his hoof to create a flow of ice cold, crystal clear water—the Hippocrene spring (from the Greek *hippo*, "horse").[66] This is reminiscent of the Old Testament story of Moses in which he strikes the rock of Meribah and water gushes forth to ease the thirst of the Israelites (Exodus 17:1–7). This water saves the wanderers in their parched state; it is the water of life.

So it is that we have come to the connection of the name with the actual elements of the earth. We have already seen the importance of holy water in various religious rituals and its presence at sacred sites. Now we can see the beginnings of a link between the site and its hero and, more specifically, the name of that hero.

In Greek, both Issa and Jeshu are rendered *Iesous*. This indicates that they spring from the same source. *Jeshu* means literally "God-man." However, "God-man" doesn't tell us much, but *Issa* or *Isa* does; it gives a clear indication of age and, as I will show, a clear indication of origins.

In the Greek and Hebrew myths, the names of particular heroes begin with *Y, I, J,* or *G.* When I first investigated the origins of these letters, much was revealed about the nature of the name, much that seems familiar. The *G,* the seventh letter of our alphabet, is derived from the letter zayin of the proto-Sinaitic alphabet (ca. 1500 BCE), where again it is the seventh letter.[67] In short, the letter zayin represents a face-to-face encounter between two people, two armies, or two situations at its highest level, a meeting between God and man.

However, it also denotes "skin pierced with a shaft," and the hero in myth is almost always pierced with an arrow, a lance, or with nails. This is borne out by the fact that in the alphabet now known as proto-Canaanite (ca. 1200 to 1050 BCE),[68] the sign was transformed into a *T.* This T is the cross or crucifix. Within the Phoenician royal house, this same letter was branded upon the forehead of the royal princes, some of whom were said to have been sacrificed as sons of god. In Greek, *zayin* becomes *zeta,* also the seventh letter of the alphabet. Seven again seems to be intimately connected to the hero. Tammuz, the "child of the seventh" was also a pierced hero.

The letters *I, J,* and *Y,* which in Latin and Greek are sometimes interchangeable, are all derived from the tenth letter of the proto-Sinaitic alphabet, yod.[69] This in turn comes from the Egyptian hieroglyph of a hand. This letter is pronounced "tot" or "dod." Dod-man again. In the Jesus story, *I, J,* and *Y* are connected to *G,* because when the hero is crucified it is the hand that is pierced by a nail. This offers another angle into the dispute over whether Jesus was nailed through the wrist or the palm of the hand.

In being the tenth letter, yod is a sign meaning "unity" or

"completeness"—I am—hence its use in the name of God. There is another interesting observation to be made. The name of Jesus is the late Latin translation of the Greek *Iesous,* which is itself a translation of the Hebrew letters yod, heh, shin, vav, and heh. This is the holy Tetragrammaton, the secret name of God—yod, heh, vav, heh, or YHWH—written around the additional letter shin, which in parts of Hebrew tradition was an expression of the feminine presence of God.

In common with other ancient scripts, written Hebrew omits the vowels and reciting Hebrew scriptures requires great dedication, since the text has virtually to be learned by heart. The vowels required for pronunciation are passed down the generations. In the case of YHWH, the proper pronunciation of the unpronounceable name of God was passed down from high priest to high priest and was believed to have been lost approximately 500 BCE. This is not surprising, since it was only offered once a year—on the day of Atonement (Yom Kippur)—when the high priest alone was allowed to enter the holy of holies in the Temple of Jerusalem. Here on the site of the old Temple of Solomon he prayed on behalf of his people.

The significance of shin, the letter added to the middle of the Tetragrammaton to make the word that eventually becomes known as Jesus, is "fire" or "light." This therefore renders JHSVH as "the fire or light of God," and could equally be pronounced *Jesus* or *Joshua.* Note, too, the closeness to *Issa.* The letter vav represents and means "nail" (its symbolism is a spiral denoting cosmic forms in motion, the Word of God spiraling outward in manifestation). What is extraordinary is that these elements come together not only in the name of Jesus but also in the name of one particular Roman patrician family.

The name *Caesar* has bequeathed us the words for emperor in a few European tongues—*kaiser* in German, and *czar* or *tsar* in Russian. Originally the Latin *C* was pronounced as a hard *k,* but it could also be pronounced as a *g.* It was this *G* that later became the seventh letter.*

---

*The letter *G* was a late introduction. It came in the Old Latin period as a variation on the letter *C* to distinguish voiced *g* from voiceless *k.* The historical originator of *G* is freedman Spurious Carvilius, who taught circa 230 BCE. At this time, *K* had fallen from favor, and *C,* which had formerly represented both *G* and *K* before open vowels, had come to express *K* in all environments.

The Caesars wished to be respected as hero-gods; indeed, Julius Caesar's divine pretensions were foisted upon all. Now we will see the ancient heritage of Caesar's name, it is really Gaesar, pronounced *geezer*. The gods are divine, they are of heaven, but they are sought after here on earth. Ancient Egypt has been shown to have been particularly intent on the bringing of heaven to earth.

In ancient times, the divine, as we have seen, was sought to be embodied in place, as at sacred sites and in people—the heroes. The heroes embody the heavenly here on earth, creating a link and setting an example. What linguistic device would be used to ground the divine attributes to earth? The prefix of the *G,* the ancient letter that denotes the divine on earth, the *G* of Geb and Gaia. *Now Isa becomes Gisa, and we see that Gisa is the embodiment of the place of Isa.*

In modern Egyptian Arabic, *Gisa* means "proximity" or "nearness."[70] If we realize that originally languages related the highest meanings in relation to the divine, "nearness to God" would be implicit. For myself, I felt that I was getting near to something approaching a source, after many years of searching. *Gisa* seems to be an expression of the child being brought forth: that child is humanity, it is an expression of "I Am," and of humanity's fragility. The similarity of the syllabary worldwide throughout history is there for all to see.

Oddly enough, while researching this thesis I came across the London slang term for a man, *geezer*. I thought that I was taking obsession to new lengths by looking it up, but much to my astonishment, I learned that this term had been brought back from the Spanish Peninsular Wars by the soldiers of the Duke of Wellington's army. It is a Basque word and it means "man" or "the man who came"—it is a reference to the hero. A *geezer* was originally a slang term for a heroic man. The "god-man who came" to set an example. The term is usually expressed as "Blinding Geezer!" A reference to the hero shining with the divine light of glory ("May the Lord make His face shine on you," Numbers 6:25). Basque is probably the oldest European language; its root is unknown and it lies outside the remit of the Indo-European language table.[71]

Further applying the Gisa key, somewhat tentatively at first, I began to follow other clues and uncovered an overall correlation pattern.

We've already encountered many heroes, including Jesus and Esus, but there are more to follow. Eshu is the messenger of god in the myths of the Nigerian Yoruba people. He, too, was associated with language and could speak them all, and like Jesus, he brought God's instructions to earth and took back to the gods the smoke and the scent of sacrifice. From the tales of the Creek people in the southeastern United States comes Esaugeteh Emisee, whose name, meaning "master of breath," is a reflection of Latin *spiritus,* also meaning "breath." He modeled people out of mud, as God did with Adam.*

We have seen how, in the ancient Middle East, these names would be prefaced with a *G* to confirm the "earthing" of the heavenly power. Thus, we would have Geshu and Gesaugeteh. It seems that this was also the case in ancient Tibet, for there we find Gesar of Ling, the early Tibetan savior god, whose second coming is much anticipated. Another version of his name is Kesar of Ling: thus demonstrating the link of *G* to *C* in its hard form of *k,* an occurrence in many languages.

Naturally, I was looking for the oldest of these correlations, and for obvious reasons my focus became firmly settled upon ancient Egypt. Through a study of names, we have come to the conclusion that Gisa is the root name of heroes, but Gisa is also a place. Usually spelled "Giza" to match the local phonetics, it is the site of the world's most fabulous monuments, the three Great Pyramids of Egypt. However, even before I discovered the Gisa/hero connection, there was one thing of which I was already certain: where there are monuments there is a hero. Gisa/Giza has been the name of the pyramid plateau since long before the arrival of the Arabs circa 600 CE.† Although it is disputed in some

---

*Also in North America, the Apache Indians have a hero figure called Herus. This figure is taken to be Jesus, and he is said to have come from their first contact with the Spanish. Their version, however, is too near to the original hero of the Egyptians, Horus, for the Spanish story to be easily acceptable. It is only when we dig deeper that we learn that in ancient times the tribe was given a book, a holy book, by a man named Herus, but that on his death it was burned as it was the custom to burn the possessions of a dead man. What can we make of this? I am inclined to the more ancient view because many Native American peoples had or still have heroes that bear striking similarities to the holy child. The Chiricahua called theirs "Child of the water"—baptism again?

†The Arabs adapted an existing place-name.

areas of Egyptology, the area has an intimate association with the god Osiris: the ultimate prince of the earth.

Osiris is the Greek translation of the Egyptian Asar. Asar is the son of the Earth, Geb. Once again, we shall preface the name with *G,* affirming that the divine is present on the Earth. It gives us *Gasar,* the same as *Gisa.* Osiris was always felt to be the nearest god to humanity: *Gisa* means precisely that, "nearness" or "proximity." Gisa can also be spelled "Gesa" or "Gesu." These are variants that do occur with myth, for reasons that we shall soon see. The name means "prince of the earth" or "blood of the earth" and is the key to unlocking a considerable amount of the world's mythology.

Osiris, to use his later and more familiar Greek name, is almost always portrayed as carrying a Djed pillar, a mysterious-looking object reminiscent of a Greek column. Osiris in his earliest incarnation was known as Djedu. *Djed* is an Old Kingdom word. The Old Kingdom is one of the oldest periods in Egyptian dynastic history. Within it there were two kings, the oddly named Djer and Djet, both meaning Horus, the son of Osiris. The syllable *dj* represents Horus in Old Kingdom hieroglyphs. When the living Horus, the regnant pharaoh, passed on, he became Osiris. Osiris, you may recall, is the Greek version of Asar. This meant that Dj became Asar* (sometimes called and spelled Wsir or User). This process would also be recognized where the hero was considered to have attained immortality while still alive; that is, to have moved from the small self to the universal Self, from "the" to "The." In other words, to have become a god-man, a god on earth.† This of course was the mark of the true hero *and could only be achieved by the rite of spiritual death and resurrection within the temple and all that it can do to affect our state of being.*

The composite term would be *Djasar* or *Djoser.* This last was

---

Dj is often translated as "Zer," this being another form of Osiris's name in archaic Egyptian—Asar.

†In Scandinavian mythology the father god, Woden (sometimes Odin), crucified himself upon the World Tree as an act of renewal. I was hardly surprised to learn that the collective name for the Scandinavian Norse pantheon of the gods was the Aesir or Aesar—Asar, and so Gisa, again.

a Third Dynasty king of Egypt, the builder of the first of the Great Pyramids—the step pyramid at Saqqara. This was the first known structure built with masoned stonework rather than rough stone.[72] Jesus was called the Alpha and the Omega; so, too, is this present in Zoser's name: Z we know is the last letter of the alphabet, whereas Asar the Egyptian name of Osiris, here present as *Oser,* means also "Prince Alpha" in Greek translation. Thus the earthly Horus, at one with a Osiris/Asar and therefore expressed as Djasar, would be the forerunner of Gasar, Gisa, and Gesu.*

De Santillana and von Dechend in *Hamlet's Mill* make the significant comment that

> once having grasped a thread going back in time, then the test of later doctrines with their own historical developments lies in their congruence with tradition preserved intact even if half understood. . . . And universality is in itself a test when coupled with a firm design. When something found, say, in China turns up also in Babylonian astrological texts, then it must be assumed to be relevant. . . .
>
> Take the origin of music. Orpheus and his harrowing death may be a poetic creation born in more than one instance in diverse places. But when characters who do not play the lyre but blow pipes get themselves flayed alive for various reasons, and their identical end is rehearsed on several continents, then we feel that we have got hold of something . . . where the Pied Piper turns up both in the medieval German myth of Hamelin and in Mexico long before Columbus, and is linked to both places with certain attributes like the color red, it can hardly be a coincidence.[73]

We find precisely these similarities in the Gisa correlation pattern. A case in point is the ancient figure of Zoroaster (sometimes Zarathustra). The founder of an ancient Persian religion, Zoroaster's teaching is a dualistic one of the continuous, eternal battle of good

---

*As a further confirmation of the Gisa-Jesus link, in *Secrets of the Great Pyramid,* Peter Tompkins tells us that *Giza* (*Gisa*) is transliterated by various authors as *Djiseh* or *Jeeseh,* the *G* being pronounced hard by the Egyptians and soft by Arabs from other regions.

against evil. Good is represented by Ahura Mazda and bad by Angra Mainyu. These are the usual elements to be found incorporated with the sacred text of this religion, known as the Avesta. The Avesta, as it is named, bears elements of the secret word that God spoke at the creation, God's own name, AWE. The good god, Ahura, is again another form of "the Lord"; it has the element of hu within it. It is Zoroaster's name that is here of particular interest. This is underlined by certain elements of his myth in which his birth is attended by marvels. He inherits the glory of Yima—in other words he is a son of Yima, judge of the dead, and, like Moses, he has the tenets of the faith revealed to him by god, in this case, Ahura Mazda.

Zoroaster's name seems in itself to be dual. Zoroastrianism as a historical phenomenon has been much disputed. The Parsees, its followers, believe it to be as much as eight thousand years old. More realistically, a conservative dating puts it at circa 1200 BCE. Needless to say, at its height in the first millennium BCE, Zoroastrianism's impact upon the development of Palestinian religion would be telling. It is essential to be aware of this when looking at Zoroaster's name: *Zr* in ancient Egyptian is *Dj*—Horus. *Aster* would be related to Astarte, goddess of love, whose symbol was the bright morning star, Venus. Looking for a root we find that *Ast* in ancient Egyptian is what the Greeks translated as "Isis." What is fascinating is the fact that Isis's original name is still alive and active today in astronomy and astrology.

Zoroaster's name then would seem to mean Horus, son of Isis, and to be based upon the theme of an ancient mythical cycle. It was while coming to terms with this aspect of the correlation that I came across other extraordinary anomalies that could only be answered, with any ease, by the formula described above. I was struck by a comment in Leonard Shlain's book, *The Alphabet versus the Goddess*: "Many historians have puzzled over the rapid and enthusiastic conversion of Egyptians to Christianity in the second century. Egypt was, after all, an immensely ancient culture whose principal characteristic was *resistance* to change. Despite having been conquered by diverse foreigners throughout its three thousand year history, Egyptians retained their fealty to Osiris, Isis."[74]

Shlain then goes on to make the interesting point that, whoever the conqueror might have been, it was *they* who were converted, not the other way around. In a similar vein, it could be that the incoming Christian religion, being the old tradition recycled, was too close to Egypt's own to be worthy of great resistance.

In the Old Testament, King David's father is called Jesse, a name quite often spelled "Isai." In the book of Isaiah 11:1–10, the first reference to "the stump of Jesse" appears, an interesting metaphor seen as a major component of some of Europe's greatest stained-glass windows, all of them inside Gothic cathedrals. In the legend of Osiris, the hero is trapped inside a coffin that, having been sealed with lead, is cast adrift and ends up at Byblos, whereupon it sprouts leaves and becomes a tree; upon its discovery by Isis, it is cut down.*

Jesse is clearly a Jesus name, and the Osiris link is no coincidence. How can it be when we compare other aspects of Osiris's myth to that of Jesus? The term used by the people of Byblos to pay homage to the wood within which Osiris is hung is rendered in the Greek *toxilon*.[75] Exactly the same term is used repeatedly in the New Testament for the cross.[76] Trojan history (ca. 3000 to 1250 BCE) also records this term, for the Trojans prayed to God and his son, an angel called *Tas* or *Tasya*, to resurrect them through the "wooden" cross, the figure of which appears in effigy on various amulets.[77]

JESUS BCE: CHRISTIANITY BEFORE CHRIST

In an image that is strikingly reminiscent of the god Osiris in the Egyptian Mysteries, Saint Paul says the following of Jesus:

In him everything in heaven and in earth was created, not only

---

*The rendering of the Hebrew text is very interesting here. In Betz and Riesner's *Jesus, Qumran and the Vatican* (180n2), the translation is given as *mig-geza*—"from the stump." Other trees that have been cut down in the same fashion feature in the myths of the Cuna Indians of Central America, wherein the sun god, Quetzalcoatl, disguises himself as a tapir and cuts down the Palluwalla tree. This tree is the World Tree that stands at the center of the universe.

things visible but also the invisible orders of thrones, sovereignties, authorities and powers. The whole universe has been created through him and for him. He exists before all things and all things are held together in him. He is the head of the body, the church. He is its origin, the first to return from the dead, to become in all things supreme. (Colossians 1:16–17)

Everything in this statement applies to Osiris, the god-man of the early Egyptian period whose death hailed the coming of the nascent Horus. *Nascent* means "being born" and comes from the Latin root *nasci,* "be born."

In the Gospel of Matthew 2:15, a prophecy is made to come true in the exile of the holy family to Egypt. Quoting Hosea 11:1, Matthew describes the words of God: "I have called my Son out of Egypt." Upon the return of the holy family from out of exile they settle in Galilee, so that again prophecy comes true: "He will be called a Nazarene." As Kamal Salibi points out, the followers of the Arabian Issa were called the community of Nasara; in other words, the Nazarenes.[78]

This particular god, Issa, had been born at Mecca in the period before the rise of Islam, which is interesting because Mecca, as well as being famous now as the central shrine of Islam, has always been famous for the Ka'aba, the cube-shaped black meteoric stone. As we have seen, the holy child is invariably the son of a mother goddess, and what can be more explicit than the fact that Cybele, whose name gives us the derivation of the word *cube,* was worshipped as a black meteoritic cube-shaped stone?

Scholars have long pointed out the textual differences between various of the Gospels and have narrowed down the file to three possible sources, Q, M, and L. It is generally assumed that Q is a lost Greek source that gave rise to the Gospels of Matthew and Luke and that additionally, Matthew drew quite separately from another source, M, while Luke drew from the source L. It has been suggested that L and M were Aramaic. I contend that it is quite possible and entirely plausible that the source gospels were derived, in part, from one original source, the pre-Islamic story of Issa, from the oral tradition. Issa's hero mythology is rooted in Arabia and predates Christianity by as much as 1,200 years. He was born of Maryam, he worked miracles, he was cruci-fied, and he rose on the third day.

It is a widely held view that the oldest of the gospel texts was the Aramaic Nazarene Gospel. The origins of this gospel would have been from within the oral tradition. I have already pointed out that, in the Islamic faith, Jesus is known as Issa. Now we see that Issa's story is very close to Jesus's story as accepted by Christians and yet maybe 1,200 years older than Christianity itself. This is a puzzle until, noting that that the story's roots are in Arabia, we learn that Saint Paul spent a mysteri-ous three years there, a point that is enlarged upon by Kamal Salibi in *Who Was Jesus?* Salibi's argument is that while in Arabia, Saint Paul had access to the original of the gospel sources, the Nazarene Gospel, and that it is from this that he gained his image of Jesus.

Early writers confirm the Nazarenes as the source of gospel knowledge.

> The introduction to Luke and the accounts quoted by Eusebius from the early second century writer Papias . . . show that all the gospels except the patently fictitious ones were based on notes taken by Greek-speaking converts from the Aramaic gospel already current among the Nazarenes; and that each evangelist, as Papias reports, "interpreted them as best he could"—that is to say, uncritically, and, in general, with studied ignorance of their historical background.[79]

Paul would therefore be the foremost evangelist of the Nazarene Gospel, working hard to bring coherence to the beliefs and practices of the fledgling church. Furthermore, it is highly likely that Paul's work strongly influenced the canonical Gospels. His own letters predate any of them and in these letters, there are few references to anything approaching a historical Jesus; as we have already seen, Saint Paul's pre-occupation is with divinity. There is also little doubt that Paul knew of at least one other Jesus, otherwise why does he refer to "another Jesus," as opposed to another saint, teacher, or Christ? "For if he that cometh preacheth another Jesus, whom we have not preached, or if ye receive another spirit, which ye have not received, or another gospel which ye have not accepted, ye might well bear with *him*" (2 Corinthians 11:4).

As we know, the problem with transferring the oral to the written is that, from then onward, its meaning becomes subject to editorial niceties and misinterpretation—and the central mystery lost. When James Joyce wrote *Ulysses,* he did so in his illegible longhand. For years, he was lauded for some of the most surreal but effective imagery in history of literature; "the man wrapped his beard in a piece of newspaper" summed up a truly absurd image—until the corrected text was published and Joyce's irregular spelling was exposed. The real text read, "the man who wrapped his bread in a piece of newspaper."

Regarding this period of transition from the oral to the written, I was intrigued by the fact that "no illustrated Christian manuscript survives from before the fourth century AD, but the tradition of sacred illustration, in the Book of the Dead style, had been firmly established among Greek Jews from pre-Christian times."[80] It is highly likely that there would have been such early Christian manuscripts and I wondered that perhaps the content was not to the later church's liking and so destroyed. I see this in the light of the existence of early manuscripts of the Egyptian Coptic church that use illustrative religious symbolism that relates both to older Jewish work and to the unfortunately named Egyptian Book of the Dead.* As a trained artist I humbly suggest that scholars should be looking at the iconography rather than the text of these works, for I venture that they may well reveal hidden links and meaning.

You will have noted here that although the theological elements of the root gospel have been dated back to 1200 BCE, it is still named the Nazarene Gospel. Significantly, there are Nazarenes (sometimes referred to as Nasoreans or Nazareans!) and Nazarites in both books of the Bible. Eusebius, in quoting from a source, now lost, relates the probability of the two being one and the same thing: an observation made by Professor Robert Eisenmann. The Nazarites are extremely ancient religious orders, whose influence upon the rise of cultural belief and

---

*Better named as "The Book of Coming Forth by Day" or "The Book of Coming Forth by Light."

religion has been lessened and to some degree diminished by later editorial incursion.

Joseph, he of the many-colored coat, is a Nazarite, as is the hero Samson. This is interesting in the sense that Joseph is described as such only *after* he had been in Egypt, not *before* he went there. Quoting from the memoirs of Hegessipus we see that, like Samson before him, James, brother of the Lord, was born "holy from his mother's womb; drank no wine, strong drink, nor ate animal food; no razor came upon his head; he neither oiled himself nor used the bath; he alone was permitted to use the holy places. . . . He was called 'the righteous.'"[81] All of these are Nazarite values, as Hegessipus makes clear, yet James was, according to the official view, the leader of the early Nazarene church in Jerusalem. The implication here is that whatever or whoever the Nazarites and Nazarenes were, they were older than the traditional founding date of the church and that they were one and the same thing.

It was at this point that I discovered another interesting anomaly. The early church father, Clement of Alexandria states that "Alexander, in his book on the Pythagorean symbols, relates that Pythagoras was a pupil of Nazaratus the Assyrian."[82] This strongly infers that Pythagoras was a Nazarene—six hundred years before Saint James, the Lord's brother.

It is another church father, Epiphanius, a very important authority on the sects within the early church, who describes the Aramaic *Natsariya*—the Nazarenes.* He also willingly admits that they are pre-Christian.[83] In the Old Testament, in the books of Ezra and Ezekiel, the Israelite heirs of King David are accorded the title of *Nasi,* meaning "prince" or "chief." In other words, sons of David—which is how Jesus is described when, in the Gospels, he is welcomed into Jerusalem. The title, however, is probably older than King David (ca. 1000 BCE) and comes from a land not too far away, Egypt.

Here I offer you two quotes. The first is a stern warning regarding Jesus from the Jewish commentary Talmud Babli Sanhedrin 103a:

---

*Significantly, showing the roots of the Nazarene knowledge, Plutarch states: "There is no difference as a matter of fact, between the texts called hieroglyphs and most of the precepts of Pythagoras." It is well known that Pythagoras spent much time in Egypt.

"That thou shalt not have a son or disciple who burns his food publicly, like Jeschu ha-Notzri." (Note the variation on Jesus's name.)

And now, Plutarch, from *Of Isis and Osiris:* "When every one of the rest of the Egyptians eats a broiled fish before his front door, the priests do not taste it, but burn their fishes to ashes before the doors [of the temple]."[84]

Placed together, not only do these two quotations identify the practices of Jesus the Nazarene and those of the Egyptian priests but also, the latter quotation shows the sacrificial offering of the fish before the entrance to the house of God. We know that for the Christian church, from the earliest times, the fish represented Jesus himself: this points to Jesus acting a role originally ritualized in Egypt. Let us therefore go with Plutarch's lead, into the land of Egypt where there is much that comes to light.

In the early church iconography Jesus was always portrayed as a king. He was never portrayed as dying upon a cross, certainly not before the medieval period. In Egyptian hieroglyphs, the word for king is spelled[85] ⳥ 𓈖.

This equates to the letters *N, S, W* in the Western alphabet.

This is the origin of related words in Arabic and Aramaic and Hebrew. As mentioned above, *Nasi* in Hebrew means "prince" or "chief," whereas in Old Semitic, *nazar* means "keep," "guard," or "protect." A quick glance at the Old Testament book of Kings reveals that the ultimate guardian of religious secrets was none other than the king himself. The first king of Egypt, after whom all the other kings followed, was Osiris. The followers of Jesus were the followers of the king.

If we now apply what we have learned about the actual identity of the mythical, theological Jesus and utilize the older version of his name, we will soon realize that the term *Nazarene* contains the original Egyptian name of Osiris—Asar. In short, to be a Nazarene was originally to be a follower of Osiris.[86] That Jesus was a Nazarene is widely accepted. That he was so called because he was of a village called Nazareth is contentious to say the least—there is little evidence to suggest that it existed in the early first century CE. If a settlement did exist, then it was a dwelling place of Nazarenes and named as such.

Their preservation of their own roots was feared by the Temple authorities, for what they preserved was the real history of the original Hebrew theology—and the fact that it emanated from within archaic Egypt.

INITIATORY TITLES

Having arrived at this position it became very necessary to step back and cast a skeptical eye over the mountain of data. However, the skeptical approach only served to yield more and more information, substantiating what I had already found. It was not as if the idea of a semblance of Jesus and Christianity existing at the time of Egyptian Old Kingdom, three thousand years before the established date, was shocking; it was just the fact that it took a considerable while to sink in. The thought had never occurred to me before.

Nevertheless, once it did sink in, more pieces began to fall into place. It was this latter process that struck me as being almost miraculous and, once my resistance was surmounted, a breath of fresh air began to blow away some of the cobwebs of my mind. The very thing that dissolved any lingering dissatisfaction was the aspect of the sacred site and its link with consciousness and the hero. I saw that an even broader view was required, broader than I had as yet applied, one that encompassed the sites, the mythos, and the linguistic angle, as well as the angle of faith and its expression within ritual.

This broader view bore fruit. Now I was beginning to think that the names of the "Gisa" correlation, give or take regional variations of culture, geography, and climate, were all somehow universal. To recap—this was a name of power. As we have seen, the sacred sites have the extraordinary ability to alter our consciousness. However, it is the method and the mode of *how* our consciousness is altered at these places that is now critical to an understanding of the nature of our quest.

It is imperative therefore that the name of the hero, the deeds of the hero, and his myth are understood as being the all-important and critical clue. The hero's name was chanted, as today the name of Jesus is sung and chanted in modern Christian ritual, which derives from antiquity. We have forgotten the power of that name, only limiting it to a kind of modified superstition, but now we can show, we can demonstrate beyond doubt that the name actually *does work as a name of power,* this is the name of God.

Seership and enlightenment are descriptions relating to consciousness. In the same way, the idea of Jesus and Christ seemed to me to be a state or quality of consciousness rather than simply an individual who did great things. This is to some extent borne out by Clement of Alexandria when he admits that the Christ familiar to him did not require to be nourished, to be fed.[87] Clement's Christ sounds distinctly metaphysical, and thus more to do with a state of consciousness or being. Now, God works in mysterious ways, but we are beginning to get to grips with the mystery. I have no doubt that the names are titles, descriptive of a state of being and given in relation to the secret sciences that we know the Ancients had and some of which are practiced today. The Gisa/Jesus names are initiatory titles, their root is the same, whereas the localized mythology—as well as pronunciation and hence, spelling—has been manipulated so that the truth has long been obscured. If you have a secret, you have power over those who do not. Religion, after Nicea, was more political than spiritual.

In Iran, the power of Mother Earth was called *Geush Urvan,* another variant upon the Gisa name; *Geush,* meaning literally "man of the earth" (but in the sense of him being the "Bull of his Father"—a titulary to be seen in Canaanite mythology and elsewhere), but when Geush was given form as a bull, the god Mithras killed him to transfer his energy to the sky.[88] We are back again to energy, to the vibrating powers of the sky, of the Earth, powers related to the Great Goddess, the Divine Mother, represented as Mary, mother of Jesus.

MARY, MARY, QUITE CONTRARY

In the books of the Bible, the name of Mary, in its various forms, occurs with regularity. In the Old Testament book of Exodus 15:20–21, mention is made of Miriam, sister of Moses and Aaron. It tells of how she led the women of Israel in a hymn of victory to Yahweh, after the

Red (or Reed) Sea crossing: "Miriam the prophetess, sister of Aaron, took the drum in her hand. All the women went out after her with drums and dances. Miriam declared to them: 'Sing to the Lord, for he has triumphed; horse and its rider he has hurled into the sea.'"

This relationship between Miriam and her brothers is now seen by scholars as an embellishment upon various early traditions, in other words, it is a piece of later propaganda, either inserted or even mistranslated in order to show her in a particular light—but also to put her firmly in her place, a lesser place in comparison to Moses and Aaron. This belies the degree of prominence that she quite probably had in the original legend.

The role of women and of goddesses is a controversial one within all areas of ancient Judeo-Christian studies, and there are very relevant questions that still have to be raised over the "masculinity" of much of our view of history. I am entirely sympathetic to the view of the Greeks that "god without goddess is spiritual insufficiency,"[89] and without a doubt one does not have to be feminist to see that women do remarkably badly between the covers of the Bible. From Eve onward there is an antipathy toward women that at its worst is positively virulent and at best downright condescending. It is with this in mind that we must approach the biblical feminine gently and with hesitancy, for there are surprises afoot.

In the disputed noncanonical book of Jasher, Miriam's story is given an entirely new slant. Here she is not just a briefly mentioned sister of Moses and Aaron, but the actual spiritual leader of the Israelites in exile. For many scholars, this understandably seems to be an exaggerated point of view to say the least, but it is not unlikely.[90]

A key to Jasher's legitimacy is that it is cited as an authority in Joshua 10:13 and 2 Samuel 1:18. Significantly, the actual theme of the book of Jasher is open to discussion; rather than being a book, it was in all likelihood originally a collection of archaic poetry, much of which has since been lost. Many commentators dismiss Jasher as having been primarily in oral from, but this was the case with all the books of the Old Testament until they were written down at a relatively late stage.

Miriam or Mary as a name has led some commentators to conclude that this woman was Egyptian in origin and of the tribe of Levi. This is reflected in the New Testament, where Jesus and his mother come out of Egypt. Furthermore, there is even a Saint Mary of Egypt whose biography is doubtful, a mark of the early church carefully covering its tracks. It was the theme of this Saint Mary of Egypt that led me to a vital aspect of the overall correlation pattern that we are now entering into; as I will reveal, the fact that Egypt had a saint called Mary is a clever subterfuge that deflects the inquirer's gaze away from a contradiction.

In many versions of her biography, this Mary lived the life of a hermit in Palestine. According to Cyril of Scythopolis, in his *Life of Cyriacus,* two of Cyriacus's disciples first came across her in the desert beyond Jordan. Upon their second visit, they found her dead and therefore they buried her. Christian tradition explains that Mary was an Egyptian who left home at the age of twelve and went to live in Alexandria, where she was a prostitute for seventeen years. It was on a subsequent pilgrimage to Jerusalem that she found herself being prevented from entering a church by a powerful and irresistible force. Upon sighting a statue of the Blessed Virgin, she was told, in a vision, to go across the Jordan where she would find rest. She spent the rest of her life there, subsisting upon a diet of dates and berries, and the divinely communicated Christian faith.[91] Significantly, the imagery of Saint Mary is often confused with that of Mary Magdalene—this is because in all likelihood, Saint Mary of Egypt and Saint Mary Magdalene are one and the same entity. The story of Saint Mary of Egypt is no mere fantasy, simply concocted as a reason for a Christian presence in Egypt, it contains a kernel of truth that is both surprising and reveals much.

However, before the revelations can begin, it is necessary to show how ancient Mary really is. The Catholic Church called Mary by the sacred title of *Theotikos,* "God Bearer," or as the more renowned, "Mother of God." Wherever one goes it is always the same, wherever there is Mary, there she is as Mother of God. In the Eleusinian mysteries the virgin-born hero is introduced with the cry, "Rejoice, rejoice, we have found our king, son of the daughter of the sea, lying in this basket among the river reeds!"

The sea, in Latin is, *mare. Mare nostrum* can mean either "our mother" or "our sea." This is a word that still thrives within modern discourse in the words *maritime, marine,* and *meridian.* Debussy's wonderful music *La Mer* is French for "the sea." The French *mercredi,* or Wednesday, is "Mary's day," a haberdasher is a *mercier* or *mercer,* and so on. Another entrance into Mary's mysteries is the narcotic that we call cannabis. It is familiar as *marijuana* and means "young Mary." That the origin of this terminology is very ancient is borne out by the guiding lights of the ancient navigator, the stars of the sea, the Stella Maris reference that is even older than those great seafarers, the Phoenicians. There was even a land called Mari, the land of the Amurrites, circa 2600 BCE. The very presence of this people demonstrates quite how archaic the name really is.

The name may also be a reference to a star, the brightest star in the sky, Sirius. Sirius was called Sothis by the ancient Egyptians and was the personification of the celestial Isis, whom they called Ast, or sometimes Aset. From this we get the prefix *ast* in astronomy, astrology, and more. It is also a component of the word *pastoral,* from which comes the word *pastor,* meaning "shepherd." Pastors are priests who busily attend their flocks.

Many within the church are devoted to Mary, but in this, are they are unknowing devotees of Isis? I felt there was possibly a connection between the two. I decided to dig a little deeper: the answer was marvelous—from the Old French *merveillos,* meaning "veil of Mary." Significantly, in the mysteries, Isis wore a veil, too. We are now about to part that veil to reveal the face of great splendor and of grace.

First, I must explain an apparent anomaly, *mer* in Egypt has many meanings, one of which is "death." This is reflected in the Latin *morte* of the same meaning. The Slavic *Mara,* sometimes *Mora,* was viewed as a destructive female spirit who actually tracked the blood of men, an activity that she pursued in the depths of night, thus inducing her own form of night*mares.* In the Old Testament we have *Marah.*[92] And as *Maranatha,* she is given her own slot in the New Testament in Saint Paul's letter to the Corinthians. The death-dealing Mary, though, must be seen in the context that she who takes life not only gives life, but can also bestow immortality.

In this light, it is all the more significant that, at the crucifixion, all of the requisite Marys are gathered. Furthermore, it is within this part of the Gospels that the story comes about full turn, for Jesus's body is anointed with myrrh, the very substance given to him by the three wise men at the nativity. Christians call Mary, "Myrrh of the sea." It was Myrrha who was the mother of another crucified hero, Adonis—at precisely the same place where Jesus was born, she, too, was delivered of her holy child. It was thorny myrrh twigs that were used as the crown of thorns. *Myrrh,* as we can see, is a variant upon Mary's name.

Mary, therefore, is older than the gospel tradition. Thus, it comes as no surprise that Mary appears within the Nazarene tradition as the mother of Issa.[93] We have seen who the Nazarenes really were, the actual identity of Jesus, and the hitherto unsuspected tradition dating back to the earliest days of ancient Egypt. However, what I uncovered next absolutely amazed me.

Mary was beloved of god. Isis, too, was "beloved," it was one of her oldest titles, possibly even more antique than her usual Egyptian name, Aset. Isis's "Beloved" title was also the very name for the land that is Egypt, Ta-Mery or Mery-ta.[94] It means simply "Land of Mery"; in other words, Isis.

Another piece of the puzzle has fallen into place and with it comes other confirmation. Concurrent with ancient Egypt we find the land of Sumer, sometimes spelled *Shumer.** Some of the kings of Sumer bore the name of Mery. A*mar*-sin of Sumer was recorded as having suffered the fate of the hero: he was stung in the heel by a scorpion in Sinai and was killed. That this is a mythic motif for death is beyond question; it is heavily symbolic and also recalls the death of the Hindu Christ, Krishna, by poison in the heel.

Within the Hindu and Buddhist traditions the divine mountain, Mount Meru, is prominent. "Hocart writes that 'the Sinhalese frequently placed inside their topes a square stone representing Meru. If they placed in the center of a tope a stone representing the center of the world it must have been that they took the tope to represent the world.'"[95]

---

*Shu is the Egyptian Heracles; thus, Sumer is the land of Heracles and Isis.

A tope is a Buddhist stupa or domed reliquary. They can be quite massive and are generally pyramidal in form. This is significant. Meru/ Mary, the beloved, we are being told, is linked to the cube. Cybele, we have already seen, is the lover of Attis and her name literally means "cube." What is interesting is that Isis's hieroglyph is a throne—and in representations of pharaoh enthroned, we see him seated on a *cube within a cube*.

Moreover, the land of Mery-ta also translates as "land of the Mr." The hieroglyph for Mr is △, a pyramid. It is Mery. In very archaic times the pyramidion illustrated by the hieroglyph may have been the sacred benben or bennu stone. The bennu bird is the Egyptian version of the phoenix, the mythical bird that immolates itself upon a pyre of its own making and then rises from the ashes, fully resurrected. The place where the rituals of the bennu bird took place was among the oldest and most sacred in the long history of Egypt. Today it is sadly a mere suburb of the increasing sprawl of Cairo. Its name is Heliopolis, known to the Greeks as On-Heliopolis, the very place where, in some of the noncanonical Gospels, as well as some of the Gnostic Gospels, the holy family came to reside on the flight into Egypt. In very ancient times On-Heliopolis was known as Annu or On. Christianity holds that the mother of Mary is Saint Anne.

DIVINE SUBSTANCE AND PLACE

The names of Mary and of Anna or Anne are the names of the spirit of place. They have been used as the names not just locations but of whole countries. The appearance of these names is no coincidence, no accident, for as we shall see, they occur worldwide with a regularity that is, to say the least, breathtaking.

In Rome, the Blessed Virgin Mary was worshipped as "the House consecrated to God," the "Tabernacle of the Holy Ghost," and as a dwelling place. The very title of the king of Egypt, *pharaoh* means "great house." Isis's hieroglyph is Mr, the pyramid. Mary carried within her the divine power, the energy of the holy child.

In 1997, I traveled to Egypt, the land of the beloved, and as I looked upon the greatest monument in the whole history of mankind, the Great Pyramid of Giza, I pondered upon what it all meant. In a metaphorical sense, if Mary is portrayed always as the expectant mother, or as the mother who has just brought forth, she has brought forth something wonderful; grace has given rise to beautiful things. The veil was beginning to part.

In a surprising amount of the world's mythology, the name of *Anna* appears. In the Middle East, *Ana* is the mother of Mari. The Dagda of Newgrange, Ireland, is married to *Boann* as well as being intimate with the Morrigan—Mary again.[96] There is *Anat* of Canaan, *Anath* of Syria, even the Romans had *Anna Perenna,* eternal Anna, Mother of the Aeons, and the Celts had *Anu,* sometimes called Danu.

In the Babylonian myth of creation, the Enuma Elish, *An* is heaven and *Anu* the underworld. This realm is called the underworld because it is the substratum that underpins and upholds our world and all worlds. Following this line of inquiry, I found out that, in the Sanskrit, *anu* means "atomic," *ana* means "to live," and *anna* itself, taken in its most subtle and mystical sense, is the food upon which the whole of existence depends. But this food or energy is not yet embodied. The embodiment is named *murti:* "Any solid body or material form, embodiment, manifestation, incarnation, personification."[97]

Sanskrit has the term *anna-maya-murti,* "body of food" or "embodiment of Anna." How close this word *murti* is to that old name for Egypt, Mery-ta. The Anna names relate to that which gives birth to and nourishes substance. Mary names relate to the substantial, the embodiments of heaven on all levels, including of course the earth itself, especially in areas of sacred power, where the finer energies may be felt, and also at a personal level, where those energies may be embodied and may give birth to something more.

Incidentally, on the planetary level, we should realize that some earth gods are not merely associated with this planet, but with all planets by virtue of association with all embodiments on that level. Mary names do not only relate to planets and people. We have seen how the hieroglyph for Mr is a pyramid and how sacred buildings are positioned at places of power and are positioned to enhance that power. If the Egyptians' beloved Mery-Isis is the pyramid, then she manifests that power. She represents that power, but what will she give birth to, what is it exactly that she carries inside her? Perhaps the clue might come from the Egyptian scriptures: "In the beginning there was Isis, Oldest of the Old. She is the neter from whom all becoming arose."[98]

Isis was, like Mary in the Catholic Church, Queen of Heaven and Theotikos, Mother of God. From the Egyptian word for a divinity or divine evanescence (not god or goddess), *neter*, we get our word *nature*. We are looking at the identity of some exceedingly ancient powers, the most powerful of which, as far as humankind is concerned, is carried inside the womb of the mother. Mery, the daughter of Annu, gives birth to Gisa. Who or what is the power of Gisa? This is the answer we will come to in due course. But first another intriguing point. After his own resurrection, Jesus ascended to heaven upon a cloud of glory, in an event testified by the apostles within the gospel texts. The Great Pyramid was known in its day, 2,500 years before Christ, as the Place of the Ascension.[99]

Here in the heart of mystery such things are not a coincidence.

THE PATTERN OF THE HERO

Jesus and the Temple

In the Gospels, Jesus says, "No one can come to the Father but *through* me" (John 14:6). This is an interesting statement in light of what we have already encountered, for it is the typical statement of the hero as arbiter between humanity and God. It also seems oddly architectural, for in Jesus's time the temple was believed to be inhabited by the very presence of God. Therefore, to get to God, via Jesus, the temple had in some way to be involved—perhaps by literally passing through it? And yet the temple is rarely mentioned in the Gospels. The central figure of Jesus is the main focus, but surely his divine aspirations, if God resided at an address in central Jerusalem, would have been temple based?

Much ink has been spilled over the figure of Jesus and his identity and even whether he existed or not, far too much to go into detail here. However, the role of the temple throughout Hebrew history is oddly muted when it comes to the official story of Jesus as given in the New Testament. The justification for this is given as the prediction of Jesus that the temple would be destroyed, that he could rebuild it in three days, and that after him it would not be needed.

And yet, he says that upon the rock called Peter, he would build his church. This could be seen as a simple metaphor, but it was one used as the foundation point of the Roman Church. However, it could also be a first-century critique of the existing temple regime and the fact that the Nasoreans saw the site as deeply polluted, ritually impure, and the priesthood as corrupt (John 2:19, Mark 14:58).

The language of the Gospels is ambiguous because, taking its lead from Hebrew script and thinking, it can mean many things—and this has obscured the original meaning of the story of Jesus. Like the English language of today, Hebrew is subtle and easy to use for punning and for altering or distorting meaning. Further to this is the fact that, in the ancient world one word could have many meanings, by contrast with today where meaning is largely imbued within a multiplicity of specific words.

The advent of Christianity is a point in question. The early Christian father, Justin Martyr (100–165 CE) attests to the fact that the Hebrew canon was altered by Jewish scribes called *Tiqune Soferim* to expunge any possible Christian references. This was easily achieved by the simple expedient of adding to or altering the square Hebrew script: a stroke here or a simple adjustment there and one word could be made to mean something completely different. The Koran mentions the fact that the Jewish authorities altered their books in the same way. Further to this is the undeniable fact that Christians in the years after Justin were responsible for an awful amount of destruction of inconvenient texts in the battle between religion and religious identity.

Overall, it is an example of how language very quickly became detached from the understanding of specific sacred sites; so, too, the original meaning of myth (of which we have already spoken in the origin of the name of Jesus in Sinai) could become changed. It was this point to which Jesus was referring in John 2:19 and Mark 14:58 as he sought the restoration of the original rites and liturgy.

In late 2007, I became involved with a startling discovery in the deserts of Northern Jordan: a series of small books or codices had been discovered in caves not far from the Sea of Galilee. What distinguishes them is the fact that they are made of metal—out of lead or, in some instances, solid gold.

Why they should have been made out of lead is a question I posed to a leading academic. Her reply was as simple as it was majestic: "So that they could not be altered."[100] The word had been set in the ore from stone.

There are many things unique about them, and one of them is the fact that they are heavily illustrated throughout—and all the illustrations are a script in themselves. What stands out most are the depictions of the temple: they are the central theme of the codices.

Unique and forbidden imagery illuminates the deep, dark interior of the temple and hints at hidden rituals that took place there, a place forbidden to all except the high priests—and then only on the day of atonement, Yom Kippur. The fact that they also bear witness to the image of Jesus surrounded by an early form of halo is the most explicit link yet of the historical figure of Jesus to the temple. What they make clear is that these codices are the meeting point of myth and history. Within them is chronicled, in ritualistic and liturgical terms, the story of Christ and his role in the temple as the meeting place of God and humanity: the very essence of the role of the classical hero.

And here is where Christianity as a faith is truly unique: Jesus personally embodied the role of the hero-king, as the arbiter between humanity and the divine, whereas in myth the role of the hero-king is very obviously a collective liturgical metaphor, a focus for the people. Not all heroes are messiahs, in the modern sense of the word (*messiah* means "Anointed," Son of God), but they are mostly the sons of kings, themselves the children of gods, and therefore born arbiters between the people and their gods. Jesus lived out the prophecies of the messiah in order to bring fulfillment and renewal, to bear witness to both aspects of human nature: he stated explicitly that we are all children, sons and daughters of god. And it can be said with some certainty that he succeeded, though perhaps not in the way he originally envisaged.

The more you look at the Gospels the more difficult it is to tell what might have been historical and what might have been mythical. This is because the language in them is heavily symbolic: "In contemporary biblical scholarship a lot of the Gospels are written with symbolism in mind. They are not setting out to be literal accounts but they are set out to be symbolic."[101] The early writers of Christianity did not see the Bible as history but instead as a coded message containing the key elements of liturgical Christianity.

The codices are clues to the Gospels and their writing. The Gospels are an already-painted canvas over which a new varnish or veneer has been overlaid to tell the new story through the old, to be underscored by its inherent power. The codices are, like the Gospels, testimonies but in the more archaic sense. A testimony is using the past to contextualize its presence, and announce the coming of the Expected One: but it is liturgy *not* a history—the history of a rite, a living myth, that opens up to us the technology of the place and helps us to activate its power. This is not to say that there are not shards of history to be found in them, but this was not their purpose.

The books, like Creation itself, are an outpouring, an ongoing act of creation to be added to, so that renewal might take place in its proper context: self-awareness within humans, and the context of the living world that is ours. The act of self-awareness on the level of the spirit is the act of renewal; awareness of it transforms us.

Jesus was a Nazarene. This much is outlined in the Gospels. And as such, Jesus is very much the embodiment of the Nazarene idea of the hero—a very ancient concept that extends all the way back beyond the time of Samson, another Nasorean figure who predates the foundation of the monarchy in Israel circa 1000 BCE. In this sense we can see that the Nazarenes were the preservers of an age-old tradition, their very name means precisely this (see "Jesus BCE: Christianity before Christ," page 174): thus in the period of the historical Jesus, as told in the Gospels, they were the restorers of what was already very ancient—it may be said for sure that Jesus did not set out to start another religion but to bring back the older traditions that had been expunged from temple practice over the previous six-hundred-year period.

As restorers we can see in the archaic legends the many similarities to the story of Jesus in the Gospels, although there are many differences, but the overall gist of the two legends in comparison are largely the same, and thus demonstrate the theme of restoration. Like Jesus, Samson is born in a cave whose entrance shines so brightly that it is obscured by the light. Samson is compared to the sun in its glory—and probably *is* the sun in origin. Like Jesus, Samson brings down a temple with his bare hands and sacrifices his own life in the process. He achieves his aim by pushing apart the two pillars supporting the temple superstructure—the precise way the temple in Jerusalem was constructed. Samson was a mythical hero. Jesus was not—he was an actual historical reality, despite attempts to show the exact opposite.[102] He was the meeting place of myth and humanity, of God and renewal. The impact of his actions was a historic first—as was his achievement. The mythos is telling us, in story form, what the liturgy truly is. Furthermore, the name of the hero is telling us of his true identity and of his quest.* The language itself has become the adventure.

Recent research into early Christian practice and faith has revealed that in the Solomonic period, at a given time—the Jubilee period—the king became the incarnate presence of God on earth: he was semi-divine. Via a process known as *theosis* he would enter the holiest part of the sanctuary, wherein God was said to reside, push apart the veil of the sanctuary, enter, and sit upon divine throne.

This was a tradition to be seen throughout the Middle East and beyond,[103] dating back to the archaic period circa 3000 BCE. In seeking to restore temple practice, away from the abstract reforms of temple priests carried out in the centuries immediately before the birth of Christ, it therefore follows that the Nazarenes sought to do the same—and when we see the claims made for the historical figure of Jesus Christ we can see their method in practice.

---

*See also Greg Taylor, "Words Are Magic—How the language You Speak (and Hear) Changes Your Reality," June 18, 2018. Accessed at DailyGrail.com (website).

In John's gospel Jesus is described as the Word ". . . and the Word was God." Jesus is become the resonant expression of the god-man as the hero, and this expression can only have taken place within the environs of the temple—the *temenos* (points on the Earth where the divine is held to be closer to humanity than elsewhere).

Jesus in the Gospels makes it very clear that the new temple is the body of man himself, of humanity. It is perhaps ironic that this was steadfastly ignored by the new catechism of the church three hundred years later—in favor of the continuous construction of further temples dedicated to his glory. However, it kept a very ancient tradition alive, and, once the mission of the historical Jesus was achieved, the aspect of self-awareness was kept alive with it.

A Comparison to a Similarity

This compendium of hero myths worldwide is a fairly general presentation that only dips its toes into the very deep waters of world mythology. Furthermore, the hero myth has a close association with the power of place, and it is essential that it should not be separated from this essential root, for if it is, context is left behind.

Each of the myths shares certain specific features with Christian belief, in the sense of the original story and the ritual surrounding it. They are not all by any means the same; geography, culture, and language seem to be the main reasons behind such differences, but the essential elements are as follows:

- The holy child is born of a virgin, in a cave or grotto.
- The birth is foreshadowed by the light of a star.
- Three wise men or shepherds pay court to the wondrous birth.
- The child is the mystical Son of God.
- The child is sought in a massacre of the innocents. The parents are necessarily forced into exile because of a certain tyrannical king who is usually, with very few exceptions, a relation of the holy child, an uncle or grandfather perhaps.
- The holy child performs superhuman deeds on the way to adulthood.

- Little is told of the childhood of the hero.
- The child appears in adult form, to be baptized by a holy man, or initiated, as in the case of Jesus or Achilles, who was schooled by a holy man called Chiron the centaur—the cross.
- The same holy man is associated with water and is born usually at the summer solstice.
- Quite often the hero is portrayed as a long-haired bearded man with blue eyes and arrayed in blue or white.
- The hero performs miracles ranging from walking on water to changing water into wine.
- He heals the sick, but he is not honored in his own town or country.
- He has twelve disciples and is transfigured before them.
- The hero celebrates in triumph and soon after is accused of a misdemeanor that leads eventually to his death.
- A last supper takes place in eucharistic fashion.
- The hero is put on trial by a man known as the pilot—the navigator who will set him on his way to the stars.
- The hero is crucified, or hung on a tree, or gored by a boar, or dies in heroic fashion and harrows hell.
- On the third day he rises again from the dead, having redeemed the sins of the world.
- He appears before his disciples and ascends to heaven in glory.
- He leaves a sign of his presence in the form of linguistic script.

These are the basics of the myth of the hero, but of course, within the worldwide telling of the myth there are variants. As I looked further afield, I noted the common similarities and also the outstanding dissimilarities. Looking closer I realized that perhaps the most important detail of all was the names themselves, their similarity not only to the type of myth but also to the type of place. The rough etymologies given* are based upon the meanings of specific syllables, which appear to be universal. Because of the very widespread presence of the names,

---

*For more on this, see appendix, "The Anatomy of the Hero."

one basic assumption that I have made is that the basic syllables concerned are intimately linked to the power of place; in other words, they are "of it," they are an emanation of all of the forces that I have written of in chapters 1, 2, and 3.

The monuments associated with the culture/civilization as its hero myth developed are tellingly built not only to a resonant purpose but are also aligned to the stars or to predict the movements of the sun. This latter is believed to have been a method of predicting the best time for sowing crops. However, those that are aligned to the stars leave us with a great mystery, one that is discussed in terms of myth. The question left behind by these monuments and their extraordinary secrets is: Could it really be a coincidence that they all share the same features, the same myths, albeit with distinctions—and even the same syllabaries?

The nature of the place varies from location to location, but again the variation is tempered by the human response. The earliest of those responses are pyramidal in structure, or approximate the same elements of pyramidal architecture within their proportion and view. In 1996 it was announced that archaeologists had discovered the world's oldest pyramids—on the Atlantic coast of southern Brazil. Dating from 3000 BCE, these pyramids predate the earliest Egyptian example by several centuries. Astonishingly, whereas the Egyptian variety is built entirely of stone, the Brazilian ones have been constructed exclusively of seashells. For this reason, these massive monuments lay undiscovered for centuries. Initial research suggests that they were originally more than 160 feet high and up to thirty-seven acres at the base. Furthermore, in terms of volume these monuments far outstripped their Egyptian counterparts. "Our new research shows that Brazil's Indians, 5,000 years ago, were more sophisticated than we had thought and were capable of producing truly monumental structures," said Professor Edna Morley, Director of the National Heritage Institute.[104]

Subsequent discoveries of pyramids in Peru, also dating to 3000 BCE, in 2019 have underscored the presence of pyramids at the outset of civilizations throughout South America. In the past thirty-five years pyramids have been located the world over, from Siberia to New Zealand, all of them very ancient, all of them a seemingly

primal response of indigenous people to the landscape around them—
a response that echoes throughout their various mythologies. It is one
of the great tragedies of world history that the conquistadores were
allowed to deliberately destroy much of indigenous Central and South
American history. They also wiped out, quite deliberately, the knowledge that would allow Mayan hieroglyphs to be deciphered. Why?

Could it have been for reasons of mythic similarity? A historical-
cum-mythical messiah/king of the Maya was a king by the name of Yax
K'uk Mo, the Lord of the West, who was interred deep inside a pyramid
within a pyramid. His name approximates to the "Blue Green Quetzal"
(or Great Sun First Quetzal Macaw). Returning to the theme much
later, I noticed that, in pronunciation, the name Yax is very similar to
the name of Jeshua, Jesus, Joshua, Yeshua, and so on. The difference
seems to fade when we look in detail at the legend of Quetzalcoatl, the
plumed serpent.

The *Qu* syllable was probably pronounced very softly, almost like
a *g*—the earth syllable. *Coatl* means "twin." We are left with *etza* or
itza, another version of *issa* or *isa.* Again, the myth is linked to the site
by the explicit fact that the hero is the son of the Earth. Pyramids pro-
liferate throughout Central and South America. Teotihuacan is a city
full of them, the grandest of which are the Pyramids of the Sun and of
the Moon.

In Uxmal in Mexico there are seven pyramids ranged in a form
approximating the seven stars of the Great Bear and so on. The list is
long. At Tucume in Peru, the Inca built pyramids and depicted alongside them a sacred stone, very much like the Egyptian benben stone, as
well as a mythical bird in association with it, much like the Egyptian
bennu bird—the phoenix. Quite probably the indigenous peoples of
South and Central America took their knowledge with them in the
migration to North America (ca. 1500 BCE), and there it remains as a
cultural memory. Could the teepee be a cultural memory of the pyramids? A portable vestige of their latent power? Other similarities, some
quite extraordinary, also occur, some of which I will raise now in a very
general manner.

The Sequorno Indians of Teso dos Bichos call their female shamans

Ameru, an obvious link to the landscape and an intriguing one given that the wider continental landmass is today called America. The idea of America being named after Amerigo Vespucci never did make sense to me. Why was it not called Vespuccia? Most people denote their surnames, not their first names, so as to be remembered.

Like the Egyptians and their obsessions with wheat and barley (see chapter 9), the Anasazi, who disappeared from history long ago, were great cultivators, in this case of maize. It is clear from archaeological evidence that they saw within this process a metaphor of the dying and rising god. The Anasazi, whose name means "the Ancient Ones," built kivas in what is now the southern United States for purposes that to this day remain unclear. However, what is known of the buildings is their extraordinary resonant properties that, when harnessed, can induce altered states. But perhaps the most astonishing things about the Native American peoples are the names that they give their gods. There is something about them that is very familiar.

The great god of the Navajo is a hero by the name of Yesha. He is one of the holy ones whom the Native Americans call the *Yei-eden-na.* The hero Hiawatha was called "the feathered one"—the syllable *hia* is another form of *Hu,* the Lord. Another great divine being of these peoples is *Ya-weh-node*—"She Whose Voice Rides the Wind." Resonance again.

There are further synchronicities that occur that bring elements of Egyptian and Middle Eastern cosmogony to mind in a way that intimately links the myth to the site, and they carry the little-known fact that there are pyramids in North America, too. In Cahokia, Illinois, there are mounds, pyramidal in form and intent, placed there because the building of them brought a closer proximity to heaven. The people wanted to be closer to their god, while still inhabiting the earth. In this sense—and the builders were quite explicit about this—the mound represents the godly power upon earth; it is the son of the Earth Mother and the Sky Father. The symbols that have been located at these sites are all symbols that have now become familiar to us: spirals, crosses, and other hieroglyphs. Wooden posts have also been located that act as markers of time, as sundials seeking out the hour of sunrise, which the Native Americans saw as symbolic with the reawakening of life, of resurrection.

Everywhere, we find the hero, the names of the hero,* the names of gods and goddesses, and everywhere monuments that revealed much about these names. They are monuments to the divine in humankind, the sound of the human voice, and on this, myth has much to say.

---

*For more on the etymology of hero names worldwide, see "Heroic Name Variations," in the appendix.

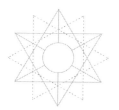

5

Salvation and the Soul

When Israel was a child, then I loved him, and called my son out of Egypt.

HOSEA 11:1

Neter *means principle of life, and the Temple is its house.*

R. A. SCHWALLER DE LUBICZ,
THE TEMPLE IN MAN

In 1922 the world was electrified by the discovery of the tomb of Tutankhamun. Up until that time, the name of Egypt was almost singularly associated with pyramids. From now on it was the magnificent gold-dust mask of the young pharaoh Tutankhamun, circa 1400 BCE, that would dominate the image of Egypt. For the first time, an intact royal tomb had been discovered. It was to lead to many surprises, some of which are only at last being understood.

In a posthumously published book, Simone Weil commented that "Ezekiel expressly compares Egypt to the tree of life of the Earthly Paradise, and Phoenicia, at least at the beginning of his story, to the cherub standing by the tree. If this view of the matter is correct, a current of perfectly pure spirituality would have flowed across antiquity from pre-historic Egypt to Christianity."[1]

Indeed, many of the contents of Tutankhamun's tomb had a direct link with Christianity, a belief system that only came along 1,400 years later. Of the items found within the tomb were personal ritual objects—a pair of gloves and a gala robe. Both bear a striking similarity to ritual items found within both the Roman Catholic Church and the various Orthodox churches.

> The two garments . . . recall official vestments of the character of priestley apparel, such as the dalmatic worn by deacons and Bishops of the Christian church, or by Kings and Emperors at coronations. . . . They take the form of a long, loose vestment, having richly ornamented tapestry-woven decoration with fringes on both sides. In addition to this ornamentation, one of them has needlework of palmette pattern, desert flora and animals over the broad hem at the bottom. The openings of the neck and at the chest are also adorned with woven pattern. One of the vestments, with field quite plain, has narrow sleeves like the tunicle; the other, with the whole field woven with coloured rosettes as well as figures of flowers and cartouches across the chest, has its collar woven in the design of a falcon with outspread wings, and it also has that titulary of the king woven down the front. . . . *Moreover, these robes may well have had the same origin as the Roman garment whence the liturgical vestment—the dalmatic—of the Christian church derives.*[2] (My italics)

Howard Carter also noted that the pair of gloves were in a much better state of preservation and were made of tapestry-woven linen. A bishop of the Roman Catholic Church wears gloves when pontificating, along with buskings, tunic, and dalmatic under the chasuble. Other objects were also found that related to later Christian beliefs and practices: "There were also a number of ostrich feathers, recalling the flabella still used at a papal procession in Rome, such as was witnessed in the Eucharistic procession of his Holiness the Pope in July 1929. These fans, like the pontifical flabella, were carried by grooms-in-waiting in Pharaonic processions, or were held beside the throne, and appear always on either side of the king or immediately behind him."[3]

These observations served to heighten my suspicions regarding further Egypto-Christian connections. They were not difficult to find.

Popes are, upon death, interred in full regalia within three coffins—as were the pharaohs. Both the papal tradition and the pharaonic tradition share the celebration of jubilee years, the wearing of similar crowns, the celebration of the same kinds of rituals, the practice of confession, and so on. The ten tablets of the law, known as the Ten Commandments, are in fact the negative confession to be found inscribed upon the walls of the pyramid of Unas, otherwise known as the Pyramid Texts.[4]

Also, a whole litany of miracles performed by Moses is attributable to Isis, sister and wife of Osiris, including the famous parting of the seas so as to walk dry shod between them.* Oddly enough, this miracle is performed by the Egyptian goddess Hathor, too, while on her way to Byblos.[5] Even the idea of the "son of God" is particularly Egyptian in flavor. Osiris as the son of Ra was an exceedingly ancient concept. Osiris was the original "Lord" and was addressed as such. Pharaoh, who was the living Horus, the son of Osiris or, if you like, the son of the son of God, had many titles, foremost among which was his Horus name and his golden Horus name. He was the mouthpiece of the gods—the ultimate truth teller.

The Egyptian word for truth is *heka*. This was also a title of the pharaoh. In hieroglyphics it is denoted by the form of the shepherd's crook, another symbol of the god Osiris—the good shepherd. The pharaoh's title as king of both Upper and Lower Egypt was "Lord of the City of the Sun." (There were two such cities, Thebes in the south and Heliopolis in the north, now a modern-day suburb of Cairo.) In its original hieroglyphic form, the title translates as the hieroglyphic symbol of Heka Iunu and Heka Iunu Shmau.

The hieroglyphic symbol of heka, the shepherd's crook, is always found upon sarcophagi belonging to the pharaohs. It can be seen clutched in the pharaoh's crossed hands. In his left hand is the crook, in his right, the flail. This association of heka with Osiris more than implies that the symbol *he* is a sign of the hero. When split into its component parts, *he* and *ka,* we come to the essence of the term. *Ka* is "spirit" in ancient

---

*In the book of Exodus, Joshua performs the incredible feat of stopping the sun in its tracks, a feat that was also originally performed by the goddess Isis.

Egyptian. *Heka* has the same meaning as *hex,* as in *hex*agram and so on. It is a term common to the perception of witches and implies the use of rite as magic. Heka is the heroic spirit. It can also mean "spirit of He," of "the Lord"—*he* meaning "Lord." This is curious, as the Hebrew letter *he* appears in the name of the Hebrew God, Yahweh.

Heka and the Pharaohs

A pharaoh was also a magus, a bearer of great authority and great wisdom. In magical tradition, *Heka* signifies the spirit of God. A magical spell in witchcraft is a hex and comes from the same root as heka (Greek *hex* or *heksa*). The hexagram is the seal of Solomon and the symbol of King David, better known as the Star of David. Heka still has a wide influence today. *Hake* in the Arabian tongue means "that which is right, that which is true." Any follower of rugby will be very familiar with the magical connotation of the term *heka*. It is found in the Maori tradition as the splendid "haka," a ceremonial war dance, performed not only by Maoris but also by members of the New Zealand rugby team at the outset of each of their matches. It is an intimidating act with a magic all its own. The hex was, of course, sacred to Hekhet (sometimes *Heqit*). *Hekau*, the plural of *heka* means "magical words of power" and almost invariably the names of the gods were names of magical power, hence the Hebrew embargo upon speaking the holy name of God, Yahweh. Thus, we can associate the pharaoh with magical words of power.

Heretics of every era have been aware of Christianity's Egyptian links. The former Dominican friar and missionary for the Hermetic movement, Giordano Bruno (1548–1600), was burned at the stake for his beliefs. He viewed his Christian cross as really being just another version of the Egyptian ankh or ansate cross.

A fellow prisoner reported him as having said that the cross on which Christ was crucified was not in the form shown on Christian altars, this form being in reality the sign which was sculptured on the breast of the goddess Isis, and which was stolen by the Christians from the Egyptians. In a reply to an inquisitional question about this, Bruno acknowledged that he had said that the form of the cross on which Christ was crucified was different from the way in which it was usually "painted," adding these significant words:

"I think I have read in Marsilio Ficino that the virtue and holiness of the character [*caraterre*, by which he means 'cross'] is much more ancient than the time of the incarnation of Our Lord, and that it was known in the time in which the religion of the Egyptians flourished, about the time of Moses, and that this sign was affixed to the breast of Serapis."[6]

In a very intriguing and important work published in 2000, *The Pharaoh's Shadow*, Anthony Sattin makes some extraordinary observations. In a conversation with Zaki Shenouda, director of Egypt's Institute of Coptic Studies, Sattin recalls the following information: "There were prophecies that the Messiah would come. Osiris is the Christ and Isis is his mother, the Virgin. There are other things we've taken. Christians talk about the fallen angel, while in ancient Egypt there was the god Seth who killed Osiris and went on to become the god of Evil. The pharaonic temple resembles our churches. In the construction of the temple and church we find high walls and no windows, because life is supposed to be like heaven."[7] This is the beginning of a host of similarities from which Sattin concludes that pharaonic culture and tradition are still very much alive, sadly encroached upon by the modern world.

Coptic churches are fascinating repositories of the deeply ancient. They have a holy of holies hidden behind a screen, away from the eyes of the profane. Above many of these hang oil lamps and beautifully crafted ostrich eggs that, as any Coptic priest will gladly point out, are a symbol of resurrection, a very ancient one going back to the Osirian era. Quite often, in Coptic iconography the god Horus is used as the figure of Jesus. Horus is the son of Osiris, called the Logos, or the Word, by the Copts. Yet again we are back to resonance, the Word of God.

In fact, there is some very interesting iconography throughout Coptic culture. There is a statue of Horus at Koptos that portrays him grasping the phallus of the god Set. This interesting image would appear to symbolize the desire of man to overcome his wild and primal self. Osiris, the higher god, was somewhat predictably canonized out of existence by the early church as Saint Onuphris, taken from Osiris's ancient title of Unnefer the Beneficent. This brings us to the origins of Saint Benedict, whose name was given to a monastic order. Monasticism began in Egypt, whereas Saint Benedict is of uncertain origin—such an individual probably never existed. However, it was a title of the Greek god Apollo, who was called, like Osiris, Benedictus. *Benedict* means "good speaker," and his legend equates him with Osiris and Apollo.

In the ninth and tenth centuries, it was believed that the Coptic Church held the secret of the pyramids.* As we shall see, this may well be true.

In ancient times, the Egyptian primordial cross was representative of Amsu or Min, the spirit of Horus the Elder (Heru-ur), who died and rose again from the dead. This was later converted by the Coptic Christians into the cross of Jesus by the simple expedient of making the upright slightly longer.

THE DENIAL OF EGYPT

The reluctance to accept an older date for Christianity can perhaps be explained in part by the way the views of scholars have changed over the past two hundred years. The emphasis has gradually shifted from the idea that the origins of civilization lay in the Middle East to the idea that they lay in northern Europe and Asia. The older view is the Ancient model, the latter is the Indo-European thesis. According to this latter view, at some stage in the second millennium BCE the Aryans invaded Europe and swept down into the south, bringing about Greek culture as a result.

This Aryan invasion overwhelmed the Aegean pre-Hellenic culture.

---

*Oddly enough, the Coptic spelling of Jesus is *Essa,* and it dates right back to the first century CE.

Plate 1. *Trethevy Quoit, ca. 3000 BCE, an example of
a Neolithic barrow, Cornwall, Britain*
Photo by Paul Broadhurst

Plate 2. *Rocky Valley, Cornwall, Britain. An ancient labyrinth, ca. 3000 BCE.*
Photo by Paul Broadhurst

Plate 3. *The amphitheatre at Delphi. The soporific atmosphere of this extraordinary place*
is due to the fact that it is densely permeated by natural earth radiation—radon gas.
Greece, ca. eighth to ninth century BCE.
Photo by Paul Broadhurst

Plate 4. The amphitheatre at Dodona, where the voice of Zeus rustled like the wind through the trees. Greece, ca. ninth to tenth century BCE.
Photo by Paul Broadhurst

Plate 5. *Bourges Cathedral, France. Twelfth century, Gothic.*
Photo by Paul Broadhurst

Plate 6. *The Tympanum at Chartres Cathedral, France;*
Christ enthroned at the beginning of all creation
Photo by Paul Broadhurst

Plate 7. *Chartres Cathedral, France. Twelfth century, Gothic. The most perfect expression not only of a style but also of an archaic theology.*
Photo by Paul Broadhurst

Plate 8. *A Lead Codex found in caves
in northern Jordan*
Photo by David Elkington

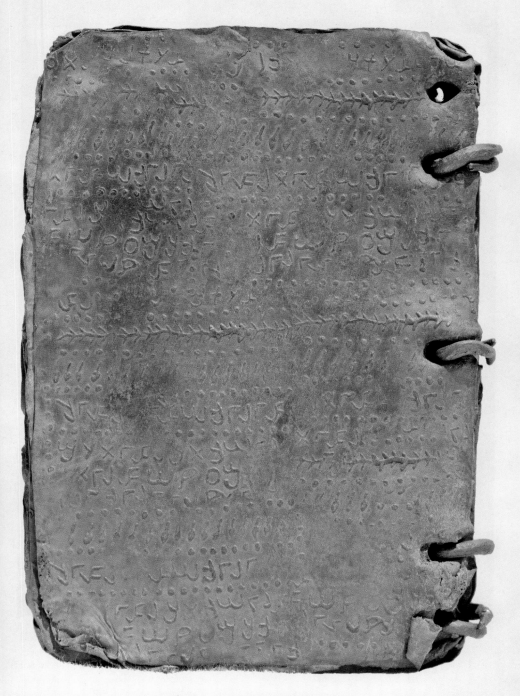

Plate 9. *Lead Codex: rear view,*
exhibiting aspects of Solomon's Temple
Photo by David Elkington

Plate 10. *Goddess figure: note how she holds the ankh, the symbol of Life, one in each hand.* Burney Relief, *British Museum.*
Photo by David Elkington

Plate 11 (facing page). *Fan vaulting at Exeter Cathedral, England. Note the extraordinary acoustic-style forms. Roger Penrose in Shadows of the Mind: A Search for the Missing Science of Consciousness posits microtubules on the quantum level as exoskeletons in a "fan-like structure" similar to fan vaulting.*
Photo by David Elkington

Plate 15. *The Great Pyramid at dawn: The cosmic birth of the hero*
Photo by Paul Broadhurst

Plate 16. *The usual internal layout of a pyramid.*
The Red Pyramid of Dahshur, Fourth Dynasty.
Photo by Paul Broadhurst

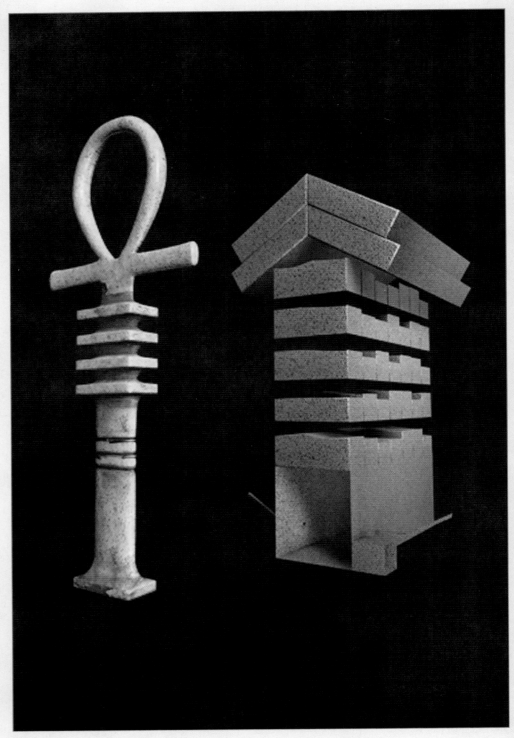

Plate 17. *The ankh and the relieving chambers*
By David Elkington with Dean Baker

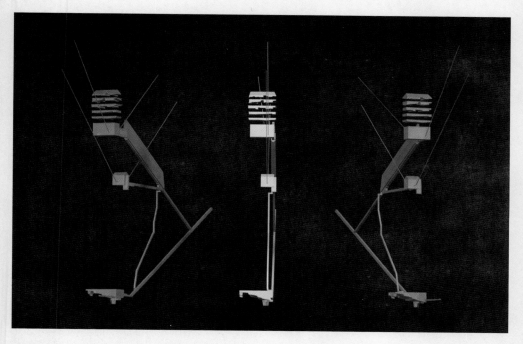

Plate 18. *The body of Osiris deep inside the Great Pyramid. Note how the form of the interior chamber layout resembles the statuary of Osiris.*
By David Elkington with Dean Baker

Plate 19. *The Neb-edjer. Twelfth Dynasty. British Museum. Photo by David Elkington*

The Ancient model was almost precisely the opposite, geographically speaking, recognizing that Greek culture, the mother culture of Europe, arose as a result of Egyptian and Phoenician colonization circa 1500 BCE. The Greeks themselves, of the periods of Pythagoras and of Socrates and Plato, circa 600 to 300 BCE, took precisely this view, as many of their writings testify. Archaeologically speaking, there is plenty of evidence for either theory; it is all a matter of interpretation.

The consequences of adopting the Aryan Indo-European thesis are significant. In effect, it sidelines the Egyptian and Phoenician cultures. They are hardly accorded the status of players in the grand scheme of things after the second millennium BCE. Egyptian settlements outside the geographical areas of Egypt and Palestine are denied, as are Phoenician settlements in later manifestations of the thesis. These manifestations, the product of the extreme edge of twentieth-century, far-right political philosophy, eventually necessitated a change of the original name of the Aryan theory, Indo-Germanic, to its more innocuous title, Indo-European.

The publication in 1987 of *Black Athena* by Professor Martin Bernal caused a storm of controversy. By highlighting the anti-Semitic angle of the Indo-Germanic view, Bernal caused annoyance in academic institutions throughout the world. In showing that the fundamental challenges to disciplines tend to come from outside, Bernal, himself an outsider, began to lay bare the Indo-European dream. He pointed out archaeological evidence that actually supported an Egypto-Afro-Asiatic model. This is not to deny that invasions from the north took place. However, these may well have been less dramatic than has been suggested, and their influence was nowhere near as great as that of the Egyptians and Phoenicians. As Bernal points out in his book, the most surprising thing about the Aryan model is that not only does it go against all of the evidence of the ancient Greeks themselves, but at heart it is also a racist theory—racist, at least, in what the extreme elements of the theory led to.

If I am right in urging the overthrow of the Aryan Model and its replacement by the Revised Ancient one, it will be necessary not only to rethink the fundamental basis of "Western Civilization" but also to recognise the penetration of racism and "continental chauvinism" in all our historiography, or philosophy of writing history. The Ancient Model had no major "internal" deficiencies or weaknesses in explanatory power. It was overthrown for external reasons. For 18th- and 19th-century Romantics and racists it was simply intolerable for Greece, which was seen not merely as the epitome of Europe but also as its pure childhood, to have been the result of the mixture of native Europeans and colonizing Africans and Semites.[8]

Bernal tentatively accepts the Aryan model's hypothesis of invasions by Indo-European speakers. However, according to this new model, the Revised Ancient Model, the earlier population were speakers of a related Indo-Hittite language, which left little trace in the rise of later Greek.

Indo-European is the favored claimant to the throne of the "master language" so long sought by scholars. They hypothesize that the earliest form of Sanskrit itself arose from a much older Vedic teaching, from which the sacred texts, the Vedas, were derived. For Western scholars and romantics Sanskrit became a kind of Ursprache, in the sense of an original or ancestral tongue.

The influence of Sanskrit upon other languages has indeed been immense. However, most of these languages are Eastern, not Western. The languages of Japan, China, Borneo, Tibet, Cambodia, the Philippines, and Java all share significant roots, both spiritually and culturally, in Sanskrit. During the eighteenth, nineteenth, and twentieth centuries the European intellectual class became collectively fascinated by all things Eastern. The rise of the Western empires and their policies of colonization saw to it that such fascination was deep and abiding. It was in the interwar period of the 1930s that this fascination was to bloom into a profoundly ugly flower, one that would leave its taint all over the world.

The non-Semitic bias of the Aryan model, together with the extreme economic conditions of a severe postwar depression in Germany, came together in a poisoned chalice held up by the Nazis as the grail of racial purity.[9] Throughout the mid- to late-1930s parties of Nazi fanatics trooped up and down the Tibetan highlands looking for the origins of the Aryan peoples and Germanic racial consciousness. What they discovered was specious to say the least, but it was put to extremely good use in propaganda against all Semites. The consequences were terrible and deeply tragic, but what must not be forgotten is the fact that the National Socialists could not have furthered any of their spurious researches without the guiding hand of elements within German academies, many of whom kept their posts after the war. It was at the end of the war that "Indo-Germanic" was simply changed to "Indo-European."

Researchers of the past decade have had a tendency to lean toward Bernal's view that before we study our ancient origins it is necessary to study the history of history. Our civilization changes and adapts the past in its own likeness. In my view, the Aryan thesis has served to obscure an essential element in the history of myth and of Christianity. That element is the spread of myth and the mystery of Christian origins. If I am correct that historical Christianity was based on very firm but very ancient origins, then there are a few observations to look at.

The manner in which language spread may have been heavily influenced by both the Egyptians and the Indo-European peoples—the history of this linguistic migration is to be found in the various mythologies of these diverse regions. *However, language itself may actually have stemmed from particular sites, as an interaction between humankind and the site,* hence the extraordinary similarity of certain primary syllables worldwide.

The apparent spread of language across Europe, Asia, and Africa may be, at least in part, an illusion. At some point in deep antiquity (ca. 30,000 BCE), there may rather have been a coming together of linked, but superficially disparate, elements. This theory dispenses with the necessarily problematic and thorny issue of racial influence, an issue that ever since the close of the Second World War, scholars have been understandably reluctant to discuss. It would also resolve a number of linguistic quandaries.

One of the striking stumbling blocks of any theory of linguistic spread is the existence of Basque. Basque is an embarrassment to many scholars because it does not fit in anywhere and thus defies convention.

It is a language to be found among the peoples of the western Pyrenees, in both Spain and France (the Basques and the Gascons). It is a region on its own, with its own culture as well as its own language, which has no Indo-European root. So, where does it come from?

Basque has as one of its ancestors the old language of Aquitaine, which the Romans called Aquitania; but before that we are in very uncertain territory. However, it is interesting that Basque is a language of a particularly mountainous area—for mountainous areas, being particularly powerful places, can have the most extraordinary effect upon humans. Furthermore, the name Pyrenees comes from the same Greek root as *pyramid,* both forms having their own particular power. Pyramids are, like Babylonian ziggurats and the mounds of North America, manmade mountains, using the same materials that make up mountain structure.

Incursions from other lands curtailed the spread of Basque and thus inhibited its use as a national language. However, another interesting thing about the Basque region is that throughout history it has been an area attractive to many people fleeing from the east. Many Solomonic Hebrews found their way there in the upheavals of the first millennium BCE, for example, and more came after.

The idea of the power of place drawing these people to it must be taken into account. Mountainous areas have commonly given refuge, but they are also recognized to be physically, emotionally, and spiritually health-giving. This is largely to do with better air quality, for not only may the air be more charged due to emissions from the rock itself and high ion counts from rapidly moving waters, but the more rapid movement of the air itself elicits clarity of mind as well.

There is a legend, prevalent over almost the entire Basque area, that certain members of the holy family either visited or were exiled to it. However, even before all of this, it was an area that legend says was frequented by Phoenician traders in search of tin and gold. No doubt they were also attracted by the sheer power of the place. We in the West have a very commercial point of view and the tendency is to see the Phoenicians solely as a trading nation traveling only for commercial gain. This is possibly a misleading view because the Ancients seem to have had a holistic attitude toward their Mother Earth, and neither she nor her inhabitants would have been looked upon simply in terms of exploitation. Also, the flat-earth view of our ancestors that is held today is a stilted one. Discoveries have shown that the Ancients traveled very far afield. Early in the twentieth century, evidence appeared that the Phoenicians may have been regular visitors to the Americas. In light of this there is no reason to doubt that they were familiar with Western Europe.*

The Phoenician tongue was closely related to Egyptian, and it is only reasonable to assume that, as the Phoenicians traveled farther and farther afield, so, too, did the language, which may have changed as it spread geographically to fit in with new experiences and ideas of the world. Thus, although the roots of Basque remain uncertain, it is known that Phoenician elements do appear both in the language and in Basque history. Although details of the Phoenician's civilization are unfortunately few and far between, enough is known of the language to see where its origins lie—in Egypt.

Although the Aryan theory of European civilization has seen the influence of both Egyptians and Phoenicians minimized and, in some respects trivialized, in the past decade things are being seen a little differently. Like Martin Bernal, Julian Baldick, a lecturer at King's College, London, sees the origins of Jewish, Christian, and Muslim traditions in Afro-Asiatic roots.[10] He has observed that elements from all these religions, such as God as Creator, circumcision, sacrifice, marriage rituals, a divine obsession with snakes, and the Ten Commandments, are not distinct—they come from an Afro-Asiatic source, which has until now been disregarded. Baldick thus places much of this heritage back into a black African context, arguing that there is a common Afro-Asiatic language base. However, while I broadly agree with Bernal and Baldick's arguments, I believe that the term "language base" should be elaborated on to become "the outward spread of a civilization to permeate, and be permeated by, the various elements of language that crop up in the

---

*More recently it has become known that even the Romans may have found their way across the Atlantic.

landscape." I believe that the origins of a language emanate from its environment, initially from the human body's subtle interaction with the planetary locale, especially its vibratory atmosphere.

The development of myth would be intrinsic within the development of language, for as language developed beyond the immediately practical into the long-term practical—for instance, to articulate a developing philosophy of life—so, too, would myth. To the Ancients, myth was the natural representation of the greater forces at work in and around them. Their philosophy of life lay within and behind their mythology. As peoples and ideas migrated, language would be removed from its natural environment and separated from its myth. It would spread outward, sometimes to be assimilated successfully, sometimes to be regarded as uncouth or perhaps inappropriate. In other words, humanity's response to its landscape differs wherever we go. Furthermore, while in linguistics it is accepted that syllabaries might remain the same or similar, something that has been completely overlooked is their punning nature.

The knowledge to be derived from understanding the punning nature of early language should not be underestimated. Wordplay is, naturally, a major feature of modern humor, but in the past, when language more clearly reflected the law of "as above, so below," its interpretation was deliberately multilayered. This in-built humor was meant to literally elicit a tickling effect, a physical realization of a greater truth in everything.* In the case of Egypt, it is quite clear that over its thousands of years as a preeminent civilization, the nation became somewhat multicultural and that its influence through language and mythology would have had a tremendous potency and spread, so much so that the effects should still be easily comprehended today.

Unfortunately, the currently accepted Indo-European language

---

*This ah-ha moment is sometimes related to a feeling of something physically falling into place in the body, but more often it's a tingling or tickling sensation in the spine or head. Such activity is also associated with kundalini. Some individuals in whom kundalini is active suffer long bouts of uncontrollable laughter. Whatever symptoms manifest during kundalini, after a period of purification—length subject to that required—feelings of discomfort give way to feelings of joy. There is much ongoing research into this subtle energy.

theory has obfuscated an honest and open search for the laws behind the origins and spread of language, and therefore obscured the truth. As we know, the Ancients were particularly interested in the power of place for practical purposes. From the earliest times, language and song were the most essential human components in the working mechanism of a site, a spiritual linguistic technology. In essence, what we have is a human-environment interaction, where language is geo-linguistic. The upward spread of this effect is thus earth based: language is where the earth bears fruit and seeds (Greek *pyren,* from the same root as both *pyramid* and *Pyrenees*). It is humankind that has gathered these fruits and spread the seed.

If we apply what I call the geo-linguistic argument to the civilization of Egypt we find that elements of early Greek and Hebrew support it very nicely, particularly in terms of the religion and mythology of those lands. Cultural and geographical differences still arise because of the influence of the power of the land, the local planetary atmosphere, as resonated within the mind.

Geo-linguistics infers that the co-joining of language, myth, and religion was, from the earliest times, complete. If the Indo-European thesis obscured the direct Egyptian roots of archaic Christianity, then it is the Greeks who must be given the credit for initiating the process. What made the Greeks so totally different from any other civilization before them was the fact that they were the first to separate myth and science in terms of language, by coining the language now commonly used for scientific description. Suddenly life became less divine in origin and more a case of varying phenomena. Physics, biology, and chemistry began their ascent to divinity, replacing the now overthrown gods.

For me, and this is entirely a personal view, the ancient Egyptian mythology was, from its outset, a language of science. The very fact that meanings of the Egyptian language have been ignored is incomprehensible. Monotheism is lauded as an intellectual breakthrough in the development of civilization. However, the ancient Egyptians had an approach that can only be described as superior to the concept of monotheism.

The Pyramid Texts are quite explicit about the worship of the

One. Those we label gods were actually *neters*—principles or forces of nature (*nature* as mentioned earlier is derived from *neter*). These same forces were scientific principles expressed as mythical and metaphorical concepts. To my mind the Greek separation of the two was actually a backward step. In Egypt everything was permeated by the divine, the divine was in all things (Greek, *pan*). If there really is an essential difference between these two cultures, it is that the Egyptians saw God in a pantheistic way. The Jewish God reflects a more objective than subjective view, in that God seems totally outside human experience and the human nous. This was eventually how things came to be seen in Greece, though unlike the Jewish text, the Greeks do not deny their Egyptian origins.

Unfortunately, the Indo-European language table seeks to minimize and even deny the importance of Egyptian religion and theology, which had a very telling influence upon the rise of Greek civilization: "Plotinus (ca. 205 to 270 AD) conceptualised Supreme Divinity as a trinity that manifests itself in three hypostases: as One, as Mind and as Soul. The prototypes of these hypostases still can be traced in the history of Egyptian thought to the first three divinities who make up the basis of the Heliopolitan Ennead: Atum-Ra, Tefnut-Mahet and Shu."[11] In many ways the philosophy of Plotinus was neo-Egyptian, to paraphrase Luckert, whose work, quoted above, demonstrates the fluidity with which Egyptian theology survived into Greek philosophy.

Plotinus was relatively late, but Luckert also makes an important observation about Plato (428–348 BCE): "Christian theologians naturally preferred the *Timaeus* to other Platonic works. In the dialogue's underlayment of Egyptian ontology, they discovered a natural affinity with their own Christology and theology. After all, Christendom was born from, and overtaken by, the same Alexandrian-Hellenistic undercurrent of Egyptian theological notions that entrapped a Porphyry and a Saint Augustine."[12]

Elements of Egyptian Christianity survived into Greece and into the rise of historical Christianity before being ignored and the least acknowledgment overridden by the racist Aryan theory. The irony of this is that in trying to purify the racial roots of the Greeks and their irresistible rise, the very foundations of such a rise are being woefully undermined to the detriment of the entire Grecian edifice by the Indo-European school.

EGYPTIAN MASONICS*

The denial of Egyptian influence by European scholars in the mid- to late-nineteenth century followed hot on the heels of the decipherment of the Egyptian script by Champollion in 1822. As a result, the mystery of Egypt began to diminish rapidly. Ironically, the hieroglyphic language had become a mystery in the first place as a consequence of Christian denial. In the fourth century CE the Christian Church banned the use of Egyptian hieroglyphics in an attempt to sever a living link with the country's pagan past. With a touch of further irony, it was to be a Jesuit priest, Athanasius Kircher, who would attempt to decipher them again 1,300 years later.

Kircher saw the pharaonic civilization as the fount of all wisdom and wrestled unsuccessfully with the language of the pharaohs for years. Modern hindsight has long since demonstrated that Kircher's approach was wrong, and unfortunately his idea that Egypt was the source of wisdom was also dismissed, a little too eagerly. In reality, the sheer power of the language and its potential are only just being discovered. The wisdom is in the application, and the application survives in forms of ritual that, over the years, have been inherited in many guises by many different cultures. To my mind, language and ritual developed together from the earliest days, the sense of ritual being akin to a divine inflection of the language at its optimum frequencies. When speaking our everyday language, we use nuances of sound to intone and differentiate and to give a sense of mood. Divine language was used and imparted in similar ways, but differently in that evidence points to it being sung, and at particular frequencies, with all of the brain wave changes this involved.

The role of language in ritual and as ritual is as ancient as the hills,

---

*With apologies to John Reid, author of *Egyptian Sonics,* for the wordplay.

as most of the world's religions still testify. However, the survival of Egyptian ritual and tradition has borne fruit in a number of surprising ways. The Coptic Church is the first and least surprising of the survivals. The other is Freemasonry.

The impressive decipherment of Egyptian hieroglyphics stems directly from the French Egypto-Masonic tradition rather than from any Romantic impulse. The length of Egypt's history far outstrips anything comparable, and therefore it is not surprising that the influence of Egypt was perceived by eighteenth-century Orientalists as conservative, a view shared by the Freemasons: "Egyptian priesthoods had in fact appealed to the conservative thinkers at least since the time when Plato had modelled his Guardians on them. In the 18th century this line of thought was taken up by the Freemasons; but even in the Middle Ages, Freemasons appear to have been especially interested in Egypt because, following ancient tradition, they believed it to be the home of geometry or Masonry."[13]

As Bernal goes on to observe: "With the formation of Speculative Masonry at the turn of the 18th century they drew on Rosicrucianism and Bruno to establish a 'twofold Philosophy.' This entailed superstitious and limited religions for the masses but, for the illuminati, a return to the natural and pure original religion of Egypt, from the debris of which all the others had been created."[14]

Freemasonry played an important role during the years of the French Revolution. Many of the leading Jacobins were Freemasons: Camille Desmoulins, Emmanuel Sieyes, and Georges Danton. More prominent than them as a high-ranking and long-established Freemason was the Marquis de Lafayette, hated by the others for his moderation. The Revolution brought down the king but raised up an emperor, Napoleon, whose attempt to conquer Egypt in 1798 was a military fiasco of major proportions, but an unqualified success on a scientific level. Egyptology was born, and immediately the race to decipher hieroglyphs was on.

The importance achieved by Masonic influence during and immediately after the French Revolution posed an acute threat to the stability of Christianity. But it was the work of one person in particular that presented the greatest menace—Charles François Dupuis, the revolutionary, anticleric, and scholar. His great work, *The Origins of All Cults,* is undeservedly obscure today. Dupuis's idea was that all mythologies and religions could be traced back to one source. Furthermore, he believed that nearly all mythologies and religions were based upon one of two principles: the miracle of sexual reproduction and the intricate movements of the stars and other heavenly bodies. He demonstrated with extraordinary and massive detail the Near Eastern mythological background of the Gospels. For Dupuis, Egypt was the source of Near Eastern, and hence world, culture.

Dupuis was active in politics during the Revolution and became the Director of Cultural Events during the Revolutionary Directory of 1795 to 1799. Later, during the Consulate under Napoleon, he was president of the legislative body. When Napoleon became the undisputed master of Italy and arbiter of papal affairs, it was Dupuis who was sent in to examine the Vatican records, which, in their entirety, were at his disposal. The Vatican has never been so exposed to scrutiny since.

In his resulting work, *Was Christ a Person or the Sun?*, Dupuis was scathing about Christianity: "If, however, it be demanded whether existed a charlatan or philosopher who called himself Christ . . . nevertheless we disbelieve the reality of Christ."[15] Dupuis was just as harsh about his contemporaries: "There are many among our readers who . . . will persist in the supposition that Christ really existed, whether wise or foolish, great legislator or impudent impostor, because before reading our book they have accepted that idea, never for a series of years so much as suspecting its falsehood. Belief with such persons is a habit difficult to get rid of."[16]

Dupuis's comments were enormously influential in the years after the publication of his work, which even two hundred years later still impresses with its shrewd eye and dangerous premise—that religion is sun centered. Dupuis and the Freemasons sent shudders of fear down the spines of those within Christianity, as well as those opposed to the Revolution—the very thing that threatened the old social order of Europe, of which Christianity was the bastion.

Christianity, it should be mentioned, had specified very precise dates for various biblical events—2200 BCE for the time of Abraham, and exactly 4004 BCE for the Creation (at precisely 9:00 a.m. on the morning of October 26, according to Archbishop James Usher of Armagh). Such was the desperation of the times that strict adherence to this date by the Christians was seen as a means of confounding the arguments of those who opposed the fundamental tenets of Catholic Christianity.

By revealing the language of Egypt, Champollion, the decoder of the Rosetta Stone, was actually thought to be helping the church's cause. His first discovery, in 1822, was that the date of the temple of Dendera was not as ancient as had been thought: it was Roman. It had been a follower of Dupuis who had first posited a speculative date of many thousands of years BCE. Edme-François Jomard had accompanied Napoleon to Egypt and been a leading scholar on the expedition. He had based many of his ideas on detailed surveys of Egypt and measurements of the pyramids. In the ancient world these measurements were cited as evidence of Egypt's great antiquity. Greek sources claimed that Egyptian measures of length were based on a knowledge of the world's circumference. Jomard published his findings at a period of passionate Hellenism, Hellenism that rejected even the ancient Greeks' own view of Egypt. Champollion's discovery delighted the Vatican. In a letter to the Holy Father, the French ambassador wrote:

> [Champollion] has . . . humbled and confounded the pride of this philosophy which claimed to have discovered in the zodiac of Dendera a chronology earlier than that of the Holy Scriptures. . . . M. Champollion establishes . . . that no monument exists from before 2200 BC, dating back, that is, to the time of Abraham, so that, in accordance with our faith, there remain approximately eighteen centuries of darkness through which interpretation of the Holy Scriptures alone can guide us.[17]

The Revolution in France had shaken Europe to its core. The cry of "All men are free" had resounded heavily within the thick walls of Saint Peter's, too, and as a frenzied spirit of emancipation was surging through the continent, the Vatican reeled from blow after blow. The sheer desperation in the French ambassador's letter is palpable. A period of enlightenment had begun, the Age of Reason had arrived, and as far as the Vatican was concerned, the blame for it all lay squarely at the door of Freemasonry. It was Freemasonry that, under the guidance of Napoleon and with the help of Dupuis, had ransacked the Vatican and exposed its archive. Freemasonry had now risen to such eminence that it threatened the very existence of the church and its influence upon the state.

Pope Clement XII had already singled out Freemasonry as an enemy of the church in 1738 when he issued a papal bull condemning and excommunicating all Freemasons, many of whom were practicing Catholics. In a letter released and published for the first time in 1962, Clement, addressing an unknown correspondent, reveals the reasons behind his condemnation of the Masons, asserting that Masonic thought rested on a heresy—the denial of Jesus's divinity. The inference was that the Masons saw all men as being divine, also a heresy. Clement also noted that the intellects behind Freemasonry were the same minds behind the Lutheran Reformation.

Two years later it was Masonic forces that had the upper hand. Whether Napoleon himself was a Freemason is uncertain. That he was deeply involved in things Masonic is undeniable; many of the higher-ranking Freemasons were in his army, and under his rule they flourished. According to Bernal, "His initial behaviour in Egypt also indicates this influence: he tried, for instance, to transcend Christianity and appear as a champion of Islam and Judaism, and he dutifully went into the great pyramid and had a mystical experience."[18]

Freemasonry comes, ostensibly, from the guilds of masons of the great cathedral-building era. The acute learning of these masons is embodied within the grand corpus of their work, and yet to a large degree they remain an enigma, as mysterious as the actual rise of the Gothic. Freemasons themselves claim a direct Egyptian influence and ancestry. The first Masonic lodge is said to have been convened by Moses at the foot of Mount Horeb. At this same meeting the Tabernacle and the Ark of the Covenant and the Sacred Tables of the Law were delivered to Moses. The second lodge was later held by Solomon, Hiram of Tyre, and Hiram Abiff, in the bosom of Mount Moriah. It was at this period that the legendary Hiram Abiff started work on the construction of Solomon's Temple. According to Masonic tradition, inscribed above the temple balustrade in the very place where the Slavonic Josephus states that the name of Jesus was inscribed, was the legend *Jah-bul-on.* This translates directly as the "Lord God of On"—*On* relates to the sun[19] and in ancient Egyptian times was the name of the place now known as Heliopolis, the City of the Sun. The god of On (or Annu) was Osiris, in his aspect as Ra. This, too, was viewed as heresy by the church.

Perhaps it is not surprising, given Freemasonry's putative origins, that elements of heresy are to be found, thanks to masons, inside many cathedrals. The long disused west door of King's College Chapel, Cambridge, contains much symbology linking the Christian and Judaic mysteries to ancient Egypt. The letters *JHVH* are carved in Hebrew within a sun symbol upon the door. Although it is known that the chapel was constructed during the reign of Henry VI (1421–1471), these symbols, as well as the proportions of the chapel, have never been satisfactorily explained—it is only by reference to Freemasonry that they can be. In fact, its plan was based upon the Temple of Solomon, and the vaulting resembles Egyptian papyrus plants, as does much of the vaulting in Gothic cathedrals for that matter.

Freemasonry, with its theme of the brotherhood of man, harks back to Gnosticism and its recurrent theme of knowing oneself. Shakespeare takes up the idea in *Henry VI,* Part 2 (act 4, scene 7): "Ignorance is the curse of God, Knowledge the wing whereby we fly to heaven." It would appear also that such knowledge came via sound and language. The term *masonic* suggests music. The term is in Old High German, *mezzo,* a musical term meaning "neither loud nor soft, half." From this comes *steinmezzo,* or stonemason. We come back to the idea of architecture as frozen music.

This terminology is interesting in relation to the Masonic obsession with the rock of Zion, the cornerstone of the temple. This is described as a particular stone or rock that is, in the many variants of the story, at first overlooked and then retrieved and used as a keystone of the temple. In the New Testament, Jesus is compared to the cornerstone, having been rejected by the people: "By the name of Jesus Christ of Nazareth, whom ye crucified. . . . This is the stone which was set at naught of you builders, which is become the head of the corner" (Acts 4:10–11).

Variations of this quotation appear throughout the New Testament, in Mathew, Acts, Romans, and Ephesians. What is remarkable is that as these references appear, they seem to become ever more explicit, stating that Jesus is *in* the rock, as well as being it. The culmination comes in 1 Peter 2:3–8.

> The Lord is gracious.
>
> To whom coming, as unto a living stone, disallowed indeed of men, but chosen of God, and precious.
>
> Ye also, as lively stones, are built up a spiritual house, an holy priesthood, to offer up spiritual sacrifices, acceptable to God by Jesus Christ.
>
> Wherefore also it is contained in the scripture, behold I lay in Sion a chief cornerstone, elect, precious: and he that believeth on him shall not be confounded.
>
> Unto you therefore which believe he is precious: but unto them which be disobedient, the stone which the builders disallowed, the same is made head of the corner.
>
> And a stone of stumbling and a rock of offence, even to them which stumble of the word, being disobedient.

The scriptures tell us that Peter is "the rock." The above texts go further in stating "by the name of Jesus Christ. . . . This is the stone." The name—the sound—is related to the stone; you will note, too, that the faithful are addressed as "lively stones." I came to realize that there is a practicality here; the text that for so long I read as metaphorical, as many others had done, was being literal. The depth of it is that the rock represents the substance, the Great Mother, the substratum of life upon which all rests. This is a clear example of "as above, so below" at work in ancient language. A hero-god would, in consciousness, be at-one with this divinity. Jesus is in the Rock; the Rock is in Jesus. The relationship is that of resonance, a synchronous vibration with the most subtle force that permeates and upholds all of life. This is not a metaphor; to the initiated, it can become a reality. As far as the physical rocks that are chosen and used in temples and cathedrals, those monuments to the human voice, again the meaning is practical. The power lies in the site, the stone, and in the building design. All of this is related to resonance, the vibratory key of life. The masons knew of this. The closing message is, of course, "If you ignore 'the stone' and 'the word,' there will be problems."

In the late eighteenth century different forms and variants of Freemasonry began to appear. It was at this time that the Oriental Rite of Memphis came into vogue among certain of these groups. According to the Oriental Rite, an Egyptian wise man, Ormus,[20] said to have lived in the first century CE, founded what was to become known as Freemasonry. This he did by amalgamating certain of the pagan and Christian mysteries; via this he founded Rosicrucianism. Not surprisingly, it is very difficult to substantiate any of these traditions, and, at first glance, they do seem rather fanciful. However, the sheer number of Freemasons and the success of the society throughout the world demands that we take at least a cursory glance.

Ormus suggests *Hor-mose,* literally "Child of Horus"—Horus being another version of *the Lord.* This suggests *Hrm,* or Hiram—the legendary king of Tyre or Hiram Abiff, legendary builder of Solomon's Temple. In the pictographs of ancient Mexico, the deity, the One God, is shown precisely as the god Ra is in Egyptian hieroglyphics: as a dot with a circle around it. This is a symbol common to world mythology, where the dot is the divine seed of infinite potential and the circle represents the infinite whole within which all exists. It is a symbol of the First Principle, the "I am."* In Persia this High God is called Ormuz. The Assyrians called the god Assur. It seems therefore that the Freemasons, in citing Ormuz/Asar as their founder, are indeed harking back to an incredibly ancient tradition—the power of the name of God.

Perhaps this is why "the Craft" is riddled with Egyptian symbolism. It was Freemasons who drafted the United States Constitution,

---

*In astronomy the symbol represents the sun, while in astrology it is used to represent the self.

and intriguingly the symbols on the dollar bill are very Masonic and very Egyptian. The adverse side to the portrait of George Washington shows the Great Seal of the United States. This seal, accepted as such by the signatories of the Declaration of Independence on June 20, 1782, features an eye in a pyramid with the legend *Novus Ordo Seculorum*—A New Secular Order. This extraordinary symbol reminded me immediately of the letter *o* of el ayin, the Eye, the all-seeing Eye of God.*

A deep vein of Egyptian symbology runs all through Freemasonic tradition. The Cleopatra's needles in London and Washington—obelisks actually dating from before Cleopatra's reign, of New Kingdom origin circa 1500 BCE—are topped with pyramid-shaped benben stones, as depicted on the dollar bill. In Masonic lodges a blazing star shines forth and is considered by Masonic authorities to represent the star Sirius. To the Egyptians the star was the embodiment of the cosmic Isis, consort of Osiris. Even the Masonic secret sign is of Egyptian origin, and is something also familiar to practicing Catholics—the tau cross.

It is difficult to ignore the Freemasons, particularly in the light of all of this. But what all these survivals demonstrate is that Egypt's influence upon the course of history, and Western civilization in particular, cannot be ignored. And yet the history, the beliefs, and the mythology have been ignored within what must surely be their proper context. History it seems, is inclined to reveal more about ourselves than the proper context of our ancestors' achievements. Indeed, the broad mass of scholars and theologians have stood together, ignoring the witnesses of the past and the brave dissenters of the present.

MASONRY AT THE CORE

Babylon, too, plays its part in the symbolism of Freemasonry. For the present, though, our attention will now focus on the philosophy behind the temple structures of Babylon. The Esa-gila, the temple of Esa or

---

*The fact that this eye is in the pyramid is all the more telling—the Freemasons were perpetuating a tradition that the church deemed heretical and wanted to be rid of. *God* was in the machine.

Asar, was the most important temple complex in ancient Babylon and was dedicated to Marduk, the supreme deity. Marduk was later superseded by the god Assur—yet another form of Asar. Assur was the supreme god of the Assyrians, who, it would appear, took their name from him.

The massive buildings were ziggurats, basically step, or unfaced, pyramids. Ziggurats were effectively graves or tombs of the suffering god, Marduk. In Sumerian, Marduk's name was Amar-utuk, "the Calf of the Sun God"; in other word, a bull calf. In Sumer he was probably a minor god, but in later times he became, as Baal, the chief god of Babylon. His other name is Bel. Marduk thus replaced the Sumerian Enlil, the "Lord of the Wind," whose other name was Kur-gal, meaning "Great Mountain." (Assur was also called the Great Mountain.) Marduk was cognate with Enlil's brother Ea, meaning "House of Water." In Hebrew, the Jews of the exile (ca. 586 BCE) called him Bel-Merodach.

Marduk has fifty names, a number of which are strikingly similar to that of the Egyptian Asar (Osiris). These are: Asaru; Asarualim; Asarualimnunna; and Asaruludu. Robert Temple comments: "We have already seen how the An of Egypt was known in Sumer not only as An but as Anu, picking up a 'u' ending. It is therefore not so senseless to see in Asaru a Sumerian form of Asar, with the same 'u' ending added. But the Egyptians themselves also had an Asaru, or more precisely, an Asar-uu, whom Wallis-Budge describes as 'a form of Osiris worshipped in lower Egypt.'"[21] Wallis-Budge also tells us that Asaruludu means "Osiris of the growing plants."[22]

The Enuma Elish, the creation myth of the Babylonians, tells the story of Marduk's victory over Tiamat (Akkadian, meaning "Sea"), the universal mother,[23] and also states that the ziggurat *is the god*.[24] In Egypt, the Great Pyramid was also accorded the same status: in other words, *the Great Pyramid is the god Asar—Osiris*. The following utterance from the Pyramid Texts, carved on the walls of the pyramid of King Unas (2375–2340 BCE), states: "O Horus, this King is Osiris, this pyramid of the King is Osiris, this construction of his is Osiris; betake yourself to it, do not be far from it in its name of Mr . . . O Horus, as your father Osiris in his name of 'Mansion of the Monarch.'"[25]

On seeing how Christian terms have been misinterpreted in the past, the assertion in the Roman Catholic catechism that *the church is the body of Christ* now becomes a reflection of the ancient viewpoint. This, at last, reveals the real purpose of Freemasonry. It also reveals its extraordinarily ancient provenance. For in Masonry, God is the temple. Jesus, the hero-god, is the cornerstone. It is a simple case of God = temple, temple = God. It is probably for this reason that the ancient Egyptian hieroglyph of the throne is the symbol for the goddess Isis and a partial symbol for the god Osiris/Asar.* The throne is the seat of power: a king might have his throne in his great palace; the gods, however, have theirs in their great mansions or temples—quite often their pyramids.

To the Ancients, the sacred site was the meeting place of heaven and earth, a place where humankind could more easily come into contact with those gigantic forces that play upon and within the earth. A temple or a church is today still called a house of God, and indeed, it is still a place wherein the consciousness of the people is meant to meet the divine consciousness of the cosmos, to meet God—although much of the means have been forgotten. To come into conscious contact with God and to remain and grow in that contact was a prime attribute of the heroic life. We have noted how Gisa is not only the hero name but also as Giza (local phonetic) it is the name of the world's most wondrous ancient sacred site. This may have been the origin of the traditional naming of Christian sites after saints, those men and women deemed heroic by the church.

The temple was effectively a point of consciousness, or a location wherein God could be experienced. In this sense the sacred place was numinous and consciousness expanding. God is everything, literally. God is both immanent and transcendent, in all things and beyond all things. By the act of communion with the divine in a prepared place, humanity could gain much. Via ritual and its attendant disciplines, one could gain control of one's senses and direct them toward the divine experience. To the Ancients this was magic, to us today what is known

---

*The Sumerian Anunnaki are also depicted seated upon their thrones.

of this is called psychology, what is unknown remains *magic*. However much we have dismissed both, they are really sciences and belong together, especially at the points on the planet where the subtle incoming and emanating energies are strongest.

Often, when the Ancients directed their senses toward the divine, they might begin by focusing upon various representations of the chosen god: a tradition that has passed into Orthodox Christianity and the theory of icons. This would help provide a bridge to the subtle, unseen realms. Provided one's intentions were pure, using the symbols or statuettes did not constitute a lapse into idolatry. In later years, with the loss of much of the wisdom that had gone before, idolatry would be a correct description, for in ignorance, the image itself would be worshipped. By that time humanity had, to a large extent, lost touch with what God truly was—the immanent, divine spark within each one of us. We give divinity a particular name and form because it helps focus our intent. We can become conscious of this divinity. Sects like the Nasoreans appear from the evidence to have had a sustained gene pool of those who were sensitives. They were born to be sensitive, trained to be of service to the divine for the benefit of humanity. The ruler was expected to be the nation's greatest servant and was associated with the sun. As above, so below: the sun was seen as the all-giving ruler of the solar system, the earthly monarch's function was the same—the all-giving ruler, embodying divinity. Thus, images of the king or queen could be used as objects of worship, not as an end in itself but rather as a psychological tool en route to the unseen Lord. In this method of the Ancients, "the solar king was no mere symbol, ornament or tax-eater, but a working instrument, a kind of electric generator or conductor of the positive current by which the spirit of the people and the fertility of the earth itself were maintained."[26] However, well before the era of King Herod, royalty had lost the plot, service was out, greed and arrogance were in.

It is apparent that to the Ancients, divine unity was seen to uphold diversity and their day-to-day welfare was best served by understanding their relationship to the whole of life—working with nature to grow food, merging with nature in order to hunt. Confirmation of this relationship with natural forces and of a king's role comes from the hand of

King Gudea of Lagash, a Sumerian who lived circa 2400 BCE: "With an astonishing directness, he [Gudea] depicts the gods as natural forces. All sources agree in presenting the king as instrumental in procuring for the community the boon of a harmonious integration with nature."[27]

The ancient mythologies were filled with relationships among divine beings and with tales of heroic mortals becoming immortals. As we have seen, the ultimate goal lay in merging with the divine, a sacred acoustic science being used for this purpose. This difference in approach, compared with that of modern times, implies a greatly lessened sense of ego on the part of the individual within their society; we have a greater sense of the collective approach. The Ancients accepted that the creative life-giving force, God, was everywhere. Their highest goal was to experience God at all times, hence humility was a prerequisite, for how can one see something's true nature, the God within, if one considers oneself above it? These people were humble before their God.

Humility enables a shaman to become his or her totem animal, taking on, for example, the attributes of an eagle and soaring over and above the Nazca lines. Such attributes would often be reflected in the shaman's name, Great Eagle or the like. Of course, it was recognized that there was a bodily difference between a human and an eagle, but what was important to these very practical people was function. Here, we can see the relationship between name, form, and function. Although the form may be different, if the function is the same, then the name will be the same or similar, hence the Gisa correlation.

Here we return to the idea of names as titles, and we can see that the function of a sacred place, as a place filled with divine atmosphere, is the same as that of the immortal hero: a body filled with divine atmosphere. For atmosphere, read *vibration*. The vibration of the temple and of the hero are the same. The body of manifestation is different, but the function is the same, they both embody divinity, and they resonate together. It should come as little surprise that, in this way, the Ancients viewed the hero and the temple as one. Again, we come to the fresh interpretation of the Roman Catholic catechism: if the hero is the temple, *then the temple is the actual body of Christ*. Further on I shall discuss just how extraordinary this concept truly is.

The point of regularly returning to a sacred place is to share in and retain its atmosphere. Here we should note the difference in the two embodiments, the temple and the hero, and the importance of the temple to the aspiring hero. The sacred building is sited, designed, constructed, and used in a very particular way to embody divinity—to reveal the atmosphere of God. The aspirant must see him or herself as a temple too, but not as an edifice that can be built from scratch. The aspirant already has an atmosphere that must be changed and a body that must be subtly restructured. The mind and body must be purified to manifest divinity. The temple and its rituals would be prime tools. We know that in these specifically sited and specially designed places, through acoustic-based ritual, the participants raised their consciousness to merge with what many have called the higher realms. I reaffirm my earlier statement that the names of Jesus and Christ, like Buddha, are titles relating to levels of consciousness. These title holders manifest higher levels of consciousness than the norm, and this benefits humankind. How one can ascribe consciousness to a building, however sacred, may require a shifting understanding. One needs to accept the understanding of the Ancients that consciousness is everywhere, embodied in different forms and hence expressed at different levels. Embodiment is expression. It is the unexpressed, universal form that sustains the expression.

God created Adam out of red clay. From the ground we come and back to it we return. This is a cycle that the temple reflects; as the human body is constructed from earth, so, too, the temple. Jesus is the cornerstone of the temple; he is the roughhewn stone that the builders almost neglected. With these metaphors we are familiar, they are all about embodiment. What we are not familiar with, though, is the fact that in the Ethiopian legends of our Lady Mary,[28] Jesus's grandmother Hannah (Anna, Annu, Anu) is described as the twenty-pillared tabernacle of Testimony. The Egyptian *Annu* also means "pillar." I have pointed out that the Babylonian *Annu* means the "underworld." So, what is the relationship between the pillar and the underworld? Simple. A pillar is a support, and the underworld is the world's support. In effect the ancient builders were endeavoring to make manifest the unseen levels of life.

This was not merely symbolism—because of site positioning, design, materials, and ritual activities—it was practical. The uplifting divine atmosphere was there to partake of, even the atmosphere of the all-pervading, all-sustaining one.

THE RITE

In the Gospels, Jesus exorcises seven "devils" from Mary Magdalene. The most ancient form of this tale appears to be about the seven spirits of the Anunnaki or Maskim, the Sumerian-Akkadian spirits of the seven planetary spheres, born of the goddess Mari.

There was a sacred drama, dating back to the third and fourth millennia BCE, that told of their multiple birth. An Akkadian tablet says of them: "They are seven! In the depths of the ocean, they are seven! In the brilliancy of the heavens, they are seven! They proceed from the ocean depths [Mary-Maria-Mer], from the hidden retreat."[29]

According to Walter Scott, in his commentary on the *Hermetica,* and to recent research, the system of planetary spheres was believed to be the body of God the Father, in which he dwelt as the human soul dwelt in the human body.[30] According to the testimony of Al-Shahrastani (died 1153 CE), the indwelling of God in the planets and in individual men

> is the personification. . . . Sometimes they say that God personifies himself by means of the celestial "habitations" [the planets or planet-spheres] in general; but that he remains one, for it is only his action (and not his essential being) that manifest itself in each of the "habitations," according to the measure of his workings on it and his personification through it. It is as if the seven "habitations" were God's seven members, and as if our seven members were God's seven "habitations," in which he manifests himself.[31]

This is reflected in something Sir Isaac Newton (1642–1727) wrote shortly before the publication of his *Principia* in 1687. Newton had developed a theory that matter was an emanation or outpouring of the will of God, a theory that the imprint of his will upon space was what we regard as impenetrability and movement: "So God may appear to our innermost consciousness to have created the world solely by an act of will alone."[32]

These interesting quotations demonstrate the philosophical influence of myth down through the ages, and the influence of the idea of the seven planetary resonances on subsequent scientific development. But what about the effect of this mythology at the actual temple site?

Seven is an interesting number, particularly in terms of music. The seventh note of an octave is the leading note, it hangs on, as if demanding to be led forth. It also reconciles, it gives a sense of something that might come after, of imminence. The seventh note is the note of completion. Furthermore, the number seven seems to be related to the body in many ways—just as the seven planetary spheres represent the body of god. Contact between man and the cosmos is said to be made via the seven chakra points that run from the base of the spine to the crown of the head. In Western medicine the existence of these areas or energy points within the body is a moot point. In contrast, in the East they are an accepted reality. Recent photographs, using super-sensitive machinery, have shown that these points do indeed exist.[33]

The aim of tantric yoga is to awaken kundalini and to bring it to fruition. Kundalini is known as "the energetic serpent" that rises up through the chakras to the crown of the head. If the kundalini energy awakens and rises to the crown—the thousand-petaled lotus—then the adept may attain a blissful union of the self with the infinite. The way of awakening this energy, and creating of the body a true temple, is held by Tibetan lamas to be through words of power, the *hekau* of the Egyptians. The most famous mantra of power is the *Om*, the word of creation; another is *Om mane padme hum*, known as the Jewel in the Lotus.

The chakras are analogous to elements of a subtle circuitry that bears an energy charge that may be activated by a word of power at the correct frequency. They are centers of subtle energy, which with the aid of subtle energy technologies, are now being measured. Each of the chakras is associated with either a major endocrine gland or a major nerve complex. For example, the throat chakra lies over the thyroid gland, the gland that controls metabolism. The others are the crown (third eye/pineal gland), heart (solar plexus), sacral, and coccygeal. Interestingly, each of these points is also "associated with a particular type of psychic perceptual functioning."[34] As Richard Gerber observes, "the chakras translate energy of a higher dimensional (or higher frequency) nature into some type of glandular-hormonal output which subsequently affects the entire physical body."[35]

This is fascinating. The idea that incoming frequencies from the outer spheres could be retuned to frequencies many octaves lower is very intriguing. And it is not as fantastical as some would have us believe, particularly in light of what we have seen in previous chapters about the influence of sun and moon on human physiology. In this sense the mantras of the Tibetan lamas and others, the extraordinary words of power that can have such a radical effect are themselves like points of frequency. They are transformers that help the spirit to ascend and also help the higher frequencies to descend to what for us are edifying levels of reattunement. Perhaps in this way the alphabet is the Ancients' legacy to us—the transmission, in symbolic terms, of innate and archetypal power. The alphabet is very much a development of earlier pictographic hieroglyphs, although the view is widely held that the letters in themselves, the individual shape-forms and characters, have no transcendent external meaning. However, this is because in translating them we have actually fixed their value and therefore their meaning. As Schwaller de Lubicz pointed out, "Each hieroglyph can have an arrested, conventional meaning for common usage, but it includes (1) all the ideas that can be connected to it, and (2) the possibility of personal comprehension."[36]

Language, in the ancient sense, is not a metaphor; it directly imparts or expresses what it wants to say. It is this that gives us a clue as to the application of language in the temple, language as frequency. Hieroglyphics were ever and always a sacred language, the language of religious philosophy and metaphysics. Hieroglyphics did, however, enter into the realms of other written languages, and were later to provide philologists with important clues to their eventual decipherment. Hieroglyphs were, of course, translated, along with the stories they tell, and the similarity of the stories to tales worldwide is remarkable. They had obviously been told and retold for many millennia before being committed to parchment.

The myth of the hero is the most prominent among the similar tales, the similarities being so clear that the differences must be due to nothing more than local cultural and linguistic interpretations. One gains the impression that the myth was there at the outset but was given a specific form by each culture according to the custom and language of those peoples. Language expresses ideas and differing languages can make them appear to be different, but ultimately the only difference may be in the language itself. The story, from the book of Genesis, of the tower of Babel and the confusion of languages makes this point.

However, having by this time encountered a broad range of facts, I was firmly of the opinion that the myth of the hero, associated with the power of the place, induced language as a physiological response committed to history in pictographic form, as hieroglyphs, literally *sacred language.*

Our bodies are the best bit of technology in our possession. We have the inbuilt technology for transformation, but in modern times, seem little prepared to use it. Or perhaps we have lost the knowledge of how to? Religion is the last vestige of an ancient spiritual technology.

There seem to be certain basic and fundamental syllables that contain all meanings within language. The names of Jesus, Mary, and Anna are cases in point. Variations of these names appear throughout religious mythology and have certain meanings, just as the words for spirit and soul seem to appear on the same basis: *ba* and *ka*. That this syllabary was used in a punning fashion I think is rather obvious, and the reader will no doubt have noticed my own use of puns within this text.

Certain of these syllables were used by the hierophant or priest as he or she entered the temple, walking sunwise—that is to say, clockwise—having made an offering at the entrance, and gradually raising the tone and volume of the voice in order to attune to the right effect. The presence of other people participating in, or merely observing, the ritual would only have added to the ambience of the site and its power.

At particular frequencies it seems that the body becomes a vessel, a receptacle for other frequencies and vibrations, a meeting place wherein opposing forces unite. Macrocosm and microcosm meet. The temple of the cosmos meets the temple of the earth to help form the temple of humankind.

> If we listen to the audible frequency of this planet, it may be possible for us to resonate and entrain with it despite the fact that the actual vibratory resonance created may be thousands of times faster or slower than the frequencies to which we are listening. This same principle applies to the frequencies of the human body which may be far removed from those sounds which we can hear but which can be affected by audible vibrations. Through the Principle of Correspondence, we may use harmonically related sounds to influence the vibrations of atoms or the stars.[37]

David Tame explains, in *The Secret Power of Music,* that "audible sound was believed to be a 'reflection,' within the world of matter, of the Cosmic Tones."[38] Music, to ancient humanity, was the carrier of a divine super-physical energy that one could draw down into oneself at certain levels of frequency. "In the spoken or intoned rituals of many of the world's religions there is again a similar concept: that the voice of the priest within the realm of time and space becomes a vehicle for the energizing Voice of the Creator to manifest its forces through."[39] As Tame goes on to observe, "the role of music and the role of religious intonation and liturgy was to release into the earth a form of cosmic energy which could keep civilisation in harmony with the heavens."[40]

In light of the revelation that temples, cathedrals, and pyramids were monuments to the male voice range, this is significant stuff. But music and chant do not only help to maintain the fabric and morals of society. They also heal.

THE DIVINE SOUND OF BIRDSONG

In October 1998, I read an article in *New Scientist* magazine that asked the question: "Were the Mayan pyramids designed to capture the quetzal's cry?"[41] In a fascinating article, acoustics engineer David Lubman, a consultant from Westminster, California, suggests that these pyramids were built to copy the cry of the quetzal, a wonderfully beautiful species of bird native to South America and sacred to the Maya.

It is a trick of local tourist guides to the Mexican pyramids to clap a pair of rocks together and listen to the resonant echo. The echo is uncanny and comes back as a "descending chirp of 'EE-OO.'" This cry is remarkably like some of the incantations that contain variant spellings of the holy, secret name of God, a name that was chanted by the hierophant or high priest. Chanting the name would drive down brain wave frequency, induce an alpha state, and open up a higher consciousness.

Research into birdsong may eventually lead to some revelations about the internal process involved when humans chant and sing. The part of a bird's brain that is responsible for song enlarges as the spring days lengthen. This process, originally thought to be controlled by testosterone, has now been shown to be controlled by melatonin. Gregory Ball and George Bentley of Johns Hopkins University, and Thomas Vant Hof, of the Max Planck Institute, have demonstrated this in experiments with starlings. The *Times* newspaper quoted Dr. Bentley: "'We would never have discovered this if we had started out trying to see what melatonin is doing in humans,' he says. 'Now, with this discovery, we can feed back the information to other scientists and perhaps help them to understand what is happening in other species.'"[42]

The parts of the human body that are stimulated most by changes in brain wave pattern and by fluctuations in the electromagnetic field are the pituitary and pineal glands. The major reason why they are more sensitive than other areas is due to the presence of a tiny quantity of magnetite within the ethmoid bone, within which both of these glands are situated. This seems to be the pattern within life in general—even bacteria have been shown to carry tiny crystals of magnetite. It is interesting that in the case of human beings, the magnetic source should be located close to the pineal gland. The pineal gland has long been associated by magicians, shamans, and diviners with the occult and with the psychic. A tiny cone-shaped gland, situated deep within the brain, the pineal is responsible for the secretion of melatonin and serotonin over each twenty-four-hour period. Melatonin is associated with relaxed states. Serotonin acts in the opposite way. Esoterically speaking, the pineal gland is associated with the third eye, el ayin, the eye of Horus.

Returning to the cry of the quetzal, what is extraordinary is that it is very like the cry of Dionysus, or even the cry of Jesus upon the cross. The Dionysian cry is *Euoii, Euoii,* whereas that of Jesus is *Eloi, Eloi.* However, the strangest and most startling correlation is to be found within the mysteries of Mithras. A portion of the Mithras Liturgy, translated by Marvin Meyer, is as follows:

> I invoke the immortal names, living and honoured, which never pass into mortal nature and are not declared in articulate speech by human tongue or mortal speech or mortal sound: EEO OEEO IOO OE EEO EEO OE EO IOO OEEE OEE OOE IE EO OO OE IEO OE OOE IEO OE IEEO EE IO OE IOE OEO EOE OEO OIE EO IO III EOE OYE EOOEE EO EIA AEA EEA EEEE EEE EEE IEO EEO OEEEOE EEO EYO OE EIO EO OE OE EE OOO YIOE.[43]

There is plenty more of this within the same text, much of it in the same repeatable binary code-type system.

The god of healing, the savior god of the Aztecs, was Quetzalcoatl, the god of spirit of life, symbolized in the breath of the wind. He was the son of the Sun and of Mother Earth. To the Maya he was Kukulcan, to the Yucatecs he was Itzamna. As we can see, the Gisa name is again present. Quetzalcoatl and Itzamna, both names mean "feathered serpent." This reminded me of the Sumerian god Ningishzida, the great healer who watches over the underworld. His symbolic creature is the horned snake. As if in confirmation of yet another global mythic connection, Quetzalcoatl rescued humanity from the depths of the underworld.

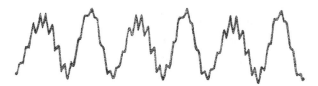

Fig. 5.1. A graphic representation of the sound *ee.*
Note the terraced pyramidal forms.

The EE-OO of the Maya is a version of their secret sacred name of God. David Lubman gained his insight into the Mayan pyramid through seeing a representation of Kukulcan in Mayan art where Kukulcan was portrayed holding a large quetzal. In comparing his recording of the echo to the cries of the quetzal, he found that not just the cry was familiar but so, too, the range of frequencies involved. Lubman has gone on to find other examples of the cry matching the site.[44] The downside is that there is little independent evidence that Mayan civilization based its ceremonies upon quetzal cries. However, as we have seen, it was not customary to reveal the secret holy names.

The name of the god was a mantra, and as such it went abroad, outward from its source. In this form it was secure, it was so entirely holy that only those who were holy in themselves could carry it. As we have seen, in Hebrew script the vowels were sacred and concealed, whereas in speech they filled in the consonant-formed words with vowels in order to give those words their real power and meaning. As Jonathan Goldman explains:

The vowels were originally very special sonics indeed, being mostly used for "God names" and other sacred purposes. Consonants gave words their bodies, but vowels put soul into them. Taken together in combination, the vowels will "spell" the Name of the Living One: I.A.O.; IEOA; HU; YAH; etc. Whichever way they are connected, they signify Divinity enlivening Existence and hence were sacred in all Magical practices. . . . Thus the vowels are to language what Life and Consciousness are to Existence.[45]

As Goldman further points out:

In traditional Hebrew Kabalah, there are certain secret names associated with specific spheres. These are names which, when chanted like a mantra, will cause the reciter to resonate to the frequency of the sphere. This is another aspect of sacred sonic entrainment similar to the practices of tantra, utilized by Tibetan monks, in which the practitioner, working with vocalization and visualization, recites a specific mantric form designed to invoke a particular deity. While visualizing this entity the practitioner unites with it.[46]

The act of chanting the vowel sounds can have a considerable effect upon states of consciousness. Each one of the vowels, chanted in a specific combination, can create particular forms of resonance. In the Western alphabet there are five vowels—*A, E, I, O, U*—whereas in ancient Egyptian and Hebrew and other ancient languages there were quite often seven, the additional vowels being a simple doubling *aa* and *oo*. William Grey has done much work on this.

> In Grey's work, the vowel A (ah) is equated with the element of Earth and the direction of the North; the vowel E (ee) is equated with the element of air and the direction of the East, the vowel I (ai) is equated with the element of fire and the direction of the South, and the vowel O (oh) is equated with the element of Water and the direction of the West and the vowel U (oo) is equated with the element of the Aether (which he called the element of Universal Truth) and the direction is "around and about."[47]

The ancient Egyptians, among others, believed that the human body houses seven souls. It is known that they used choirs within the pyramids and that they also used seven vowels to enliven states of consciousness within the body. Could the seven bodies and seven vowels be related to the seven musical notes? I believe that we should realize that there is a direct correlation between the various states of energy around us, music, consciousness, and the divine.

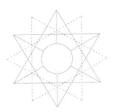

6

In the Realm of the Mer

I am the eye with which the Universe
Beholds itself and knows itself divine;
All harmony of instrument or verse,
All prophecy, all medicine is mine,
All light of art or nature;—to my song
Victory and praise in its own right belong.

PERCY BYSSHE SHELLEY, "HYMN OF APOLLO"

TA MERY

The history of Egypt is the history of a long, slow decline. Curiously, it could be argued the height of Egyptian civilization* took place during the building of the pyramids—early on in Egypt's history.

The pyramids were natural developments of the mastaba form that predated it. A mastaba is a "house of eternity"; they are flat-roofed, rectangular structures and were used from the earliest dynastic times onward (ca. 3500 BCE).

The revolution in building design started in the reign of King Djoser of the Third Dynasty in approximately 2686 BCE, and its author was the legendary, soon-to-be divine figure of Imhotep, the royal architect.

---

*The word *civilization* comes from the Latin *civitas*, meaning "city" or "city dwellers" (in the sense of the body politic).

His step pyramid at Saqqara is an extraordinary feat of engineering and led directly to the construction of the Great Pyramids in, and around, Giza and at Dahshur. This was the period of the Old Kingdom, 2575 to 2150 BCE, after which Egypt went into one of her periodic declines.

Egypt was called the Black Land by her people because of the silts washed down from the Ethiopian highlands, just over 1,500 miles upstream. This annual inundation took place during the month of June and fertilized the strip of land on either side of the river. It was a narrow strip that, less than two miles out on either side of the river, turns immediately to desert. It is in this sense that the Nile was Egypt. Egyptian life revolved around the Nile, and the Nile was the mother of her people.

Unlike the land of the two rivers, farther east in Mesopotamia, which flooded at irregular intervals causing chaos, death, and destruction, the Nile was reliable, as reliable, in fact, as any modern timepiece. If the Nile rose, it must be June. This image, this sense of constancy, pervaded all things Egyptian. To her people, Egypt was the beloved land as well as the land of the beloved, a reference to the goddess Isis: hence the oldest of the names for Egypt—Ta-Mery. (This appears to be the name mentioned in the Old Testament story of Judah and Tamar; namely, Egypt.)

Literally, *Ta-Mery* means "Place of the Mr"[1] and *Mr* has been understood as meaning "canal" or "waterway." However, given the definite article, *the* Mr, it must translate as something more specific. Its hieroglyph is a pyramid, $\triangle$ meaning a tomb.[2] It is also the name of certain specific royal pyramids, most notably those of the Giza Plateau. This hieroglyph encompasses the concept of the divine power of the Mr; in other words, Mary carrying the divine child. Mr is another hieroglyph for the goddess Isis, although rarely used outside the Old Kingdom period.* It simply means "beloved," the title of Isis, whose name in Egyptian is Ast (sometimes Aust). This is represented by the hieroglyph for the throne,[3] ⌐⌐. Isis was therefore the Egyptian throne, and the pharaohs were seated upon her lap, protected by her outstretched wings.

---

*Old Kingdom language usage became obsolete to a degree by the time of the Middle Kingdom.

The power of the Beloved was the underlying principle behind the throne. It is significant that the man sitting in Isis's lap, the pharaoh, is in fact a "great house"—the precise meaning of the word *pharaoh*. In other words, the power of Isis runs through that same house, and pharaoh was the living embodiment or incarnation of Horus, the child of Isis and Osiris. Pharaoh was the incarnation or manifestation of the divine child—the same divine child that the Mr—Mery—carries in her womb.

This is beautifully summed up by Timothy Freke and Peter Gandy: "To initiates of the Mysteries, a human being consisted of a material body and a spiritual soul. Our divine 'father' is God, who gives us our immortal soul, our material 'mother' is the Earth (matter), who gives us a mortal body. Matter cannot give birth alone, but is mysteriously impregnated by invisible Spirit to produce Life, and so is portrayed as a perpetual virgin."[4] The matter is the rock, the feminine, the substance that is impregnated by that which moves, the life stream, also symbolized as blood, that which streams throughout the body to give life. As we can see, the language of the ancient Egyptians can seem very complex, but it need not be.

The first thing that struck me about hieroglyphs was how unconscious they seemed. Now this may be an odd thing to say about these enigmatic symbols, but something about them goes straight to the mind, as if something primordial were trying to get through. In a word, the experience is *magical*. In contrast, in the modern world—the Western world of the twenty-first century—*magic* is a word largely disapproved of. It is almost unthinkable to use it when trying to explain the phenomenal world. This is where the tragedy of the separation of magic, religion, and science truly becomes apparent.

For the Egyptians, hieroglyphs were most certainly a living link between the cosmic mind and the human mind. The key to myth lay in language and also its representation in fundamental form: hieroglyphics—the representation of actual images as a means of thought transference, the same kind of symbolic language that speaks to us in dreams and in visions, and that is the basis of all religious thought and inspiration. Language is an expression of that which is conscious, and consciousness is expressed in myth. The hero myth, about birth, life, death, and regeneration, about the cycle of the seasons and the crops and their fertility, as conveyed in the purest linguistic terms, actually represents the divine power of the earth and the cosmos. Hieroglyphs are an emotional expression of a universe built upon patterns of sound.

Egypt is perhaps mistakenly called the land of the *gods,* implying that there is no One God. However, to the ancient Egyptians the concept of God was so vast that a human could not encompass it, except perhaps in the abstract, and therefore God is expressed as the sum of every divine part—in all of the gods. Thus, "for learned ancient Egyptians, there has persistently been a single God, who has staged the entire combined polytheistic show. No less than nine divine names were fused at Heliopolis into a single Ennead, a Ninefoldness. On what basis could a Christian scholar classify his own theological 'Trinity' as monotheism and keep insisting that the Heliopolitan 'Ennead' belongs to polytheism?"[5]

To say that the Egyptians had gods is an exaggeration, for the many aspects of the One God were not referred to as "gods," but as *neterw* (pronounced net-er-oo). The singular of this is *neter;* a neter was a divine principle. The fact that many of these neters were portrayed as animal headed is not a sign of a primitive society but rather a sophisticated one (*soph* meaning "wisdom"), for animals themselves represented principles within nature, principles that were divine and that served as functions. The falcon-headed god Horus represents, for example, the principle of resurrection, as does the scarab beetle. The stork, which in fairy tales is the bringer of babies, is actually the symbol of the migrating soul, and so it goes on—there are many examples. This in itself is a wonderful illustration of the complexities of meaning that, if properly understood, can reveal some amazing things.

In the ancient world one word could mean many things: it could have a philosophical, numerical, and linguistic meaning. In the world of the twenty-first century we have many words, with the result that understanding has become conditioned by the inability to see wider meanings in symbolism and the alphabet is symbolic only in its primary purpose.[6]

CREATION, RESONANCE, AND AWARENESS

The creation of matter ex nihilo—out of nothing—ranks as the greatest of mysteries, not only for the ancient mind but for the modern mind, too. For physicists it has become a holy grail, as they seek a theory that unites the very large—relativity and gravity—with the very small, the quantum world of subatomic particles, so far unsuccessfully. Perhaps a clue to the unraveling of the mystery can be found, after all, in the ancient Egyptian Pyramid Texts, collectively, but misleadingly, known as the Book of the Dead. What they seem to reveal is an idea that has only recently begun to appeal to physics, the idea of consciousness manifesting as energy—an idea that finds expression in the proper title of the Book of the Dead—"The Book of Coming Forth by Light."

"Energy" expresses the principle of something active. It comes ultimately from the Greek *energeia,* meaning "activity," this being two syllables: *en* and *ergon,* meaning "in" or "work." Energy as a concept thus expresses a kind of movement or motion. This is interesting because physicists are beginning to theorize that even before the hypothesized big bang that gave rise to the universe, there must have been some kind of movement within the pre-primordial energy fields whereby creation was brought about.

However, if we now extend this concept to accommodate the ideas of the ancient Egyptian creation mythos, we will see that at the beginning of all things the god Tum became aware of himself, and in doing so created Atum, the creator god, out of Nun, the primeval waters. In this sense, Tum is transcendent cause: in becoming self-aware he not only regards himself but the entire universe, too. This is expressed by the primal vowel, *a,* hence, *A*tum. Furthermore, Tum, in becoming self-aware, becomes self-conscious. Reality is born. If all this sounds rather tricky, bear with me, because when I express it differently the penny will drop.

This basic concept of the creation bears a striking resemblance to the Gnostic idea of matter and creation. Tum is the unknowable, ineffable, unspeakable godhead, who in becoming self-aware creates, or becomes, the Gnostic demiurge. Required in all of this is the image of the feminine, of Isis the goddess, who was called, even in the earliest days, the Oldest of the Old. From a cursory look at the creation texts, Isis would appear to have been the concept of the whole, just as Gnosticism informs us. However, the whole needs to stir, to vibrate, something needs to move across the waters ("the waters" being a common term for the cosmic expanse). An explanation would be that consciousness would be there before creation, for it is the "I Am," the ever-still, knowing observer. The creation then arises through a duality represented by male and female. The female is the substance, the male is that which brings movement. Another analogy is the receptacle and that which fills it. Both of these explanations are reflected in the sexual act. Through the motion, relativity begins, worlds are created, and the I Am observes itself in the myriad forms that manifest.

In Christianity we have seen how the rock is related to the substratum and the substance, but in Christian ritual, the substance is usually represented by bread and the movement by wine, which as I have mentioned, represents the lifeblood of the creation reflected in the ever-moving life stream of the human body. Wine is used because it also represents the divine intoxication that can be experienced by the individual when these godly forces are at work. Thus the concept of a great god and great goddess in motion brings us to another important expression referred to earlier—emotion.

Emotion is from the Latin *emovere,* which consists of two elements: *e* and *movere,* meaning "to move." The *e* is taken as meaning *ex,* or "out of." So, emotion is, or arises, *out of movement.* However, the involvement of the letter *e* does not end there, because it is also an expression of the third note of the diatonic scale of C major. This note corresponds to *mi* in the do-re-mi notation. Furthermore, *mi* was, according to the *Oxford English Dictionary,* originally the first syllable of the Latin *mira.* We are back to the pyramids again, to the ancient Egyptian Mr.

The exploration goes yet further, for mi is the third note of the hexachord, a row of six notes (hex, from Egyptian *heqa*—"lord of"). Do-re-mi (et cetera) originally gave their names to the six notes of the hexachord. This was a formulation brought about by an eleventh-century monk by the name of Guido d'Arezzo (995–1050 CE). What intrigues me about Guido was that he was no ordinary monk but a Benedictine, a follower of Saint Benedict, who as we have seen, was in all probability, none other than Apollo Benedictus—Apollo also known as "the good speaker." Resonance again—and here the voice of the God and of gods in general becomes exceedingly interesting.

The names of the six notes of the hexachord may be found in the hymn for Saint John the Baptist's Day (June 24), which happens to coincide with the annual flooding of the Nile. Oddly enough, it was Saint John's father, Zacharias, who upon hearing of his wife's pregnancy, sang the Benedictus as a hymn of thanksgiving. The hymn to Saint John features all of the musical syllables and key encoded words so far revealed within this text: "*Ut queant laxis resonare fibris Mira gestorum famuli tuorum, Solve polluti labii reatum, Sancte Ioannes.*"* A translation would be, "So that servants may sing with loose strings and can sing of your wonderful deeds, Saint John, loosen the charge of defiled lips." A request for purification before the singing of praises?

It seems though, that the outer meaning is secondary to the sounds and the effects they may have. Within these lines we not only have the six notes of the hexachord but also various expressions of the names of the gods. We have here *resonare,* resonance; *Mira,* the Egyptian Mr, the beloved of God, Mary; and directly after that *gestorum,* which, as it relates to the Holy Ghost—the divine in the body—is also used as another expression of the name of Jesus. So what is it that I am getting at?

In the beginning God is said to have uttered the Word. The movement of energy needed to utter that Word, no matter how infinitesimally small, would have created motion, energy motion: e-motion. And out of (Latin *e* or *ex*) this would have come the expression of the Word and Creation. To the ancient mind this was and is an ever-continuing process. Plato expressed it thus: "Pure being was confronted with non-being, and the result was becoming."[7]

This is the fundamental ancient Egyptian concept.

---

*Note that te and do, adopted later, are present here in Sancte Iohannes.

SOMETHING FROM NOTHING

Entering the tomb of Pharaoh Rameses VI (reigned ca. 1141 to 1133 BCE) in the Valley of the Kings, you begin a slow descent into darkness. In the dim light you will be able to make out murals, these murals give a picture of Rameses's soul in the afterlife and his perilous ascent into the Duat, guarded and guided by various incantations from the Book of the Dead.

The hieroglyphs signaling the beginning of this journey are sited very close to the entrance, and among the first of them you will notice the symbols of the neterw, looking rather like the flags on a golf course. There were seven of them in a row 𓊖𓊖𓊖𓊖𓊖𓊖𓊖.

These are preceded and followed by the royal crowns of Upper and Lower Egypt. The crowns mark the beginning and ending of a cycle, so that their combined value would be a whole, a unity. This opening hieroglyphic sequence is expressing, when taken as a whole, the cosmic octave—the hieroglyphs in this tomb were meant to be sung or chanted. Hieroglyphs were known as "the words of the God."[8] What this says is quite revelatory—that all sound and all language is but a further emanation of the very first primordial Word uttered by the divine self-consciousness: God.

The American composer Edgar Varese (1885–1965) had a strong belief that sound had an intelligence of its own. This is reflected in the Egyptian myth of creation and its expression in hieroglyphic symbolism. The organizing powers of nature, the gods, are sound and subtle vibration. Everything begins to fall into place.

The hieroglyphic symbol for unity, for the figure 1, is a vesica piscis ⬯. It represents the open mouth and the creation of all from out of nothing by God. Its meaning is primordial scission, being the division and separation from out of the ineffable, and is philosophically unfathomable and incomprehensible. For 1 is the absolute whole—from out of the one, multiplicity comes forth, all matter, all words, all numbers.

The ancient Egyptians held that before all things there was a primeval abyss of waters, everywhere and infinite, endless, with no boundary and no direction, a limitless void, the infinite, as yet unmoving, substance. This sounds amazingly like recent scientific speculation about the nature of the universe. It is estimated that a vast quantity of the universe, known as *dark matter,* which has been sought out over the decades using a whole gamut of experimental equipment and data, might actually be made up of hydrogen, the chief component of water. On May 8, 2000, it was announced that the Hubble Space Telescope had at last located some of this missing water.[9] The hydrogen was found to be in the spaces between clusters of galaxies, and it accounts for nearly half the known matter in the universe.

Now, interestingly, water is one of the greatest carriers of energy, if not the greatest, and the universe is a sea of energy. Within this there is a phenomenon known as zero-point energy, which is contained within a zero-point field. This is an all-pervading electromagnetic field that exists within a quantum vacuum. Zero-point energy is basically the vibrational energy associated with matter as the matter itself is defined all the way down to zero. In effect a zero-point is a point of stillness in a universe that is totally in motion. Now, if all of this seems complex, let's get to the point: that zero is defined as *neuter,* meaning "neither masculine nor feminine." *Neuter* is from the Latin, and beyond that its etymology is uncertain; but to me it seems obvious—*neter.*

It is believed that the historical peoples of the Middle East did not have a concept of zero: but this is based upon a dearth of proper evidence. No evidence does not mean to say that there wasn't any. The idea that perhaps it was a theological and mathematical concept must now be taken seriously. *Zero* was considered so secret, so dangerous even, that the understanding of it was limited to a select few, the initiates of the mysteries. To understand a point of zero is to come close to a greater understanding of creation, for Nothing defines Everything.

As we have already discovered, water, called *hydras* by the ancient Greeks, was associated with strange qualities and was a prime ingredient in the construction of the sacred site. This brings us to probably one of the most ancient temples in Egypt—the Osireion. Quite literally, it is deluged in water, so much so that little of it can be seen. This seems to be a part of its actual purpose; the Egyptians were very good at pointing out the completely obvious. The primeval hill in the center of the great hall is totally covered with water. This reflects the myth of the creation, for in the beginning, God created a primordial mound or bennu that rose up out of the waters. It was from this first dry spot that God went on to polish off the task.

As we can see, the hieroglyph of unity ⬭ (the open mouth) resembles a vesica piscis: exactly the same as the cathedral tympanum, in the center of which sits Christ enthroned. Christ is again being implied as *sound*. Furthermore, the glyph looks like a vibrating string. It has recently been speculated that the universe is nothing but a gigantic standing-wave pattern, vibrating to the tune of billions upon billions of frequencies. It is this vibrational energy that is set in train from the finest to the gross level, and everywhere this energy continues to be moved by the initial breath of God—the Word. Another thing that catches the eye is the similarity of this glyph to an eye. In fact, what is really striking is its resemblance to the perceived shape of a UFO: we are back in the territory of el ayin—when we adjust the word, we end up with *alien,* meaning "stranger."

The Hebrew-Greek *ayin* signifies an eye, the third eye—in this case the eye of God. If we place within the vesica the hieroglyph for the sun god, Ra, the determinative that means both light and time, it looks exactly like an eye ⬭⊙.

We now have, from the initial Word of Creation, *light,* and its measure, *time**—entangled just as Einstein said they would be.

If we now look at the texts in the order of the story that they are telling, the first thing that must be corrected, from the standpoint of their subsequent influence, is the opening line to the book of Genesis, which has been translated as: "In the beginning God created the heaven and the earth." The more direct translation is "The Lord of the beginning created . . ."

This is more sophisticated; Tum, the transcendent, becomes aware as Atum, this is the beginning. From the self-aware Lord, come all of the pantheon. The divine breath is facilitated or given form by the god

---

*If one compares the two versions of the Creation from the Old Testament and the New, the version in Saint John's gospel is much more related to the very ancient tradition of creation mythology, particularly the Egyptian.

Ptah. And this divine breath is the sound of the Om, sometimes spelled *Aum*—this seems to be a variation of Atum's name, with a silent *t*. Atum is *the Word*. Ptah is the creative power immanent in Atum—the demiurge. Atum appears in the Bible as Adam, whose son is Cain, the Hebrew equivalent of Ptah. Atum is the god who fell to earth, whose name means "the complete one." Ptah's facilitation of the Word is reflected in the sheer power of the prologue to Saint John's gospel: "In the beginning was the Word, and the Word was with God, and the Word was God. The same was in the beginning with God" (John 1:1–2).

According to Egyptian belief, the heart and the tongue represented thought and speech; attributes of the Creator who would be deified as Horus and Thoth. By bringing forth thought and speech, Ptah brings forth order from out of chaos, and the gods come into existence.[10] This is well expressed in the idea that the Word, *vibration,* facilitates self-luminosity as exemplified by the sun. This is where Ra, the self-luminous one, comes into his own. The disk hieroglyph representing Ra, who at the Creation is at Atum-Ra, is also the eye of *Khepera,* the Egyptian verb "to be." Ra sails around the heavens in his boat of millions of years, with all the other gods in tow. He is the pilot, the navigator of the celestial waters. The boat of Ra was sometimes called Ur or Uru, or even Makaa, which means "great protector"—literally calling self-luminosity the great AA, the double vowel of creation.

So, in this sense, creation is a process of quantum sound expanding ever outward. It is interesting that the later word *sound* is related to that for sun. In Anglo-Norman the word is *sun,* whereas sun in the English of the Middle Ages is *son* or *sonne.* Here we see the relationship between all three: *sound* is the child of God, so is the sun, and so is a son or a daughter.

In the *Corpus Hermeticum,* Asclepius warns King Ammon not to translate the Egyptian Mysteries into Greek. The quotation is a good analysis of the nature of words: "For the Greeks, O King, who make logical demonstrations, use words emptied of power, and this very activity is what constitutes their philosophy, a mere noise of words. But we [Egyptians] do not use 'words' [*logoi*] but 'sounds' [*phonai*] which are full of effects."[11]

The power of the Word is expressed in another version of the creation myth wherein a heronlike bennu bird flies over the waters of the primeval abyss and lights upon a stone or rock. It then emits a piercing cry to break the silence of the abyss. The bennu bird is the bird recognized by the Greeks as the mythical phoenix, also identified as Osiris, who is said to have uttered his name at the Creation. The Edfu texts are explicit in stating that it took seven words or sounds to create the world. In the Old Testament, God takes six days to create all things and on the seventh rests.

Oddly enough the seven words of the Edfu texts became trapped in a chest, rather like Osiris, who, in his myth, is also trapped in a chest. Is this a reference to God's secret powerful name being made up of the seven primal notes? Is this the original name of Osiris? This brought me back to Guido d'Arezzo's hexachord, which, as the *Oxford Companion to Music* points out, is in reality a heptachord—seven notes, not six—do-re-mi-fa-so-la-te. There seems to be a long religious tradition of preserving the sacred words of God, a tradition that tells that all is music.

This struck a chord, which in turn brought something else to mind. In Greek the word *chord* also reveals a certain amount about the mechanics behind creation. Spelled, χορδε and transliterated as *chorde,* meaning "of the emotions," the Greek word contains the letters chi, omicron, rho, delta, and epsilon.

The Chi-Rho was the symbol of Emperor Constantine's conversion to Christianity; but here the two letters are separated by omicron, O. This is symbolic of the pineal gland, the ayin of Hebrew and Egyptian. The cross is symbolic of matter. Rho is the creative female principle: it is the power of fertility, of reproduction. This aspect of *chord* suggests *awareness,* in the form of the O, of the two creative principles, chi and rho. The Chi-Rho represents the cross of the hero and the O our awareness of it—God within us. The philosopher Iamblichus (second century CE) stated that the names of the gods were impressed on souls before birth and that theurgic chants awaken them. Theurgy was the magical system practiced by the Egyptians for the invocation of beneficent spirits.

At the beginning of time it all came down to words and music.

When I looked at the words, I could see why: God, whose name means "voice" expressed the Word—*ex* meaning "out of" and *press* as in "pressure." This gave rise to the universe, literally *uni,* or "one," and *verse,* or "word"; *one word.* Thus the Word reverberated, from *verb,* or "word," and *re,* or "back." In other words, the word bounced back and forth like an echo.

The Word, the thought of God musing upon the state of self-awareness, became music, from *muse,* "to reflect in silence." This music shattered the silence. The music was a reflection of the *experience* of awareness, in Latin *expriri,* or "out of trying, effort." This is related to *expire,* in Latin *expirare,* "to breathe out." That which was breathed out was the spirit, literally the breath of God. It is this, in its form of music, that *resounded* around the newly formed universe—resonance.

Now, at last we understand the enthusiasm of the Ancients for all things creative: it comes from *en theos,* "from out of God." This is exhilarating, literally *ex hilare,* from "out of laughter." God had a sense of *humor: Hu,* "the Lord" and *mour,* "love": the Love of the Lord. There was an old Roman festival that celebrated God's son, it was called the Hilaria, where the Romans made *merry* (from the name *Mary,* the beloved of God.)

Something had emerged from out of nothing—ex nihilo. And it was very funny.

THE POWER OF THE WORD

The Word is all around us, within the universe. Within one word—universe—there are billions of vibratory permutations that give the universe its momentum. As these words spin around the cosmos, they form little vortices, giving rise to matter, to mass, volume, weight, gravity, and quanta (the very small). And all of these cross-correlations of frequency make the universe what it is—home.

Science and religion are, it appears, speaking the same language. The universe, both will admit, is a mass of eternal creation. The recent research of Professor Andre Linde of Stanford University, California, has suggested that the big bang is a bit of an exaggeration. Rather, it was a seething foam of creation that, he suggests, will refresh and renew itself for eternity. Professor Linde believes that there was more than one bubble and that even though our section of the universe is created, other parts are still going through the process of creation.[12] This view accords well with the ancient conceptions of the Creation—that it is an ever-expanding, ongoing event.

Another thing that accords is the structure of the hydrogen atom, which has a single proton and a single electron in orbit about a nucleus. The hieroglyph for the god Ra is startlingly reminiscent of this: ⊙. Without hydrogen, there would be no water, and thus no life.

Electrons, protons, and nuclei have even smaller particles, subatomic particles, at their core. These bundles of quanta are in fact bundles of energy, vibrating at different but incredibly high frequencies. Frequency describes the rate of energy flow or the amount of energy in a particle or a bundle. It has also been suggested that gravity is an effect due to standing waves, moving in toward the ultimate structures of matter—in other words, the interaction of different bands of frequency.[13] What seems to have happened throughout history is that these ideas* have been redefined, as language and understanding have changed their perspective over the course of millennia.

"Being is change, motion and rhythm, the irresistible circle of time, the incidence of the 'right movement,'" say de Santillana and von Dechend.[14] It seems to me that resonance, the resonance of the Word, defines each moment by a deft act of consciousness. The breath, *spiritus,* is released, and the primal substance moves, giving rise to cause and effect; the eternal is in motion, and time is born as a consequence. This description has barely changed through the centuries.† The ancient discovery is our rediscovery.

Even particle theory existed all those years ago. According to Sextus Empiricus (second century), it was not Democritus who first formulated

---

*Ideas that I believe are forms of inherent memory, visions of consciousness reaching out from the human standpoint to a point of infinite memory, godhead.

†Other descriptions that have already been explained within the text include analogies wherein the breath, *spiritus,* is replaced by the blood, the wine, or the life stream; and primal substance by the substratum, the rock, or bread of life.

atomic theory, but Moschus, a Phoenician who, unlike Democritus, rightly believed the atom to be divisible. This is exceptional stuff given that Democritus is always given the credit for the discovery. The idea that even atoms are divisible is too often thought to be a modern discovery. Similar scientific observations were not rare. The philosophers Thales and Anaximenes both saw the Milky Way as being made up of an infinite number of stars and planets—just as astronomers, now discovering planets beyond our solar system, have realized.

There are many other instances where it is being revealed that the Ancients had more depth than we previously thought. The Kabbalah, for example, is not just pure mysticism but may have been describing complex mathematical concepts in a philosophical vein, the origin of today's scientific paradigm. The ten sephiroth of the Kabbalah, the ten branches of the Tree of Life, mean numerical emanations. They are, as is now being discovered, "nothing less than a mathematical blueprint of the cosmos."[15] These branches are linked by twenty-two paths, which represent the psychological states of experience that are encountered as the logos—the Word—descends into matter, or as the monad (soul) ascends to higher worlds. The sephiroth have been described as follows:

> As can be done with most complex symbolism, the same glyph of ten sephiroth can be used to represent Dionysus' nine celestial hierarchies (plus a tenth for mankind), with seraphim or lords of love in Kother (the Crown), cherubim, the lords of harmony, in Chokmah (Wisdom), and thrones, or spirits of will in Binah (Intelligence). The rest follow in descending order into ever greater density . . . which, according to Phillips, denotes equally the outer organic form of the Cosmic logos, the entire universe, this solar system, a human body, or a single evanescent subatomic particle.[16]

Perhaps deeper even than this is the principle of the spin of the electron. Without it life and matter would be quite different—nonexistent even. The rotation of the sun gives off energy in the form of the solar wind—positive and negative particles that take about forty-eight hours to reach the Earth. As the Earth travels around the sun, she receives this fertilizing seed, which in many ways correlates closely to the breath of life. The sun has two very distinct magnetic fields—the equatorial and the polar. The equatorial field consists of alternating negative and positive polarities and rotates at a different rate from the polar field. It has been calculated that every seven days the sun's radiation alternates in polarity. This correlates nicely with the description of God making the world in six days and then taking the seventh day off. This led me to the idea that perhaps, here again, the Ancients expressed the same concepts as us, but in a different, more poetic language. It follows then that the ancient language of religion may well be an old language of science, a science based upon pure observation with an inherent subjective element.

In the temple of the Pharaoh Seti I (reigned 1291–1278 BCE), the pharaoh is shown receiving the water of life from the gods. What is interesting about this is that the water is not portrayed as one might be inclined to think, as a torrent, but as a series of tiny ankhs. The ankh is the Egyptian cross, symbolic of life, and of matter. As can be seen ☥ the ankh has a strange looping shape, reminiscent of a teardrop. Strange looping figures like this also appear upon the pyramid capstone of Amenemhet III, a Middle Kingdom pharaoh (reigned 1842–1797 BCE). They appear right next to the figure of the celestial Osiris, holding a star in his hand. The meaning of this glyph varies, but in this context, it means "in the neighborhood of" (m-s ht): it is a variant spelling of Sahu, the celestial Osiris. The glyph is telling us that he is near.

This reminded me of the later Hebrew tradition regarding the origin of the double vowels. These seven double vowels, the original elements of the primordial Word, were specifically related to the seven directions of space, which they are also believed to have formed.

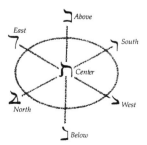

Fig. 6.1. The kabbalistic Tree of Life

If we exclude the center, the six directions correspond to the six types of quark, the fundamental particles of matter. Scientists call the six types, arbitrarily, *strange, top, up, bottom, down,* and *charm.* Quarks combine in twos or threes to form other particles, including electrons; in other words, they decay into larger particles, as all particles do.[17] However, it was when I stood back to look at the Hebrew idea of the double vowels that I noticed something really striking. Here we seem to have all the elements of the spin ratio of a particle as it manifests in space. The particle concerned is the electron, and electrons, as far as matter and life are concerned, are fundamental for the negative charge. Electrons will give off radiation, unless they are in a stable orbit. When giving off radiation, they are moving from one orbit to another. Electrons, however, also absorb radiation; that means they move farther out from the nucleus. It is this that defines the makeup of the elements. This behavior of electrons gives rise to an easily observed set of spectral lines that can be used to identify the atoms of different chemical elements. When electrons absorb energy, their frequency changes, thus changing the chemical that they form, to a greater or lesser degree. Of all the particles so far defined, it would seem that the electron is the one most responsive to other frequencies and prone to change on its own.

However, it is the spin ratio of the electron that is of most interest here, for all particles spin in various orbitals. It is the nature of these electron orbitals that gives a substance most of its physical and chemical properties. Electrons, unlike most, do not follow a definite path around the nucleus. Instead they can be found within orbitals, and these are arranged around the nucleus of an atom like shells. There are S orbitals, P orbitals, and D orbitals. It is the D orbital that is the one of interest to us (see figure 6.2), because it and the Egyptian ankh are similar in

Fig. 6.2. The D orbital electron spin

shape, especially if you imagine that the orbital is edge-on. Significantly, they both refer to the same thing—life.

The D orbital consists of a bubble that is actually a torus, a doughnut-shaped ring that has a rapid rotation. From within this rises a teardrop shape, through which flow spinning particles, sub-elements of the electron itself. Energy does not enter in, but rises up from the inside of this orbital. Without the D orbital spin of an electron the spark of life would not exist, for this orbital defines the wavelength of the electron, as well as its frequency.

The other intriguing phenomenon that the shape of the ankh resembles is the teardrop shape of the Earth's magnetic field, which extends outward toward the sun as the magnetosphere. The lines of magnetic force flow downward at the North Pole and upward at the South Pole. Also part of the magnetosphere are the doughnut-shaped Van Allen radiation belts, which consist of highly energetic charged particles trapped within the Earth's magnetic field. This most closely resembles one of the many variations on the theme of the ankh, in form and in motion, the variation from which, it is thought, the fleur-de-lys originated.

There were other cultures that used symbolism as a precise way of describing life. The Druidic and Celtic cultures spring to mind. Although seemingly less sophisticated than the Egyptians, they used a symbol, a "cross of matter," very similar to the ansate cross or cross of Osiris. As a representation of creative power, as if from the cross, they used a trinity of rays, this was the AWE, /|\.

I find it quite extraordinary that modern theory should have been so accurately foreshadowed by the philosophical and also the alphabetical

Fig. 6.3. The fleur-de-lys

symbolism of the Ancients. Both Krishna and Christ declared that they were the first letter of the alphabet, the letter *A*. Jesus stated, "I am the Alpha and the Omega" (Revelation 1:8). Omega is the sound of a long *o,* as in "eau." *O* symbolizes unity, the boundary of the universe, the world egg of the Orphic Mysteries. *Eau* is the French for "water," the watery abyss of the primordial universe, until it was broken by the newly self-aware God, uttering the double vowel *AA*, the beginning. This *AA* coming out of the toroidal *O* again assimilates electron spin.

What Jesus and Krishna are both stating is they are at-one with the Lord of the beginning—God.

PRINCE ALPHA AND OMEGA

The name of Osiris in its later native Egyptian form—Middle Egyptian—is Asar. *Sar* means "blood" or "prince" (princes being "of the blood"). Thus, Osiris's name translates as Prince Alpha. Osiris's name in hieroglyphs is a throne above an eye, signs representing the syllables *as-ar.* Sometimes this is transcribed as Usir, or more commonly Wsir. There are many variations of the name, one has a neter symbol ⌐, another has the symbol of Ra, ☉ and yet another shows the crown of Upper Egypt, but all contain the throne and eye symbols.

Wallis-Budge concedes that the meaning of the name in its original sense is not known and doubts that it is ever likely to be.[18] However, as I have already revealed, the myth and its location at a place of power offer us more than enough clues. The site itself would have inspired the initial human response in terms of a language with a mythic value that sprang from both the site and its direct experience. Hence the early hieroglyphic form of language. The key to the language was cross-correlation, at the site and beyond. This would have resulted in a lordly and inspired reaction—humor. Puns would have been the order of the day, without the barriers created by one word having only one particular meaning.

The key glyph in Osiris's name is the eye—*ar.* The throne—*as*—is symbolic of the center of all things, the axis. In hieroglyphs it is a determinative sign that embraces the concept of equilibrium, support, and exaltation. In Asiatic symbolism the throne stands midway between the mountain and palace on the one hand and the headdress on the other.[19] If anything symbolized the pharaonic history of Egypt it was the crown—the headdress, known as the nemes.

The pun of Osiris's name Asar is that it translates as Prince Alpha, yet in its form *Wsir,* it also translates as Prince Omega. In Greek the *w* is a double *u,* which became the Western *w.* In Hebrew it is the letter vav, represented by the glyph for the nail, ٦, with which Jesus and other heroes are crucified. In being the Alpha and Omega, Osiris is all things; it is his name in its seven-voweled form that gave rise to all creation, and still does, for it is an ongoing process. In other words, it is a name of most considerable power.

There is a reflection of this in other mythologies of the world. In Mayan mythology, the god of death (Osiris, too, was a god of death), Yum Cimih, was also symbolized by the hieroglyph meaning completion. There seem to be many striking parallels between Egyptian and South American culture, even though the two are separated not only by an ocean but also by several millennia, with South American culture dating to a much later era. The Maya believed, like the Egyptians, that when they died their souls traveled to the underworld, facing trials along the way. Furthermore, rather like the Gaelic-Arabian linguistic similarities, Mayan and Egyptian language shared extraordinary resemblances, even though both have many dialects. For example, the word *bak* in both languages means "chickpea." There are other correlations, too, that are mythological, architectural (pyramids), and linguistic.*

Alphabetically speaking, the shape of a pyramid is similar to that of an A. The great German Rosicrucian Jacob Boehme proposed the idea that God could be represented by the letter *A.* In German, *A* is *auge,* meaning "eye." This is the all-seeing eye of God that reflects the divine expression throughout the cosmos and is the key to consciousness. The

---

*For example, in Central America the Mexican King Can had a son Prince Aac, meaning "A the great." This figure slew Prince Coh, meaning "the great O." The biblical brothers, Cain and Abel, mean, respectively, "the great A" and "the Lord O" (Abel in Hebrew is spelled with an *O*). In the Bible story, Cain slew Abel. To my mind, this twist does not negate the connections, especially in the light of geo-linguistic theory.

A of *auge* would reflect its own A, which, as Boehme is quick to point out, gives a diamond-shaped glyph.

For Boehme, this was the image of the interpenetration of cosmos and humankind, of macrocosm and microcosm—the very large and the very small. This is an image to be found in many kabbalistic texts, as well as in Rosicrucianism and also in Freemasonry.

You may recall that Jesus, known to us as a carpenter, was actually a *tektos*—a builder. In Masonic imagery, God may be seen as the Great Architect, the builder of the "great house" (in Egyptian, pharaoh). In his hand are the callipers with which he sets out the measure of all. In Freemasonic symbolism there is a reflection of the imagery used by Boehme. The compass and the T-square of Freemasonry are shaped as in figure 1 and figure 2, respectively. When placed together they form the ideogram in figure 3. This is one of the heretical symbols of the early medieval Cathars. This symbol appears in many watermarks of the period. It means "Thy kingdom come" and was expressed as an aspiration, literally the desire "to rise up on the breath of God." This drew my attention to its resemblance to a pyramid. The connection seemed obvious in view of the significance of architecture within Freemasonry and also in view of the pyramid name of Mr—Mary, the beloved of Osiris.

1.

2.

3.

ISIS AND OSIRIS, JESUS AND MARY, TOO

The legend of Isis and Osiris is among the most curious and the most beautiful in world mythology, as well as being among the most archaic. The earliest references to the legend appear in the Fifth Dynasty Pyramid Texts (2498–2345 BCE), which are so named because they are inscribed upon the interior walls of the pyramid of the pharaoh Unas (reigned 2375–2345 BCE). Until recently it was doubted whether the religion of Osiris had developed in archaic times. George Hart stated that the earliest known mention of Osiris is dated to the Fifth Dynasty.[20] However, in contradiction to this, Professor Walter Emery said in 1961 that Osiris's religion was already well advanced by the inception of what today we recognize as ancient Egypt.[21] The discovery at Helwan of a girdle with the name of Isis was supplemented by the discovery of a hieroglyphic symbol of the god Osiris, a Djed pillar, both of which dated to the First Dynasty. These symbols in themselves confirm the already well-developed character of the religion.

It is in the Pyramid Texts that the story of Isis and Osiris is given its most complete airing. Isis and Osiris were the offspring of the earth god Geb and the sky goddess Nut. Geb and Nut were themselves the children of Shu, meaning "air" or "space," and Tefnut, the goddess of moisture. These two sprang from out of the Creator, Atum. Thus, Osiris was a great-grandchild of the Creator and was destined to become the lord of the underworld. So human is the story of Isis and Osiris that some commentators have suggested that it may in fact be based upon real historical figures. However, this element of humanity had a higher purpose—to draw humankind nearer to the gods and, ultimately, God. Osiris was the human aspect of God. After all, humankind was made in God's own image, and Osiris was seen as the full realization of that image. As the eldest son of Geb and Nut, Osiris inherited the right to the throne of Egypt. Osiris was a prototypical messiah, establishing the tradition of all heroes and all messiahs that came after.

The ancient texts tell us that, as the first pharaoh of Egypt, after whom all other pharaohs were heirs or "sons of Osiris" (Horus-kings),

Osiris set about civilizing the barbarian* world by becoming its king. He taught the world how to grow and cultivate corn, how to worship, and how to live in general. More importantly, he also educated the world, giving it wisdom and the alphabet—the hieroglyphic language. For these deeds he was called *wennefer,* or "eternally good."

Because of these things, Osiris's brother Set (Greek *Typhon*) became insanely jealous and plotted Osiris's downfall. Set gathered a gang of conspirators and they hatched a plan. A magnificent wooden chest was constructed to Osiris's specific dimensions, and he was duly asked to come and view it. Set and his companions tried to fit themselves into the chest, but each, in turn, was unsuccessful. At last it was Osiris's turn, and fitting into the box perfectly, he fell victim to the ploy. Set and his conspirators nailed down the lid of the chest and, with Osiris inside, threw it into the Nile. Osiris drowned, but the box made its way past the Nile Delta and across the sea to Byblos (modern day Lebanon), where, mysteriously, it sprouted leaves and turned into a palm tree.† This palm tree was subsequently cut down and taken to the royal palace, where it was used as a pillar.

Fig. 6.4. Byblian palm tree

Barbarian means literally "hairy man," a fitting description of various of the heroes; in the Old Testament, Esau is described as a "red and hairy man."
†Sometimes it is a sycamore tree. (See Hooke, *Middle Eastern Mythology,* 69.)

In the meantime, Osiris's beloved wife, Isis, distraught at the disappearance of her husband, went in search of him and, after much travail, ended up at Byblos, where she saw the pillar—and what it truly was. Isis disguised herself, became a handmaiden to the local queen and in token of the queen's generosity, immortalized her children. However, just as she was finishing off the process of the last child, she was caught in the act—holding it by its heel over a sacred flame. To the total astonishment of the royal court, Isis was forced to reveal herself. Then, taking the pillar with her, she returned to Egypt.

However, the angry and murderous Set once again managed to get his hands on the body of his dead brother and tore it into pieces, fourteen in all, and scattered them all over Egypt. In great sorrow, Isis and her sister Nepthys, the wife of Set, gathered up all the pieces and put them together again, as the first mummy.* Isis hovered over the body as a kite, and with the help of Thoth and Anubis (the guide of souls to the underworld), she breathed life, *spiritus,* into the body. Unfortunately, though, the body was incomplete, Osiris's phallus had been swallowed by a fish. This did not stop Isis, who manufactured another, and Osiris, resurrected as the ithyphallic god Min (sometimes Men or Menu), the equivalent of the Greek Pan, was able to impregnate Isis, who conceived the divine child Horus. Horus would later avenge his father, who resurrected in spirit, became lord of the underworld, the Duat, and the judge of the dead. This then, in brief, is the legend of Isis and Osiris, and at first glance it seems most unlike other myths—until we take a closer look.

The first comparison that came into my mind was one that must ultimately remain speculative. At the beginning of the myth, Geb the earth god and his wife, Nut the sky goddess, had a curse placed on them by Ra, who feared usurpation by any of the children in Nut's pregnant womb. Ra forbade her to give birth on any day of the 360-day year. Thoth subsequently played checkers with the moon and won sufficient light for five extra days, during which Nut gave birth to Osiris, Isis, Set, Nepthys, and Horus (called Apollo by the Greeks). This is very similar to the story of the birth of Jesus. Herod, as the king, represents the sun;

---

*Mummy, from the Persian *mum,* meaning "wax," in this case preserving wax.

Fig. 6.5. The Djed pillar

Jesus is his heir. Herod, in his final days, tried to find the prophesied child and have it killed. He recognized it as a probable grandson, and suspicion fell on his eldest son, Antipater. Herod was thought to have died and Antipater was killed. Herod, though, managed to live on for a further five days. It is likely that during these five days Jesus was born. This, too, would be the birth of the hero-god during the intercalary period. As I have already stated, this is speculative, but probable, given what we have already seen of the broader evidence.

The other thing that struck me was that Osiris became a pillar at Byblos. I could not help but notice that Josephus mentioned, in the Slavonic version of his *Antiquities of the Jews,* that Jesus's name was carved above the temple balustrade in Jerusalem: a balustrade that was itself supported by pillars, the most famous of which had names—Jachin and Boaz—sacred within Freemasonry. In an early form, Osiris was known as Andjeti, or was at least identified with the god of that name, taking from Andjeti the symbols of the shepherd's crook and the flail. Actually, I believe that from the outset, the two were the same, for the Egyptian texts claim that, originally, Osiris came from Djedu, the capital of the ninth nome (administrative district) of Lower Egypt. The Greeks called this town Busiris, from its Egyptian name, Per-Asar, literally "house of Osiris." (Later on, during the First and Second Dynasties, Osiris's cult moved to Abydos.) The only difference between Osiris and Andjeti is that the latter was always represented as a living king, whereas Osiris was always dead. However, Osiris was not really dead but resurrected, and as such represented a living force both within the body of his own son, the Horus-king, as well as within the land and the cosmos. This is reflected in the name Andjeti. Literally "pillar of the underworld, of Annu." The very word is talking of these forces immanent within the earth. The Egyptian word for pillar is *djed,* and this is a component of Thoth's name in its hieroglyphic form. *D* and *t* are cognate, hence the Greek form Dod or Tot—Thoth. In Egyptian the name is Djehuti.* The component *hu*—the Lord—is within the name, within the pillar, the *djed.* Djehuti is a pun upon Osiris's title reflecting Isis's search—My Lord is in the pillar.

This reflects the legend of Adonis, whose name means "the Lord," and who was said to have been born from a myrrh tree—*myrrh* being another form of Mary.† Thus Adonis's mother was Myrrh,[22] the very substance gifted to Jesus at the Nativity.

What we have here is an age-old myth, one that can be found the length and breadth of the Middle East and beyond, such as the myth of Aleyin and Mot, the rival twins of Canaanite mythology. Aleyin we have encountered before as El Ayin; another of his names is Baal—or "the Lord." *Mot* simply means "death." Aleyin, like Adonis and Osiris, is life, while Set and Mot are death.

Going back to the tree/pillar, at the base of it, or curled around it, is a serpent, the Egyptian hieroglyph for "goddess," the goddess concerned being Isis. Could it be that this became the story of the temptation of Adam and Eve? Adam is Atum, the Creator, and Eve is Hebe, darkness, mother goddess, and mistress of the Tree of Life. Any icon displaying these elements could easily have been misinterpreted in later years.

---

*Recognizable in the Indian Vedic literature as *Jyoti.* This word and its progeny relate to illumination and all that shines.
†The myrrh tree was known as "the tear of Horus."

Another thing that struck me was the story of Osiris's missing phallus. Apparently it had been swallowed by a fish—a letos fish. The letos fish is a curious thing, and certain parts of it, particularly its liver, were used in ritual purifications. The fish is of course a symbol of Christianity, but the name of the fish suggests that it had something to do with the god Apollo, for Leto was Apollo's mother. As we have seen, Horus was equated with Apollo. We are back to "the Lord" again, the meaning of Horus's name.

The very archaic Egyptian mythos would appear to be the forerunner of them all: recognizable features of it appear in quite a few mythologies from other countries. Isis's search for Osiris and her later anguish over her dead son Horus (who in some versions of the myth is, as an infant, stung by Set, disguised as a scorpion) resemble the Eleusinian Mystery of Demeter and her search for the lost Persephone[23] in that Isis's anguish halts Ra in his tracks, thus causing the world to become dark and infertile. The similarities are almost limitless. However, it would be unfair to label Egypt as unquestionably the oldest source. It is just that so many thousands of years later Egypt has the most intact remains, and what came after appears to be but a remnant of an older, more complex theology, out of which emerged what today we term *religion*. Egypt as she stands now, allows us to penetrate deep into the mysteries of belief, into her way of life, which was so totally permeated by the gods.

We have approached the holy of holies and are about to enter in. Having drawn apart the veil of Christianity, a greater mystery loomed before me, one closely related to church dogma and doctrine, which in its youth had risen like a phoenix from out of the ashes of ancient Egypt.

We have seen how Osiris's name in Egypt is Asar, represented in hieroglyphs as a throne and an eye. The former means "to have power," and given the nature of Osiris's myth, the implication appears to be power within. Osiris/Asar, is the prototype of Jesus, the names being the same— Gisa or Gesu.* Also, like the Indo-Aryans, the Assyrians, and others, the Egyptians had a solar dynasty of kings represented by hawks or falcons.

---

*This is confirmed by the name of the Vedic god of the dead, Yama, who was also called Asu or Esu-niti, and who performed the same function as Osiris/Gesu as a guide of spirit. (See Frawley, *Gods, Sages and Kings*, 271.)

Also common to these mythologies is the myth of the phoenix, which Jesus is often compared to, as are Osiris and a plethora of other gods worldwide. The myth of the phoenix, the bird that sets itself alight and then arises reborn from its own ashes, is a potent one. It is a motif that recurs to this day, a rare survival of ancient thinking, possibly more than five thousand years old. The word *phoenix* comes from the Greek *phoinikos* (generative). There is some dispute as to whether this word is related to *phoinos,* also a Greek word, meaning "red." Even if it is only a pun, I think that it is related, for in Phoenician *phoinix* means "purple," from the dye of the mollusk, *Murex.* This same dye also produced the red hues that may be seen in the robes of modern and medieval cardinals of the Roman Catholic Church. The dye was a very expensive commodity in the ancient world.

What is interesting is that *phoenix* is a masculine form of *phoenissa,* "the red one" or "the bloody one."[24] This would appear to be a reference to the hero, whose fate is indeed bloody. However, *phoenix* also refers to the date palm, which brings us back to Osiris and his sojourn at Byblos. The Persian form of the mythical bird is called the huma, and it is quite probable that this comes from the same root as *human* (Latin *homo*).* The incarnation of god into human form is symbolized in the Egyptian glyph for the phoenix—a stork or heron, the bennu bird. The bennu bird was the carrier of the soul of Osiris,† it represents the divine Logos—the Word of God. The bennu bird also initiates a new world age, where order is saved from cosmic entropy. As if to underscore these wide mythic connections, it was believed by the ancient Egyptians that the phoenix wrapped up its father's ashes in an egg of myrrh—the tree from which Adonis/Osiris was born. Furthermore, the phoenix utters a cry from its mythic pyre, the cry of the soul; just as Jesus, Dionysus, and Quetzalcoatl, too, utter a death cry. Heroes die to rise again, like the sun and the phoenix.

Osiris was originally the Good God, whose "thrones are the stars

---

*This brings us back to Thoth, who was also the herald of childbirth, a role taken in Christian tradition by the Archangel Gabriel.
†Osiris is sometimes referred to as Herm-Anubis (Heru-em-Anpu). Cocks of a saffron color were sacrificed to him—reminiscent of the Buddhist colors.

which never rest."[25] He was the King of Kings and Lord of Lords and God of Gods.[26] More than this, he was the Resurrection and the Life, the Good Shepherd, Lord of Eternity and the Everlasting, who made men and women to be born again. According to the ancient Egyptian scriptures, "As truly as Osiris lives, so truly shall his followers live; as truly as Osiris is not dead, he shall die no more." The great scholar Sir Edwin Wallis-Budge wrote, "From First to Last, Osiris was to the Egyptians the god-man who suffered, and died, and rose again."[27] For the devout Christian unaware of things Egyptian, this is profoundly surprising stuff, for it is also deeply familiar—as the life of Jesus from the Gospels.

For the ancient peoples, of at least the second and third millennia BCE, there was no distinction to be made between myth and history. Together, with immanence and numinousness, they provided the ongoing sense, the perception of creation, the everlasting now, the universe unfolding to the story of the gods. In a Jungian sense these gods represented archetypes, and the greatest and most human of them all was Osiris, whom today we called Jesus.

The story of Osiris/Jesus was about the burden of life, the burden of getting to know oneself. To the Egyptians, the journey of the soul in death was a reflection of self-knowledge in life. The journey to heaven of the pharaoh's soul was actually a reflection of his philosophical journey through life's experience. These similarities are not limited to Christianity, but of all the world's religions and mythologies, it is Christianity in doctrine and theology that most approximates to the ancient Egyptian mode of the sacred. In fact, so very similar are the two that the more ancient way should be labeled "ancient Egyptian Christianity," for that is precisely what it was.

Osiris was the Redeemer at the center of the ancient Egyptian belief system. As in the modern church, hymns cry out to his glory, exalting him beyond all compare. Here is an example:

> *He is born! He is born! O come and adore Him!*
> *Life-giving mothers, the mothers who bore Him,*
> *Stars of the heavens the daybreak adorning,*

Ancestors, ye of the Star of the Morning,
Women and Men, O come and adore him,
Child who is born in the night.[28]

There are other verses, and the whole thing reads like a Christmas carol.

Osiris's nativity was announced by three wise men, who were represented by the three belt stars of the constellation Orion. In the Greek mysteries of Adonis, a cry of celebration went out that the Star of Salvation had dawned in the East. In myth, the nativity of the savior is often marked by the dawning of a new star. This motif also occurs in the legends of Krishna, Pythagoras, Yu, and even Caesar. Zoroastrianism, too, shares the same legend. This motif appears in South American tradition too, in the legend of the hero savior-god, Quetzalcoatl, who appears as the morning star.

The three wise men, having found the child, pay homage to it and its mother. In the Egyptian mythos this divine child is the reborn Osiris, the "hawk of his father" (as in Canaan/Phoenicia, Baal is the "bull of his father"), called Horus. Another comparison is that the newborn child's father is dead before the child is born, a point made by Wallis-Budge in *Legends of Our Lady Mary, the Perpetual Virgin, and her Mother Hanna.*[29]

These Christian similarities are underscored by Pilate's accusation to Christ, "Chrestos ei!" which, translated from the Latin, reads, "What a simpleton you are!" *Chrestos* means "simple," as in innocent and gentle, whereas *Christos* means "anointed." It is thought that the two words became confused in the early Greek texts. Chrestos is one of the traditional epithets of Osiris, and on the island of Delos, sacred to Apollo, there is an inscription to Chreste Isis. Pilate may well have known these things.[30]

But what is the name Pilate? It makes one think of pilot—the navigator of a boat. The sun god Ra sails through the heavens in his boat of millions of years, and in this boat are assembled all of the gods. It is Ra who navigates it, Ra who is the pilot, Ra whose son Osiris was the Word made flesh, and Ra who decreed that the flesh should go through the wheel of suffering, just as Pilate condemned Jesus. There are various maritime references in the New Testament. The Gospel of John contains the following: "And he said unto them, 'cast the net on the right side of the ship, and ye shall find.' They cast therefore, and now they were not able to draw it for the multitude of fishes" (John 21:6).

Needless to say, the fishing motif is one that occurs regularly within the Egyptian sacred iconography, scenes that are trivialized as "everyday" by some Egyptologists who fail to see the symbolism. The modern idea of Jesus represents two different aspects of Egyptian thinking: the dead and the resurrected Osiris, and the living Horus, the son of Osiris. The pharaoh was the incarnation of Horus, and when he died, he would become Osiris and his heir the new Horus.

Thus it was that the mythic cycle lived on in the pharaohs and the pharaonic bloodline. Horus

> represented the soul of life, which came by water to a dried up world, upon the verge of perishing with hunger and with thirst. Here the fish, as the first fruit of the earth, was a sign of his incorporation in matter; hence the typical shoot, the green ear of corn, or the branch that were imagined in Child Horus—the saviour who came by water.
>
> The Saviour who came in fruit as a product of the tree was the Natzar [Nazarene]. The Saviour who came by spirit was the soul of the sun. This was the earliest rendering of the incorporation of Horus, as the primary of life and light of the world.[31]

The Egyptian element that might seem out of place is Osiris's wild brother, Set. While it is of note that there are some stories of Jesus having a twin,* it is interesting to discover that the pharaoh was actually the incarnation of them both. He was at once the living Osiris—Horus—and he was Set. During the Nineteenth Dynasty there were two kings named after this wild god. This raises an interesting point,

---

*Thomas Didymus or Judas. This is another theme that runs throughout the hero myth.

one often overlooked by researchers, and one that leads to innumerable misunderstandings.

Set is generally held to be the god of, or an incarnation of, evil. Quite simply, this is only half true. The only evil about Set was his wild nature and thus his ignorance. Kings of Egypt would hardly have named themselves after evil! In calling themselves, or associating themselves with, Set, they are acknowledging their all-too-human characteristics. Set was the man of the wilderness, wild, red and hairy, unkempt and vicious—Osiris as man in his lowest form. Even before a pharaoh died, he longed to rise to the heavens, homeward bound as the higher Osiris—knowledgeable, cultured, meek and mild, saintly and godlike.

If ever there was a god of evil, it is more likely to have been in the form of the serpent who opposed the passage of the boat of Ra in the underworld, the Duat. For Ra the sun god, to rise and fill humankind with his divine breath, he had to overcome this serpent, called Apophis. However, to the Ancients, the serpent's obstruction was functional, for they understood that wisdom is gained through dealing with adversity. Hence the serpent's traditional association with wisdom.

Apophis was chaos, disorder from order, the swallower of light. As if to prove the point, as the boat of Ra plows on through the cosmos, there at the prow is a marksman with a spear, the spear destined to kill and overcome the serpent. And who is the marksman? None other than Set. The very act of spearing the serpent illustrates how Set is destined to overcome his own chaos, his dark side, his ignorance—and to ascend as one with his brother, Osiris. That is why Set is in the divine boat of Ra, who as Atum fills the universe with the sound of his holy voice.

There is a red hairy wild man in the Old Testament, too. He is also one of two brothers. His name is Esau, brother of Jacob. It is Esau who is tricked out of his rightful inheritance through the gift of a bowl of red lentil potage. Red lentils were always ritually offered to Horus, and as if to confirm the Egyptian origin of Jacob and Esau, Set, too, is tricked of his rightful inheritance of the land of Upper Egypt, even after Geb, the earth god, adjudicates and decides that he should have it. Horus gains it at Set's expense.*

Set, who probably gives his name, his red-hairy name, to sunset, the red dying sun, was the god of primal regenerative power. From *Set* is derived *Satan,* known to Hebrews as God's adversary, and also *satyr,* an early Greek variation on the theme of Set, portrayed as part man, part goat. And it was the goat, the scapegoat in Hebrew ritual that is associated with the adversary, with Set.

However, Set's name is the only one of the Egyptian divine names that, although male, is actually spelled in the feminine form—with a *t* at its end. The female inside Set is *Aset*—Isis. This would be the primal expression of the letter *a,* as in *A*set, giving us the alchemical base matter, the matter that needs to be transmuted. Is this the reason why in Christian iconography the devil is sometimes portrayed as a woman, a woman with the hoof of an ass, the animal that sometimes represents Set? Is this why on the Christian Palm Sunday (palms were sacred to Osiris), Jesus is set on an ass? To represent that he has overcome his bestial nature?

The myths and scriptures were there to encourage everyone to overcome their lower nature, but how might the process be represented in myth? It is back to the cross, but not in the way that the church sees it today. There is a good example in the Greek Chiron, the boatman of the River Styx in Greek mythology. Chiron received a wound and although he was able to heal others, he was unable to heal himself. He therefore bequeathed his own immortality to someone else and then died. However, by virtue of his actions, his wound was healed and he became immortal. His action was one of complete selflessness, offered without thought of personal gain, one without guile. Chiron is, of course, our old friend the Chi-Rho, the cross. In the ancient myths the hero-gods are nailed to the center of the cross.

---

*In the myth of Osiris and Set it is Set who is crucified (R. Graves, *Greek Myths* 1:283–84, and Campbell, *The Mythic Image,* 29), for Set was a king, and a ritually crucified one at that. This crucifixion is rather like that of Woden, who crucified himself so that he might gain higher knowledge of the Mysteries. Thus once again we have a motif of Set metamorphosing through his own suffering and becoming his higher self. Ass-eared Set ruled Edom—the country of the Herods.

The center represents the still point from which all movement proceeds. This aspect of the cross therefore represents innocence in action. Further to this there is, in the Valley of the Kings, in the tomb of Tuthmosis III, a hieroglyph of a strange ram-headed god, which looks nothing like the only other ram-headed god of my acquaintance, Amon. It is in fact Lwf[32]—Ra in the form of a lamb, the gentle creature that, in its newborn innocent activity, represents the Word made flesh. This seems to confirm that Lwf was the soul of Ra, the Word of consciousness. Innocence in action.

The ancient symbols and gods not only represent all of the powers and attributes of nature, but in the form of the hero-gods, they also supply us with instructions for the betterment of our estate.

7

Epiphany

Science must begin with myths, and with the criticism of myths.

KARL POPPER, *CONJECTURES AND REFUTATIONS*

Any sufficiently advanced technology is indistinguishable from magic.

ARTHUR C. CLARKE

THE SCIENCE OF THE SOUL

All good things in civilization begin with inspiration—the highest of the high; and from that point on it is as if humanity begins its long, slow descent and decline from the metaphorical into the literal.

We have inherited the components of an ancient and sacred spiritual technology, one that seems, like the gods, to be somehow inherent within language—a language that the sensitive Ancients translated as an expression of the power of place and their understanding of the evolutionary process: evolution of the spirit that is.

Human beings as we have seen, seem able to tune in like radio receivers to certain frequencies and thus gain inspiration (literally, *in spirit*). And yet in the modern world we tune in to that which is separate from us, including religion, and diminish the past in the process. To ancient humanity we were the technology, as sure as the sun went around the Earth.

The hero's association with nature was not just with the daily round of the setting and rising sun, but also the seasonal solar aspects. Ancient Egyptians would beat their breasts and sing a lament to the god as they thrashed the corn. The god was dying, this was the season of the harvest, and the life force was withdrawing back into the earth, leaving only seed heads behind, seeds that carried the hope of the future. Earthen corn dollies would be made and buried at this time; when excavated later they would be found to have sprouted—an omen of the resurgence of the spirit of life, immanent within the earth. Hieroglyphics are quite explicit about Osiris/Gesu as life itself, and as such he was defined by death, a concept familiar within Christianity. The science of the ancient Egyptians has become the faith of today: "He that eateth my flesh, and drinketh my blood dwelleth in me, and I in him" (John 6:56).

In Mithraism the same passage reads: "He who will not eat of my body and drink of my blood, so that he will be made one with me and I with him, the same shall not know salvation."[1]

Both the ancient Egyptians and the followers of Mithras used communion wafers. The Mithraic wafers even bore the sign of the cross. The followers of Mithraism had a deeper understanding of the workings of nature as represented in the communion, the celebration of the Eucharist, where the knowledge that the active principle and the substantive principle were at work within was well understood.

A question that one may ask at this stage is: In what sense were the Ancients aware of this as science? The answer is that science was looked upon differently by the Ancients. It was not removed from life in the objective sense. It was inseparable from daily living and personal, subjective experience. This does not mean that the philosophers and priests were unaware of the finest realms of nature; the evidence to the contrary is plain to see. *Science* as a term means "knowledge," from the Latin *scientia* (from the present participle of the verb *scire*, or "to know"). For these peoples knowledge came from within—*gnosis* (Greek). And ultimately, knowledge in this form was held to be of divine origins: information was transmitted by those cosmic powers, the gods who permeated all things.* The very names and symbols of the gods expressed the scientific principles that we see everywhere.

The hieroglyph denoting the name of the Egyptian god of wisdom, Thoth, is referred to by Egyptologists as an ink jar. This is a reasonable assumption, for Thoth was the scribe of the gods.† However, the ink jar is also the hieroglyph for the heart, symbolic of the conscious mind, the very thing that Thoth represents.[2]

Egyptian water jars were quite frequently molded in the form of the heart hieroglyph, for reasons that until recently remained unknown. Until, that is, Bruce Mace, an American enthusiast of things Egyptian, got his hands on some ancient Egyptian pottery and subjected it to an analysis. Having analyzed it, he blended certain clays into the right mixture and consistency and made a water jar, heart shaped, of course. What he discovered was remarkable. Tainted water, when poured into the jar, was rendered clean by an extraordinary process of reverse osmosis. Inside the jar the water spun at a frequency counter to its various polluting agents. These were absorbed by the clay and expelled as waste through its outer shell. The water inside was of such purity that it became almost sweet to the taste.

---

*In ancient Mesopotamia, the powers or properties of the gods were called *me* (pronounced "may"), and they enabled a whole host of activities central to human life, especially religion—the binding back to the source. The names given to the powers were the names of the gods, and thus the gods denoted the nature of these powers: in ancient Egyptian, *neter*—literally, the power of god in whatever variant aspect. The ancient Mesopotamian term that denoted how these powers really ought to be was, interestingly, *Gishur*. This term means "play," or "design," in the sense of how God brings things about in the grand scheme of things. Therefore, it is most fitting that *Gishur* is so similar to *Gisa*. This sense of the word *design* is the theme that runs through not just the ancient Egyptian Book of the Dead but through scriptures and mystery teachings worldwide. The ancient Mesopotamian me are the "powers which make possible the powers of the Gishur and which ensure the continuation of civilised life" (Black and Green, *Gods, Demons and Symbols of Ancient Mesopotamia*, 130).

†*Thoth* to *tot* is interesting. *D* and *t* are cognate; thus we may derive *Dod* from the name. This is the ancient form of David. Jesus/Asar was of the line of David. *Thoth* is Greek; in Egyptian it is *Djehuti*, meaning literally "Dje, the Lord, has come"—Dje is Osiris. Thoth was brought to Britain possibly as early as 4,000 years ago: his name survives in the many place-names beginning with "Toot"—Tooting, Toothill, and the like.

This reflects a discovery about blood flow, announced in May 1998. Surgeons have realized that as the heart pumps blood around the body, the blood forms swirls as it rushes through the arteries. Furthermore, nature has designed arteries with a helical "twist" to encourage this swirl. A further observation to be made is that the Earth's core spins at a much faster rate than the rest of the planet, while generating the planet's magnetic field. The same observation can be made of the sun, whose inner core spins at a rate phenomenally faster than its surface. We have already seen the importance of the spiral to the Ancients. Here, through the symbolism of the heart, the inkpot, and the rediscovery of an ancient technology, we see that the Egyptian god Thoth, while relating to the conscious mind at the heart of the matter, relates to the spiraling force that directs the blood in microcosm—and the galaxies in macrocosm: e-motional intelligence.

This is very remarkable. It is the kind of knowledge that reaffirms the nature of the journey. In the words of Teilhard de Chardin, "A truth once seen, even by a single mind, always ends up by imposing itself on the totality of human consciousness."[3] Surely, it is now time for this ancient knowledge to be accepted as a spiritual science and to be respectfully researched as such.

In the mid-1970s, at the annual conference of the world education fellowship, Professor Sir Alistair Harding (1896–1985) made

> a powerful appeal for the building of what he called a scientific theology for the era that lies ahead of us, and on three separate occasions in the course of his address he reminded his audience of the dangers of dogmas of materialism which were supposedly built on secure scientific foundations. He remarked upon the fact that our civilisation had been built upon a spiritual interpretation of the world. . . . His concluding words stressed the need for "an experimental faith which can regenerate the spiritual that has been the driving force of all the great civilisations of the past."[4]

This recognition of past civilizations by an eminent scientist, preeminent even, gives a novel perspective on the issue of our origins and a rare accreditation of the spiritual intelligence of the Ancients, an intelligence not based purely upon the rational elements of the mind, but upon the emotions, too.

The seat of emotional intelligence may have been the heart as presided over by Thoth, but the seat of spiritual intelligence was marked upon the forehead by the uraeus of the pharaoh's crown. One of the finest examples of the uraeus may be seen upon the golden likeness of King Tutankhamun in the Cairo Museum. It protrudes from the forehead part of his nemes headdress. The uraeus, from the Greek *ouraios,* or "cobra," was exactly that—the cobra goddess Wadjet. The uraeus was the archetypal serpent image of kingship and was often seen in tandem with the vulture goddess Nekhbet. Together Wadjet and Nekhbet offered the pharaoh protection against all comers. Significantly, Wadjet was sometimes portrayed in leonine form and as such was known as the Eye of Horus.

Fig. 7.1. (left) The united crowns of Upper and Lower Egypt;
(right) the red crown of Lower Egypt.

Jutting out from the crown of Lower Egypt, the red crown, is a proboscis. There were four main crowns of Egypt: the blue *khepresh* crown;* the ostrich-featherlike *atef* crown; the crown of Upper Egypt, called the *hedjet;* and the crown of Lower Egypt, the red crown, or *deshret.* From the First Dynasty onward the crowns of Upper and Lower Egypt were united, but still had the strange proboscis. The crown of the emperor of Japan is almost exactly the same as the pharaonic crown of Lower Egypt, and as the emperor himself originally and mythically represents the mysterious force of fertility, the creative force of life itself. This is intriguing and fascinating it is only when we look at Tutankhamun's famous golden mask in profile that we see an answer to this question: side on, the uraeus cobra has a similar line of flow to the old Egyptian letter (hieratic) rendered as *r, l,* or *n* in Semitic scripts. Its shape infers the movement of energy: ꝫ.

The exact sound of the letter in old Egyptian is not known. However, it does have a marked resemblance to the Hebrew ayin, the *O* (the Greek Omicron).[5] Its sound probably varies with the different meanings of the glyph, though the Semitic rendering *R* gives a broad hint that its divine meaning was the double vowel *AA,* representing the breath of God. (Just say the letter *r* and you will see what I mean). The confusion over which letter it represents may be solved by the application of a little Egyptian theology—the fact that the holy, inexpressible name of God was made up solely of vowels. Ancient languages can only be written as consonants, but the very sound of these is what gives a clue.

The letter *A* when doubled means "primal energy" and "possibility" in terms of *potential.* These are the very things that are said to have spurred on the creation and to have been behind it.[6] The uraeus appears to be protecting the *ba,* or soul, of the pharaoh. Its position upon the head is over the spot approximating to the third eye or pineal gland, the fount of consciousness. Oddly enough, Heliopolis was called *Ayin Esh Shems*—"the Fountain of the Sun"—by the Arabs.

Resh in khep*resh* is the letter *R* in Egyptian and Hebrew. It is the pronunciation of the double *AA*—the Word of God.

Heliopolis was, from the very outset of Egyptian history, the greatest cult center of them all, and the most ancient. Ayin is also the fountain of the soul.

In ancient Indian tradition the gland associated with the pineal, the pituitary, was called the "cave of Brahma." In ancient Egypt it was the "cave" or "grotto of Ra." In researching the connection between the cobra/uraeus and the pineal gland, there is the following statement by Manly P. Hall, written at the beginning of the twentieth century: "Did the Egyptians know that reptiles present the highest development of this gland and for this reason coiled the serpent upon their foreheads where the third eye of the Orientals is placed by symbolic license? Was not the uraeus the symbol of wisdom and is not the pineal gland the organ of a method of acquiring knowledge?" This is followed up by a further intriguing comment.

> As an emblem of divinity, the pineal gland would naturally be associated with royalty, for the kings were the shadows of the gods upon earth. The crown of Lower Egypt and also the Pschent, or crown of the Double Empire [consolidated Egypt], were surmounted by a curious antenna, feeler, or very thin curved horn, which is most reminiscent of the descriptions of the structure of the third eye that have descended from the first ages.[7]

From what we have seen so far regarding the pineal gland and its symbolism, it is very obvious that the Ancients had a detailed knowledge of its workings and abilities, and knew it to be the seat of the soul. To quote Manly P. Hall once again concerning the occult knowledge of ancient man: "The pineal gland is regarded as a link between the objective and subjective states of consciousness: or, in exoteric terminology, the visible and invisible worlds of nature."[8]

This knowledge, gained firsthand via the ancient technology of the sacred site, was put to supremely good use and preserved via the mythology that has come down to us many thousands of years later. We have described the various aspects of those mythologies in order to demonstrate their extraordinary similarities across the world and we have only skimmed the surface of these uncharted waters. That Christianity emerged out of Egypt many thousands of years ago seems to be the only obvious conclusion that one can draw. It is also obvious that many and various aspects of it were drawn directly from the power of place—no matter where that place might be.

It is now time to name names and point the finger in a more definite direction. Reflecting upon the pharaohs of Egypt and the symbolic nature of their power, suddenly, after many years of research and many questions, the answers begin to come thick and fast.

NUTS ABOUT MARY

Mary Magdalene is one of the most prominent characters within the New Testament. A follower of Jesus, she is famous for being a reformed prostitute, though the text that describes her is not at all explicit. The passage, in the Gospel of Luke (8:2) merely states: "And certain women, which had been healed of evil spirits and infirmities, Mary called Magdalene, out of whom went seven devils."

The name of the Magdalene does not appear in any text earlier than the Gospels, and no historical documents of the period mentioned her either, not even Josephus. However, this does not mean that she was not historical. As usual, we need to examine in what sense she was historical. The Magdalene was said to be a prostitute.

The first real clue comes from ancient temple practices, by way of a remotely related term. A prostitute is a "tart." The *Oxford English Dictionary of Slang* states that *tart* is probably short for "raspberry tart," the rhyming slang for "sweetheart." However, it is also reminiscent of Astarte, who appears in the Old Testament as Ishtar, the Babylonian goddess of love and sexuality. Among the Greeks she was identified with Aphrodite, and another heavenly goddess of love called Venus by the Romans.

Astarte is one of the oldest forms of the Great Goddess in the Middle East. Interestingly, she was known as the Lady of Byblos, the place where Osiris was transformed into a palm tree and where Isis went in search of him. In Hebrew, Astarte was translated into *Esther*, meaning "star" or "soul."* *Esther* is the word from which we derive *Easter,* the pivotal time for our dying and rising god.† But why was Mary Magdalene described as a whore?

It seems that some priestesses gave themselves over to sexual activities of a tantric nature in the service of the goddess, before going on perhaps, to other things. Initially they were temple virgins, and it is at this point that we come to a very delicate comparison, because Mary, Mother of God, was a temple virgin, at least according to some accounts. But this is where things begin to get a little interesting.

The word *whore* comes from a variety of European sources: Low German *hore;* Middle Dutch *hoere;* Old Norse *hora,* and so on. These probably come from the Greek term *choron.* The nearest in meaning to the gospel term (which is from the Greek) is the Gothic *hors,* meaning "adulterer," which is what the Magdalene is implied as having been. However, the argument is sealed by the Latin term *carus* (from an Indo-European base, *qar*). *Car, q're, kar,* and *qar* are all early forms of the Greek *kore.* (There was even a nation named after her, the Carians, who were formidable seafarers.) The male version of this name makes the answer entirely obvious—Horus. A whore was an Egyptian temple woman and would be a temple dancer at that: it is Christianity that has made the term one of abuse. Remember the dance of the seven veils by Herod Antipas's niece Salome? Whore has the same etymology as *hour,* and it was the Hourae who performed the dance of the hours—and also acted as midwives to the gods.

It then becomes significant that Salome was lusted after by Herod Antipas and was, in thought at least, an adulteress. Was Salome the historical Magdalene? Is this the reason why a person named Salome is portrayed in the Gospel of Mark 15:40 at the foot of the cross? To delve

---

*"Astarte" seems to be a reference to Ursa Major, the Great Bear, *Ast-Arte* meaning "star of the bear." This is a constellation of some significance, as it may have played an important part in the funerary ceremony surrounding the dead pharaoh.
†Esther had a consort Mordecai, who sacrificed the god Hammon. What this story is really about is the goddess Ishtar and her consort Marduk. Hammon would seem to be Amon. This was a myth current in the civilization of Elam, ca. 2000 BCE, and seems to be yet another mythic retelling of the story of Isis, consort of Osiris.

into the question of a historical Magdalene is beyond the scope of this book, but who or what was the mythical Magdalene?

In the early Christian treatise, the *Shepherd* [or Pastor] *of Hermas* (second century CE), Christ is compared to a cube-shaped mountain, upon which a tower is erected.[9] The cube-shaped mountain's name is Mount Meru. The origin of the word *cube* is *Kubaba,* another name of the goddess Cybele. Cybele is a traditionally associated with the pine tree and the pineal gland. Cybele is also known as the hairy woman. This is interesting, for this is precisely what Mary Magdalene is said to have become, on vacating the Holy Land after the crucifixion. Of course, the legends are apocryphal, but relevant nonetheless.

In the tomb of Tutankhamun, a wonderful representation of his birth—his ritual birth—was discovered by Howard Carter and his team. It portrayed the newborn Tutankhamun emerging, headfirst, out of a lotus flower. This was representative of the god Osiris arising out of the primordial state. As a piece of statuary, it is quite breathtaking, but it also has great implications. It was believed that the lotus seed and the almond seed were containers of the spirit of the unborn divine child. I thought back to the Festival of Cybele, which fell on April 4, the Megalensia—a word not far removed from *Magdalene*—and the "coincidental" similarities between Cybele and the Magdalene. Was there more to it than this?

In the brain the seat of the emotions is the amygdala,* and it has been suggested that the amygdala may use the Earth's ELF frequencies in association with the environment of the power of place. Electrical stimulation of the amygdala and hippocampus can induce a whole range of paranormal sensations, which include out-of-body experiences. These experiences often involve the consciousness moving through a tunnel or

---

*"Neuroscientists using positron emission tomography (PET) have demonstrated that long-term memory for emotionally arousing events in healthy humans is highly associated with increased brain activity in the amygdaloid complex. The finding provides dramatic graphic evidence that the amygdalae—a pair of walnut-shaped structures near the hippocampus—play a critical role in the formation and modulation of emotionally influenced memory in both animals and humans." (University of California, Irvine, July 22, 1996, Susan Menning, Press Release.)

tower of light toward a brighter light. This is especially so in near-death experiences.* In the Old Testament book of Micah 4:8, the following can be read:

> *And thou, O tower of the flock,*
> *the stronghold of the daughter of Zion,*
> *unto thee shall it come, even the first dominion;*
> *the kingdom shall come to thee daughter of Jerusalem*

In a motif common within mythology, when the hero wishes to leave the earth and "join his fathers" he does so by entering a tower. This could be a reference to a monument where the dead king is entombed, but the near-death evidence and the Ancients' focus on the otherworldly should encourage us to look a little deeper.

In Hebrew, "tower of the flock" is *magdal-eder*. The epithet *magdala* means literally "tower." It also means "elevated, great, magnificent, high." *Amygdale* is the Greek word for almond, from which we derive *amygdala*. In French the word for almond is *amande*—the origin of the Christian name Amanda. This means, appropriately, bearing in mind the Magdalene connection, "fit to be loved." The very similar Sanskrit word *ananda* means "bliss." The almond is curiously shaped, like a vesica or standing-wave pattern, with nodes and antinode. Recall the story of Jesus driving seven devils out of the Magdalene? Seven is the number of music, the notes of the scale, and music heightens the emotions, identified within the brain with the amygdala. The book of Micah, chapter 4, relates to the healing of afflictions and the return of all, particularly the feminine, the substance of life, to god. In other words, the translation of emotion from pain to bliss, a process that can be stimulated by ritual and all that ritual entails.

In pharaonic ritual, when a pharaoh was crowned, the spirit of the hero-god entered into him. The pharaoh became Horus—the living Osiris/Gesu. His wife became the symbolic Isis/Mary. In the coronation

---

*This is now common knowledge since the publication of *On Death and Dying* by Dr. Elisabeth Kübler-Ross in 1969 and subsequent research.

ceremony, he was anointed with almond oil and his marriage became a covenant with the beloved goddess, for that was who the queen was.

Interestingly, the eating of almonds can increase the production of semen (called the *water of life* by ancient peoples). Almonds were not only understood to be an aphrodisiac but also associated with consciousness and the powers of the mind. They were the fruit of immortality and, as such, their use in ritual was paramount. In some of the noncanonical gospel texts, the young Mary was given an almond seed or sometimes a lotus seed to swallow—the seed was the newly conceived fetus. The myth was fairly widespread. In some texts it is Mary's mother, Hannah who swallows the seed in order to give rise to Mary.

The Virgin Mary, being pure, gives birth to the hero who must attain the heavenly state despite the adversity of this material world. The *fallen* aspect of Mary as attributed to the Magdalene may well be due to a fall into materialism and the subsequent loss of the knowledge of the use of el ayin, the third eye, by humanity, for Mary Magdalene is also Mary "Amygdala," who falls and then reclaims her divinity through devotion to the way of the hero-god.

Through this, the pathway to higher consciousness, the tower to the heavens, is reinstated. Based upon near-death experience research, this tower may not be an analogy, but a real attribute, perhaps available through the correct functioning of the amygdala and the pineal gland. Mary relates to the substance of life. The story of Mary Magdalene could well be urging the end of gross materialism and the recognition, rejuvenation, and proper integration of practical spiritual attributes into human life. The tower that rises to the divine would be operative. But what is the tower? Could it not also be spoken of as a column or pillar?

IN THE CITY OF THE DEAD

Although no overall description of the pharaoh's coronation ceremony has survived into the modern era, enough research has been done to show that a procession around "the walls" and then inward toward the temple would take place. The walls referred to may relate to the long-disappeared boundary of the city of Memphis, but it is more likely to be the boundary of the sacred Giza complex. The idea of sacred site being processed in an ever-decreasing spiral is worldwide and still operative.

Round and Round

The word *gilgal* means "spiral" and is part of the name of the Sumerian mythical hero Gilgamesh, said to have been an actual historical king, the fifth king of the Second Dynasty of Erech. *Gilgamesh* means "man of spirals." A gilgal is also a circle of stones. In aboriginal, *gilgal* means "saucer-shaped depression"—one that has formed a natural reservoir. Gilgal is the name of a place where Joshua, Moses's heir, spent the night—it was within sight of Jericho, where "the walls came tumbling down." The proper reading of the story would seem to intimate that it was the name of Joshua/Jesus that was used or invoked to such a devastating effect—a great blast of air perhaps? Or is the entire story merely the coded description of a temple ritual? The fact that Joshua, who is cognate with all the heroes so far described in this book, stayed the night is interesting. In relation to this ritual timing, there is an interesting letter by the Roman governor of the Bithynia in Asia Minor, Pliny the Younger, writing of the Christians ca. 111 CE, that states: "They meet on a certain fixed day before sunrise and sing an antiphonal hymn to Christ as a god, and bind themselves with an oath."

The practice of antiphonal singing, a kind of alternate chanting sung by two parties, is indeed very ancient and in ritual at certain sites, as we have seen, is very powerful. At the culmination of the *heiros gamos* within the Santa Sanctorum, the name of God was ritually chanted in a rite known as the epiclesis. The epiclesis was the invocation of the spirit of God into the Eucharist—in Greek the indwelling of God was called the *eutheos*. In the heavens the term is *eutheoi*. God was called upon to send the Holy Spirit upon the bread and wine, thus making them the Body and Blood of the Christos, his Holy Son.

According to the Gnostic group known as the Sabians, the three Giza pyramids were the three kings come to pay homage to the holy child. In Osirian legend, Osiris's coming was announced by the three wise men, the three stars of Orion's belt, Mintaka, Al-Nitak, and Al-Nilam. The terrestrial positions of the pyramids at Giza reflect the stars' positions,[10] but not fully. As Herold and Lawson point out, the magnitude of the stars does not match the size of the pyramids.[11] This does not mean that the synchronicity of their sighting would have been lost on the builders.

There is, however, a distinctive mathematical correlation to the siting that has a clear association with the tradition of spiral procession. Rostau,* the very ancient name of the area, means both "the place of the cross" and "symmetry." The symmetry is this: the apexes of the three pyramids of the Giza Plateau lie along the curve of a Fibonacci spiral.†

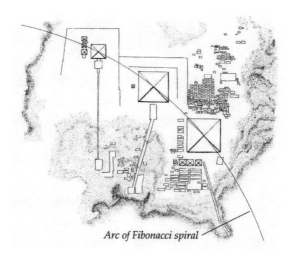

Arc of Fibonacci spiral

Fig. 7.2. The arc of a Fibonacci spiral
over the Giza pyramids

---

*In the Shabaka texts (line 18), carved into a single slab of solid black granite that now resides in the British Museum, the following description of Rostau appears: "This is the land . . . [of] . . . the burial place of Osiris in the House of Sokar."

†Named after Leonardo Fibonacci, also known as Leonardo of Pisa (ca. 1175–1250). His work, *Liber Abaci* helped introduce Arabic notation into Europe. In their simplest form, Fibonacci numbers form a sequence where, beginning with 1, each number is the sum of its two predecessors: 1, 2, 3, 5, 8, 13 etc. There are clear applications in botany, psychology, and astronomy.

The Fibonacci spiral is based upon the 5:8 ratio, a ratio that is found widely in nature, especially in relation to growth. The siting of the pyramids in this way seems to confirm that in being houses of life, they follow the patterns of life, exude it even. Almond and lotus seeds, too, the seeds of the immaculate conception, reflect the pattern of the Fibonacci spiral as it flows ever outward. It is as if the ancient Egyptians are saying or even telling us that we are attuned to the powers of nature.

Oddly enough, the Fibonacci proportion of 5:8 also betrays the proportions of cereal grain. It also relates to the workings of the harmonic overtones of a vibrating string as well as the golden proportions of a piano keyboard, all of this illustrating wonderfully well the outward flow and proportions of life itself.*

The word *church* comes from the Greek *kyriakon,* "belonging to the Lord," and if anything belonged to the Lord, the Giza Plateau did. In fact, the entire plateau is nothing but a massive church, a "church of the horizon." Apart from being "the Lord," Horus, of whom the Sphinx is a representation, is also known in a different aspect as Ra- Horakhty,[12] "Horus of the Horizon." In Greek *horos* means "limit," and this is the root of *horizon.* However, horizon also implies a circle—one's circle of vision into the distant landscape, as far as the eye can see. Horizon, therefore, suggests the wheel of Horus—the god is often called Horus of the Double Horizon.[13] The symbolic horizon represents the ultimate limit of creation, the limit of all that is godly, the limit of Atum-Ra. The beginning and the end of the circle are the Alpha and Omega, and within this boundary Atum-Ra is manifest as Osiris, and as Osiris's son, Horus—the Lord.

In Greek mythology it was the father god Zeus who was the guardian of boundaries and limits, just as the Greek word for horizon is

---

*There is too much contrary evidence for one to suggest that the Egyptians were not aware of these things: in fact, some of the imagery has come down to us, after many thousands of years, in the form of the pointy-hatted wizards wearing cloaks with stars and spirals and carrying long staves. I. E. S. Edwards observes that this was precisely the apparel worn by the Heliopolitan high priests, the greatest of whom was known as Chief of the Astronomers. His staff even had a five-pointed star on it. (See Edwards, *The Pyramids of Egypt,* 286.) Even the image of a magician's hat bears similarity to a pyramidal-shaped structure.

from *horos,* "limits," and *kuklos* or *kirkos,* "circle." *Choros* is, according to Liddell and Scott's *Greek-English Lexicon,* a word meaning "circular" or "cyclic chorus," in the sense of "any which were danced around a ring, round an altar."[14] This is interesting because the English word *orison* (from the old French *orison,* French *oraison*) means "a prayer." The root of this is the Latin *orare,* "to pray." Hence oratories and musical oratorios.

Everything to do with this word implies the secret. In Egyptian hieroglyphs this is underscored by the sign for horizon: the sun rising over a mountain, properly, "the place in the sky where the Sun rises."[15] In Greek, *ori* a means "mountain." The fact that *choros* means "cyclic chorus" brings together two concepts here: one is entirely to do with the idea of measure and the other with music within that measure. We are back at resonance again.

The church is a house of God, and we resonate within it, both on a physiological level and by chant and by the singing of hymns. Church in Gaelic is *kirk,* in German it is *kirche,* in Dutch *kert,* and so on. In Old English the terms are *circe* and *cyrice.* Circe was a Greek goddess. She was the daughter of the Sun. Recalling the religions of the dying and rising gods—all associated with the sun—and the fact that Jesus, too, was a dying and rising god, the term *church* is well chosen. Cyrice comes direct from the Greek *kuriakon,* which pertains to "the Lord" and his house. However, all of the former terms—*circe, kirk,* and so on—imply a circle, and indeed many of the oldest churches were circular, as were medieval Templar churches. The Greek word for circle, *kuklos,* is more in the sense of a cycle—hence we now have the aspect of continuation, of time. There is much about of these words that fascinates and thrills, for indeed all of them have mythical implications.

Horizon is the circle of Horus, the Lord: it is horizon of the circle of the church. In Hebrew, *hug* means "circle." *Hug* of course, contains the lordly syllable *hu.*

Saint Peter was the legendary holder of the keys to the church. Peter is possibly a variation on the theme of Petosiris, meaning "Osiris within the rock." Petosiris is likely to have been the root of the Latin *petra.* The Roman god of the underworld is Dis-Pater (*Dis* is cognate with *deus* and *divus,* or "divinity.") *Dis-pater* means "divine-father," or "divine Peter." Dis was represented by a dot within a circle: ⊙. This is precisely the same as the Egyptian hieroglyph for the father god, Ra. This brings us immediately full circle to Circe and the church.

Now, if this sounds convoluted, let me sum up, and thus come to the point: we have a circular boundary, *horos,* meaning "dance around an altar"—which altar? We have orison—where prayers were sung? We have the Egyptian glyph of the sun rising over a mountain—but where on the Giza Plateau will we find a mountain? We have a church within a circle—where? And finally, we have Dis-pater, the power within—within what?

Of course, they were all in front of me on the Giza Plateau: the altar, the oratory, the mountain, the church, and the circle. The power? Certainly, one of attraction: I turned from the Sphinx and stepped into the circle and began to walk through the site.

When Plutarch writes in *On Isis and Osiris* of the visible and calls it Isis, he is pointing to a particular spot within the circle, the Mr, the womb of Mery—the pyramid, whose holy child is within. This would seem to be pointing to an *actual body:* the body of the Son of God. All of the indications are that the whole of the Giza Plateau was a temple to this man. It was earlier in the year, while I was perusing a dictionary of Basque, Europe's oldest and oddest language, older even than Latin, Greek, and Hebrew, that I discovered another synchronicity. For most Europeans, the word *church* is a variation of *kirk;* but not for the Basques. Their word for church is *eliza;* in its syllabic form it is literally, "the God, Isa."

IN THE WOMB OF THE MER

The Great Pyramid of Khufu is probably the most extraordinary place on the planet. Photographs cannot do it justice, they only serve to minimize both the size and the impact of this incredible monument. Standing before it, cupping your hands over your brow, looking up at layer upon layer of stone until at its peak it meets the sky, is quite simply a breathtaking experience. Although the white Tura limestone casing has long since been stripped off, the pyramid stands firm, the ultimate enigma of the Ancients, whose construction all those thousands of years ago we can only guess at, still.

It is astonishing to consider that London's St. Paul's cathedral, Saint Peter's in Rome, and a whole host of Gothic cathedrals could easily be incorporated into the Great Pyramid, such is its volume. Sir William Matthew Flinders-Petrie, the father of modern archaeology, estimated that this immense building consisted of 2.3 million stones, each weighing approximately 2.5 tons each. The stepped exterior is made of limestone, while the interior chambers are made of both limestone and granite.

How this gigantic monument was completed in approximately twenty years is a problem that has vexed historians and architects alike. This figure comes from the *Histories* of Herodotus, who was informed of the figure by the Egyptians themselves, when he visited the region in approximately 450 BCE. There is also further historical evidence from the Fourth Dynasty that seems to confirm this number of twenty years. The logistics of completing this task in a mere twenty years without modern technology are staggering. It has been calculated that even in a period of thirty to thirty-two years, one block would've had to have been put in place every two or three minutes in a ten-hour working day.

The term *pyramid* is Greek. Individually, each of the ancient Egyptian pyramids had a different name and, in short, seem to be "horizons" of their respective pharaohs. We have seen that the pyramid *is the god to whom it was dedicated*. Therefore, by being dedicated to a priest-king, the pharaoh, the pyramid represented the meeting place on earth of humanity and the divine. The Great Pyramid was called Akhet Khnum-Khuf, or Akhet Khufu, for short. This translates as "the horizon of Khufu." However, the other pyramids of the Giza Plateau, those of Khafre and Menkaure, are simply referred to as Mr, in their original glyphs: $\triangle$. In its day, Khafre's monument was called the Great Pyramid, whereas Menkaure's was the Divine Pyramid. Khufu's was actually *the Pyramid that is the Place of the Sunrise and Sunset,* in effect, the place of the dying and rising god. All of the pyramids of Egypt still have names, most of them very beautiful: the Pyramid which is Flourishing of Places; the Pyramid of the Ba Spirit; the Purified Pyramid, and so on.[16]

In the *Hermetica,* reference is made to the pyramids. Ibn Zulaq (died 997 CE) states: "Among the wonders of Egypt are the two great pyramids . . . of which one is the grave of Hermes, i.e., Idris, and the other is that of Agathadaimon, the pupil of Hermes; and the Sabians used to go on pilgrimage to these pyramids."[17] A little earlier (about 987 CE), An-Nadim writes:

According to some, Hermes is said to have been one of the seven temple-priests who were appointed to have charge of the seven temples [of the planet-gods]; and it is said that the charge of the temple of [the planet] Mercury was given to him, and that he was thence named Hermes. . . . They say that he was a wise man, and that after his death he was buried in the building which is known in Cairo under the name of Abu-Hermes, and which the people call "the two Hermae" [pyramids].[18]

To the Egyptians, theirs was a sacred landscape, highlighted by equally sacred landmarks, their additions to a land given them by the gods. It so happens that Osiris was entombed upon or within this landscape, in the House of Sokar, a realm that lay within the fifth division of the earthly Duat, or underworld. According to the Shabaka texts, the House of Sokar was in the realm of Sokar, a land that Sir E. A. Wallace-Budge recognized as "a large extent of territory . . . in the deserts around Memphis."[19] According to R. T. Rundle-Clarke, this area was the modern Giza, which is the burial place of Memphis and the home of a form of Osiris known as Sokar. This area is also known as Rostau. This point is confirmed by various texts and stelae, but perhaps the most telling of all is the stele that stands between the paws of the Sphinx. This identifies Giza as "the Splendid Place of the First Time" and demonstrates that the Sphinx stood beside "the house of Sokar."

Sokar was a falcon god of the dead, indigenous to Memphis. Apparently, according to some Egyptologists, he was taken over by Osiris at some time in the past. The Egyptologists' conclusions remain hypothetical, but, needless to say, the house beside the Sphinx is the Great Pyramid of Khufu.

The pyramid *is the god,* but were the pyramids tombs? It seems that the theory of pyramids as tombs is all down to a misconception.* While the Giza site was certainly a necropolis, all of its pyramids have been discovered to be devoid of human bodies. The same is true for other pyramids, too—perhaps the odd bone or two, but nothing complete. Even when untouched, fully intact sarcophagi have been uncovered, they proved empty when opened. The pyramids were empty tombs.†

The step-pyramid ziggurats of Mesopotamia were not used as tombs, either. So, how to explain this riddle? Perhaps the answer may lay in the fact that tomb in Middle English is *tumbe,* sometimes *taumbe,* in Old French it is *tambe,* and then there is the Latin *tumba,* hence *tumulus* and *tumuli.* Tomb in its original Greek, *tumbos* from *temenos,* was a place dedicated to a god. This is the key. In the original creation myth of the Egyptians it is the god Tum, who brings forth all things as an act of self-awareness. The pyramid was also known as the Place of Atum. The distinction between these meanings is an important one in that the most obvious ritualistic use of the pyramid was as a cultic center, a place where the spirit of pharaoh ascended—it was a place of ascension. It might be that in later traditions all of these hidden facets have been co-joined.

All of this is to say that the place was a *temenos*—a sacred enclosure or precinct. In other words, a sacred place of power. If we look at the Giza necropolis from the sky, we can see the answer very clearly: recent infrared imagery taken of the plateau has demonstrated that it is one vast graveyard—but none of the all-too-human tombs or graves encroaches upon the sacred space of the pyramids. There appears to be a zone, obvious to the naked eye, that is virgin territory: it has never been touched. The presence of the graves marks the approach to the

---

*This is reflected in Mexico, where the pyramids of the area are known as *teocallis,* from *teo calli,* meaning "house of god." The Mexican word *teo* is very like the Greek *theo,* also meaning "god."

†Jesus, like the pharaohs, was entombed and after the Resurrection his tomb was found to be empty; but even then it was Jewish custom to entomb bodies for a statutory period before re-entering and collecting the bones of the decayed corpse and depositing them in ossuaries.

divine, but it is also a statement that the place is not a tomb: except in the metaphorical or even sacred sense.

There is an observation to made here: the Sphinx, in this respect, is acting like the gargoyles on later Gothic cathedrals, it is guarding the sacred space. Warning off those who are impure in heart. Furthermore, the "Place of the First Time" is a holy of holies: as God inhabits this place of extreme holiness time cannot exist *in it*. God is beyond time. Therefore, the boundary marking the entrance (*en-trance*) point *is the end of time:* it would also be the beginning of the "First Time"—the eternal sense of *now* wherein the Creation flows forth.

At present the common consensus is that the Great Pyramid was erected during the reign of Khufu. This dates it to approximately 2589 to 2566 BCE. However, an assessment of the site reveals that it was busy hundreds of years before Khufu. There is even the possibility that Khufu's pyramid itself may have been constructed over an older site. The Giza Plateau was an important site from at least the First Dynasty, circa 3050 BCE, and possibly before. Several archaeological finds have confirmed this.[20] Several of the mastaba tombs found at Giza, dating to this period, are known to have influenced certain aspects of design in the later Fourth Dynasty pyramid-building era.

There are undoubtedly other archaic elements, such as the boat pits, and the remains of several recently discovered pyramids, in the style of the Third Dynasty but believed to have been constructed in the Fourth. However, what is of interest is the fact that the entire necropolis is exactly that—a city of the dead, littered with tombs, graves, and mastabas, right back to the earliest period.

Geological analysis has revealed that at a short distance beneath the sands there are a few rocky outcrops that, if they had been built upon, would have provided a massively unstable foundation. It is in the space between these outcrops that the three great pyramids have been built.*

---

*However, the Great Pyramid is sited directly over a huge mound of bedrock. Jean Kerisel argues that topography is not a significant answer to the siting of the pyramid. Furthermore, the bedrock of the Giza Plateau could not take the weight of a vast pyramidal structure. Nor were the three Giza pyramids built in succession—after Khufu, Djedefre built his at Dahshur.

The mystery of their building is a debate that rages and remains an, as, yet, unanswered question and beyond the scope of this work to go into in detail.* However, one of the most accepted theories is that a causeway was constructed allowing the building materials to be delivered during construction of the pyramid.[21] In ancient Egyptian there is a word used in association with the construction: *pdwt,* which means "something that stretches," like a rope or line. The Westcar Papyrus contains a story about King Khufu and a magician called Djedi. The story revolves around Khufu's search for the secret chambers of Thoth. The Egyptologist Sir Alan Gardiner made a translation of the papyrus in the 1920s, in which he suggests that Khufu was looking for the number of the chambers of Thoth. This, however, has been countered by the Egyptologist F. W. Green, who observes that Gardiner may have confused two words: *pdwt* and *ipwt,* meaning "account," "archive," or "plan." The two words have a similar sound, and *ipwt* can mean a stretched string for marking a wall.

This is interesting because from time immemorial, hemp ropes were used to join the sacred sites to their associated settlements, if these were any distance away. It was believed that the hemp aided the conduction of the site's energy into the settlement.† This brings to mind the pyramid causeways that link each pyramid to its associated mortuary temple. The causeways are composed of very conductive quartz—surely a medium used for precisely this purpose.

Standing now before Khufu's great pyramid, I looked up toward its peak, recalling that in the nineteenth century its amazing conductivity was demonstrated by Sir William Siemens (1823–1883). Siemens was a British electrical engineer who visited Egypt and climbed the pyramid. While he was standing on the summit, "an Arab guide called his attention to the fact that whenever he raised his hand with his fingers outspread an acute ringing noise was heard. Raising his index, Siemens

---

*Due to years of sensationalist claims there is a reluctance at present to investigate the deeper mysteries of the Giza Plateau.

†There are many examples worldwide of straight roads and paths of this kind, complete with threads or ropes. In fact, to impede the progress of particular types of spirit, shamans would often tie knots in the rope.

felt a distinct prickling in it. When he tried to drink from a wine bottle he had brought along he noted a slight electric shock."[22]

Siemens moistened a newspaper and converted the bottle into a Leyden jar by wrapping the newspaper around it. "It became increasingly charged with electricity simply by being held above his head. When sparks began to issue from the wine bottle, Siemens's Arab guides became distrustful and accused him of practicing witchcraft. One of the guides tried to seize Siemens's companion, but Siemens lowered the bottle towards him and gave the Arab such a jolt that he was knocked senseless to the ground."[23]

While a similar effect can be observed at a host of domed structures and even cathedrals worldwide, given what I had already uncovered about the technology of the Ancients, I can only suspect that this column of energy atop of the pyramid was a goal of the designers. Both the Egyptians and the Greeks believed that power—real power—resided in the pillar or column. To the Greeks the actual place of power was the space above the column, called the attic. It is interesting that in classical mythology, the hero-god Attis was also strongly linked to the earth, in the form of the goddess Cybele. Columns of the Attic Order were square, and it was believed that the column co-joined the earth's power to anything constructed above it, perhaps because static electricity would accumulate on top of the columns, either from the earth or to be absorbed by it.

I wondered if this was a symbolic co-joining of the earth and sky, or perhaps something related to the Tower of Heaven often featured in near-death experiences. As the evening approached, the Giza air was cooling and I took a walk around the pyramid that is the Place of the Sunrise and Sunset and then returned to my hotel, thinking that there was a pillar nearby: an unseen column that must have accumulated vast power in these terms.

I would step *into it* the next day.

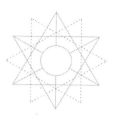

8
Communion

He is made one with Nature; there is heard
His voice in all her music, from the moan
Of thunder, to the song of night's sweet bird;
He is a presence to be felt and known
In darkness and in light, from herb and stone.
Spreading itself where'er that Power may move
Which has withdrawn his being to its own;
Which wields the world with never-wearied love
Sustains it from beneath, and kindles it above.

THEODOR H. GASTER, "'BA'AL IS RISEN . . .':
AN ANCIENT HEBREW PASSION-PLAY FROM
RAS SHAMRA-UGARIT"

There is a story of King Khufu and his search for the secret chambers of Thoth, and the deeds of the magician Djedi. Intriguingly, in the *Antiquities of the Jews*,[1] Flavius Josephus tells us of how Herod the Great had sought out the tomb of David to take money from it so that he might pay for the rebuilding of the Jerusalem Temple. Money he found not, but apparently, he decided to go farther into the tomb when suddenly two of his guards were slain by a mysterious burst of flame, and Herod, frightened, turned tail and ran.

This story is spurious, and William Whiston (1667–1752), in his translation, makes it abundantly clear that he thinks so, too. So, where did Josephus get it from? Well, if we look at the stories of Khufu and Herod, they bear remarkable comparison. Khufu consulted a magician called Djedi, wanting to know about the secret chambers of Thoth. Djedi responded by saying that he did not have the information himself but that he knew where it could be found—inside a box in the Room of Revision in Heliopolis. As Sir Alan Gardiner writes, "What ambition could have fired Cheops (Khufu) more than to possess in his own pyramid a replica of the mysterious chambers in the hoary sanctuary of the god of Wisdom?"[2]

At the time of this supposed raid on the tomb of David, Herod was rebuilding the Temple of Jerusalem, the home of Yahweh. *Thoth* is often transcribed as *Dwd* and is the root of the name David. However, *Thoth* is the Greek transcription of the Egyptian name Djehuti. Another form of this root name of David* is *Djedi*—a foreshortening of Djehuti. For the tombs of Thoth and David, read the same thing; it is the same story. Furthermore, no remains of any Davidic tomb have ever been located at the temple site in Jerusalem, nor is it by any means certain that the historical David of the Old Testament ever existed.

Khufu's pyramid took twenty years to build, so, too, the inner temple at Jerusalem by Herod's builders. Both kings raided a tomb involving a chap called David. Khufu asked the help of a magician. Herod has magic occur unaided. Khufu wanted to seek information about Thoth's sanctuary, so, too, Herod, who wanted to complete the magnificence of his own refurbishment of David's tomb, his monument being constructed around it, mythical though its existence seems to have been. The same, too, of Khufu, whose pyramid, it is suspected, was erected over the mythical chambers of Thoth. Khufu was a Horus king. Herod's name is not too far from the same spelling.

All of this should make us wonder if there was, during the first century, an ancient tradition of sorts permeating the religious culture of Jerusalem. Was Josephus perhaps responding to an older Israelite tradition, one with its origins in Egypt? This made me think about the story

---

*David as a name means "beloved"—the title of Osiris's sister/wife, Isis. Osiris may originally have been known as Djedi.

of Jesus's name over the temple balustrade in the Slavonic Josephus—a reference to Osiris of the Great Pyramid, perhaps? Was it the continuation of an age-old tradition, hence the name of Jesus in Israelite tradition?

Josephus had links to far older traditions in first century Judea, traditions that went back many hundreds if not thousands of years before Judaism, traditions that he is probably hinting of in his writings. He, too, had hidden a story with a mythic cipher: "The Djedi will lead you to the secrets of Thoth."

THE RETURN OF THE DJED AND I

Heliopolis, the City of the Sun, was known in deep antiquity as *Iunu* or *Annu*. Literally, it was the City of the Pillar. In full, it was Annu Mehret, the City of the Northern Pillar. Tradition has it that atop this pillar was the benben, one of the most sacred and powerful objects in all of the realms of Egyptian belief. This benben, with its pillar, is very intimately associated with Osiris, whose pillar is called the *Djed*. The bennu atop the pillar symbolizes, in this case, the power of the phoenix within the pillar.

The story of Khufu and the magician Djedi brings us back to this point—the secret of the tale is in the magician's name. The djed itself is heavily associated with magic. Quite often in Egyptian religious iconography the djed is portrayed linked to the key of life, the ankh, thus underlining life's magical qualities, but also underscoring its mystery. On a cosmic level the djed represented the backbone of the universe, the axis around which everything spun. So, in a way the djed also symbolizes stillness, the only point of no motion in the entire cosmos. This equates with a point of perfect balance at the center of the cross. The word *djed* also means "to endure" in Egyptian—the enduring one, everlasting, immortal: all synonyms of Osiris. Recall that in the myth of Byblos it was Osiris who was sought by Isis, the beloved, and that he, the immortal one, was in the pillar. Extending this point further, this brings us to the role of the column in Greek, Roman, and Egyptian architecture. It did not merely hold up the higher sections of the building but extended the power of the earth, the numinous, throughout it.

The following line appears in the King James Bible—it is part of the Lord's Prayer, part of the very same text that is to be found inscribed on the interior walls of the pyramid of Unas: "Thy will be done in Earth as it is in Heaven." In the Aramaic tongue the text reads: *Nehwey, tzevyanach aykanna d'bwashmaya aph b'arha.*

The point of introducing this now is to demonstrate again the fact that, in many ancient languages, words had multiple interpretations, which followed a meaning from subtle to gross, from microcosm to macrocosm. Unfortunately, not only is the reader offered just one interpretation in the King James version, but it has come to be looked upon in a very literal sense. The Earth spoken of is taken solely as the title of our planet. The text is very dry and does not reflect the poetry and multileveled approach of the ancient wordsmiths. Alternative Aramaic meanings include: "Your one desire then acts with ours; as in all light, so in all forms" and "Let all wills move together in your vortex, as stars and planets swirl through the sky."[3]

The Arabic word for earth is *arha*, and its Hebrew roots carry the meaning of all nature, all gatherings of mass and form produced by the universal force. The ancient traditions associate this force with the feminine. "Mother" and "earth," as we have seen, have the same connotations and share the same essence. Thus we follow the feminine path from microcosm to macrocosm: earth is the firmament on which we live, in turn the Eearth lives within the cosmos, which lives through the universal force, which as the ultimate Mother, gives birth to all, and, as the ultimate firmament, nurtures all. This firmament is the substratum, the primal substance that upholds all. As if to underline this, Sanskrit has the word *artha*, meaning "substance." We find the same root, *ar*, here as we do in the midst of the word *Mary*. How does the Mother give birth to all? Through the Father, the movement, the life stream, equated with that oft-used term *spirit*. It is this that is the essence of the myth of Isis and her search for Osiris, the life stream that arises within the djed, within all things. Isis and Osiris, the two together as substance and spirit, bring about life.

When we see the word *djed* and its pictorial symbolism, we have to understand that its meaning is very different and very male, but that it has been interpreted quite differently in a number of contexts, particularly as the symbol has passed into history.

The Osirian Djed pillar was essential to kingship. There are many representations of it depicting it as the backbone of Osiris/Gesu, with human arms at each side holding the royal regalia. The kingly ritual of the raising of the djed was symbolic of the royal power, or *puissance,* as well as the resurrection of Osiris.

The prophet Jacob raised a pillar at Mizpah (Genesis 31:49) and the first king of Israel, Saul, was acclaimed at this very same place. So, we can see a continuation of the tradition beyond Egypt.

The architectural element and its close connection with the life-giving part of the Djed pillar is intriguing: the force in cosmic terms went beyond life and death, and implied consciousness and rebirth, in the form of Sokar, the cosmic Osiris. This was also associated with the god Ptah, who was sometimes referred to as "the noble djed." The fascinating thing about this association is that Ptah was a god associated with creation, whose instruments were his heart and tongue—through the power of the Word he created the world, but the world could only come about through the djed, hence Ptah's connection with it.*

THE BACKBONE OF GOD

For those who do not know the Great Pyramid, there are three accessible chambers of three distinct levels. The lowest is a crypt or grotto that is directly accessible from the pyramid entrance via a descending passageway. It lies beneath the base level of the building and is therefore named the Subterranean Chamber. In myth, the grotto is often where the divine child is born. This subterranean grotto is in a rough and unfinished state. I wondered if it had been used as an abaton.

In the era of the pyramids the Earth Mother was held to be the source of dreams. A process of dream incubation was used at many

---

*The whole Osirian mythos and its symbolism is complex but fascinating. Through his death comes life anew—the fundamental mystery. Osiris, as the god of death, is portrayed mummiform, wrapped in a white shroud—a swaddling band—the symbol of a dead pharaoh. And here we have another connection with the historical Jesus.

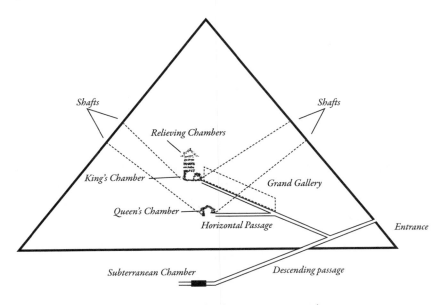

Fig. 8.1. Inside the Great Pyramid

temple sites as a means of escaping from the conscious mind and receiving guidance from the spirit of the place. The abaton was the pit or dream chamber where the dream was incubated. The initiate would be sleeping a womblike sleep within the realm of the Earth Mother. This was a means of seeking remedy for physical, emotional, and mental distress. Release from these burdens would be a purification, a kind of birth or initiation: but one might go through a personal hell first.

In Hebrew and medieval demonology, the lowest pit of hell is called Abaddon. Literally, this means the "heart of the Lord" (*ab-Adon*). As *d* and *t* are cognate, Abaddon and abaton are the same word. The Subterranean Chamber is the deepest pit in the Great Pyramid and we should remember that the pyramid is the Lord—Osiris. We should also not forget that hell is associated with transformation.

The Queen's Chamber comes next in ascending order. To get to it one moves up the ascending passageway to the foot of the Grand Gallery and from there along the horizontal passageway that leads directly to the chamber. The name "Queen's Chamber" does not seem to have been based upon any particular evidence. It is said that the incumbent Arabs named it so because their women were buried in places with sloping roofs, while their men were laid to rest beneath flat roofs. The Queen's Chamber has a sloping roof and the King's Chamber has a flat roof and this simple reason is given as the origins of their names. If this is the case, these designations may be accidentally correct. The masons used limestone in the Queen's Chamber and granite in the King's Chamber. It is likely that they saw limestone as feminine compared to granite as masculine. Limestone is relatively permeable, being the prehistoric residue of great seabeds and therefore associated with water, recognized as a feminine element by ancient peoples worldwide. On the other hand, granite is harder and is created as a consequence of the Earth's heat, the fiery element being associated with masculinity. Furthermore, whereas granite seems to have been used for its overtly radioactive properties and was therefore a stone of the spirit, limestone has no such qualities, it was everyday building stone and as such probably represented material things.

Before entering the King's Chamber, with its silent ever-haunting sense of presence, the Grand Gallery has to be ascended. This great corbeled hall is a breathtaking site. Such is it scale, design, and position that the pressures bearing down upon the masonry must be immense. At the top of the Grand Gallery is a high step and then one must duck and dive before entering the King's Chamber, a granite-clad chamber of striking simplicity and no adornments. However, it is above the King's Chamber that the mystery deepens. Constructed, it was thought, to relieve the enormous pressures bearing down from on top, the so-called relieving chambers are five in number, and from the bottom up are named Davison's, Wellington's, Nelson's, Lady Arbuthnot's, and Campbell's Chambers. But to call them relieving chambers is a misnomer. Relieve pressure they inevitably do, but the relief only comes from the presence of the upper chamber—Campbell's Chamber. Here there occurs the inconsistency of the ceiling being formed of sloping limestone slabs—everything else in these remarkable chambers, including the King's Chamber itself, is composed of granite. Furthermore, as has been pointed out,[4] the architect of this pyramid need not have incorporated five relieving chambers. To relieve the pressure bearing down upon the flat ceiling of the King's Chamber it would only have been necessary to insert pairs of angled limestone beams above it, forming a pitched roof that would withstand the pressure. However, in the King's Chamber there seems to be a purpose at work, one that suggests the symbolic and the acoustic. Other chambers in other pyramids have been constructed using local limestone, whereas the granite used for the King's Chamber was brought from Aswan, which is 500 miles up the Nile. Granite, as we have established, tends to be saturated with fairly high doses of the Earth's natural radioactivity in the form of the gas, radon, which has been implied as the "spirit" in the stone; this is perhaps reflected in the glyph for granite, ▭, whereas ordinary rock is ▥. The glyph implies a sense of greater solidity and something within making it so. The pitched limestone ceiling of the Queen's Chamber is enough to sustain the pressure bearing down from above. These pitched beams remain uncracked, whereas the granite beams in the King's Chamber are all cracked.

So far, the reasons for constructing the King's Chamber in granite remain unclear, as does the reasoning behind the relieving chambers. The bafflement continues when we address the King's Chamber sarcophagus, also made of granite. Limestone and alabaster, used in other pyramids, are far easier to fashion. Granite, in contrast, must have been very difficult to work with in this period, and yet the task has been completed with seeming and accomplished ease. The choice of granite is significant—in all areas of the King's Chamber. Having seen Newgrange, Chartres, and a host of other sites, we can see the reason why: the quartz content and the radon emissions.

The only thing that left me baffled was the interior plan; why were the relieving chambers present, if in reality they were not relieving anything? The answer may well lie in the Pyramid Texts themselves, which state (utterance no. 600) that the pyramid *is the god Osiris*.

If we could zoom in on the magnificent edifice with x-ray eyes, would we see the King's Chamber, and the chambers above, forming a significant pattern perhaps? In a mystical sense the Great Pyramid was constructed as the body of the god and whatever the god represented. In the Christian Church the church *is the body of Christ*. This pyramid was the house of life, a symbol of living sovereignty. And what

Fig. 8.2. A cross-section
of the King's Chamber
and relieving chambers

representative symbol may be seen in the hands of both gods and pha-raohs? A Djed column—the backbone of the god Osiris. The King's Chamber and its associated relieving columns are in fact a massive Djed column. The interior superstructure of the Great Pyramid is a vast great pillar—a design taken directly from the step pyramid of Djoser. What also strikes the eye at this point, although it must remain purely in the realms of conjecture, is that Khufu's quest to find the number of the secret chambers of Thoth may have borne fruit in the form of these very same chambers.

The djed concerned is one with five crossbars. Normally the num-ber of crossbars is four, though sometimes there are three. Though five elemented djeds exist, they are a rare sight indeed. The only one that I know of resides in the British Museum, and it is of a late date. Thinking further about this correlation, and pondering upon the function of the relieving chamber/djed complex, there was, I realize, another answer: the djed-ankh co-joining "the key of life and heaven." It is the actual connection point of the hero; his "backbone" is the point at which his life force attains the heavenly—connecting humanity to the higher dimension, to heaven.

The uppermost chamber, Campbell's Chamber, with its sloping roof, would represent the ankh—life itself. If granite is spirit in the stone, then limestone (being representative of that which is material and being the component of the Campbell's Chamber ceiling) would seem to be symbolic of the divine manifesting into the material world. Radiation readings taken inside the King's Chamber are high and are similar to those taken at Europe's granite-built sacred sites. In terms of the living sovereignty of the Great Pyramid, the internal cladding of granite, heavily crystalline in nature, is a living, breathing thing—breath in Latin being *spirit*. Many scientists in recent years have drawn parallels between crystals and living organisms—we must not forget that it is being suggested that at the outset of evolution, life itself was crystalline. Granite, being full of semiconducting crystals, is a container of the breath of the god, and as we have seen, people were and still are, capable of interaction within this environment; hence the reason and the realization that the architectural djed is not unique to the Great Pyramid.

The Djed Pillar of Osiris

The Djed pillar, with which Osiris is intimately associated, was also called the "palm of Byblos" and, more intriguingly "the spine of the god." Ultimately the Djed pillar is the symbol of the resurrection of Osiris in ancient Egypt.

Thus, it is interesting that in relation to quote 600 from the Pyramid Texts, the King's Chamber complex bears a striking similarity to the Djed pillar of Osiris. In the plan view on page 301, the five so-called relieving chambers above the King's Chamber resemble the intermediary spaces in between the cross beams of the column. In the Pyramid Texts there is a reference to Osiris that states: "You have been crowned Lord of Djedu." It seems to refer to the same.

The Setting Up of the Djed pillar was the most sacred of all Osirian rituals and involved Osiris's final resurrection. Osiris was called variously Asar-em-het-Djedet, which translates as "Osiris in the House of the Djed pillar," and Asar-Neb-Djed, "Osiris, Lord of the pillar." However, perhaps the most interesting reference is the following: "I am He of the Djed pillar, the son of He of the Djed pillar. I was conceived in Djedu. I was born in Djedu." This last line is particularly interesting because the structure of the pyramid itself was said to represent the pregnant womb of Isis, whose name of Mer (Mr—meaning, "beloved") is precisely what the Egyptians called these monuments. The Great Pyramid is not the first pyramid to incorporate the form of a pillar in its design—the very first of the pyramids, the step design of King Djoser (Third Dynasty, 2686 BCE), has a more orthodox square-shaped pillar at its heart. What is striking about the design of Khufu's pyramid is that the internal design is so very accurately that of Osirian symbolism.

In various of the Chinese languages there is a glyph similar to that of the djed hieroglyph, and this led me to realize that Buddhist stupas are very djedlike in structure, as are Chinese pagodas, some being five-layered and others seven-layered. I also arrived back at the pineal gland, whose root is the same as *pine,* for in seeing the djed I could also see that it was a stylized version of the pine tree. The same symbol occurs in the realms of Native American creation myths as the "Tree of Great Peace," also a fir tree.* Again, with the Native Americans, the Navajo creation myth tells of four-columned clouds,† reminiscent of the djed symbol.

It was while looking at the Native American traditions, which have much in common with ancient Egyptian thinking and culture, that I came across petroglyphs depicting the shaman and his rattle—the

Fig. 8.3. Shaman and rattle petroglyph

---

*In the Christian world, of course, we use fir trees as part of the Christmas festivity.
†My thanks to Rod Bearcloud for this and other helpful suggestions.

depiction that is at once pinelike and djedlike, associated as it is with wisdom. In Egypt the word for wisdom was Thoth's name, *Djehuti*—literally "the Lord within the Pillar."

In east and central Arizona these glyphs are particularly vivid. Can it be a coincidence that the shaman's rattle was used for healing or that many of the hero-gods in Native American tradition bear the names Yeshe, Esa, or Esu? What is more striking is that the name of many a father god is Yahwe, like the Jewish Yahweh.

The resonance of the shaman's rattle made me think more about the acoustic connection between sound and healing. The healing hero-god who used words of power to bring about transformation had fascinated me throughout this journey. The fact that he had occurred almost everywhere was astonishing. It was often in the most unlikely places. The Australasian Aborigines provide one of the most important of these correlations. Taking a wider view of things, though, that should not be in the least bit surprising, given their closeness to the earth and their incredibly well-tuned gifts of mind, body, and spirit.

The Aborigines have always been of great interest; they are masters of an instrument called the *didgeridoo*. When I first heard the diminished form of this name, I was astonished. The word *didg* is almost exactly the same as *djed:* the didgeridoo is nothing more than a hollowed-out *pillar* of wood, used as a wind instrument of very low resonance.

When the body of Osiris found its way to Byblos (sometimes Gaza) it was in the form of a tree, which was then cut down and used as a pillar in the royal palace. The Aboriginal didgeridoo appears to be yet another version of this legend.

The holy sycamore tree at Mataria in Cairo, Egypt, was the Shrine of the Tree, sacred to Isis-Hathor at Dendera. It was also the place where the holy family are said to have stayed when they first entered Egypt. This tree was said to be the receptacle for the spirits of the dead, especially Osiris, for this reason mummy cases—coffins—were made out of it.* One of the didgeridoo's ceremonial uses is in fact as a means

---

*Sycamore is *sycos* in Greek, meaning "fig"—the original forbidden tree in the Garden of Eden, because it was the Tree of Death, the very thing that Adam and Eve reaped when they ate of the Tree of Knowledge.

of connecting with the ancestral beings of the Australian Aboriginal Dreamtime. The didgeridoo is, in effect, a column of the earth, a pillar or djed. It is the backbone of the hero, through which the hero connects heaven and earth.

The name of the dreamtime deity is sounded by the didgeridoo master down the hollow tube, to low resonant effect. What this can do for brain wave patterns we have already seen. Musical pillars like these are used the world over in varying shapes and sizes, in Tibet, Switzerland, South America, and so on. Like the Osirian column at Byblos, the god is *within* the pillar—present as resonance.

THE BODY AND THE BLOOD OF THE ANOINTED OSIRIS

I will now tell the story of a set of very remarkable analyses carried out by my good friend John Reid deep inside the Great Pyramid in 1998.

John is an engineer, businessman, and independent academic who I met in Egypt many years ago. Like me he believes that the pyramids of Egypt were designed to be resonant at specific frequencies. I immediately warmed to John's friendly, relaxed, and easygoing manner and tried, in the brief time available when we first met in Egypt, to tell him a little of my own way of thinking. I described to him my vision that the Great Pyramid was the finest example of a building constructed to aid the alteration of individual states of consciousness.

I explained how I even felt that the building itself was in some way conscious, mainly through the building materials used, and how the idea of the gods was a description of the powers within.* I told him that I believed that the Great Pyramid was the spiritual technology of the

---

*Beyond the presence of a high radon content, some studies had revealed differing levels of magnetism. The upper chambers had very much lower readings of magnetism than the lower chambers, while at the same time the upper chambers were significantly higher in terms of frequency.

gods at its peak.* However, I did not tell John of the Osiris connection. I felt that perhaps I might seem a trifle crazy if I did. I need not have worried—that day he would undertake the first of two, prearranged, acoustic experiments in the pyramid.

What he was soon to discover would actually have seemed incredible to me, had I not known the Great Pyramid so well. John is a sound engineer who has spent a lifetime analyzing, as a part of his profession, the acoustics of cathedrals, churches, and other structures of like use. When he returned to our hotel after his first analytical session, he confirmed to me that the prime resonant frequencies of the Queen's Chamber, the King's Chamber, *and* its sarcophagus were all close to 125 Hz, well within the male voice range. He is the first to admit that, given the disparate dimensions of the two chambers and the much smaller proportions of the sarcophagus, this cannot be a coincidence.

John also carried out a test for prime resonance in the Grand Gallery and discovered this to be 250 Hz, exactly *double* the frequency of the King's and Queen's Chambers. The acoustic link between the chambers could hardly be firmer.

The Grand Gallery is a layered structure of seven corbels of four feet each. It has all the appearance and acoustics of having been built to reflect the diatonic ratio. I have never had the opportunity to experiment, but I have often wondered what would happen to the brain if the pentatonic chant was sung therein. I pondered over this Fibonacci dimension to the resonance within—5:8. Is this what happens inside the King's Chamber and relieving chambers? My hunch is an educated one but, until recently, with the introduction of air-conditioning, salt would accumulate on the ceiling of the King's Chamber—crystalline salt, which grows following the 5:8 ratio.

In any complex tone, harmonics can be heard as part of the overall frequency. They are an integral multiple of the fundamental tone so, for

---

*Strange things had happened there, peculiar "doublings" of energy, which generally is supposed to diminish with output—not so in the King's Chamber. But there was something else peculiar, too: people with broken bones would visit the place and walk away significantly further along the path to recovery—healing that took place by coming into contact with the dying and rising god Osiris.

example, with a total of 100 Hz, you may hear a harmonic frequency of 200 Hz and 300 Hz, et cetera. There is a strong relation to resonance here. If you pick up an acoustic guitar and intone a note into the sound box, you will find that, subject to the frequency intoned, sympathetic tones come from the strings—some higher than the notes given. These are harmonics.

Of more interest regarding the King's Chamber are subharmonics. A subharmonic is officially described as an integral submultiple of the fundamental. So, in the case of 100 Hz, that number can be divided by whole numbers—two, three, and so on—in order to find the subharmonic. Actually, subharmonics are not recognized in science as being implicit to a complex tone.

John Reid published his figures in the winter of 1999/2000,[5] and subsequent to discussions between us, he provided me with figures that include important predictions regarding subharmonics. We thought that until we returned to Giza, these would remain predictions. However, subsequently published details of experiments in the King's Chamber by Tom Danley* and his team have helped confirm figures given in the charts that follow.[6] Tom Danley describes the moment dramatically: "What really made everyone get up and run to the exit was the resonance near 30 Hz. At that moment, I aborted the test. This was a good resonance, it got nice and strong and scared the wits out of several crew members."[7]

Now, if we extrapolate some of the subharmonics from the frequencies that John discovered, we obtain the following results:

Area	Prime Resonance*	1st Sub-Harmonic 2:1	2nd Sub-Harmonic 4:1	3rd Sub-Harmonic 8:1	4th Sub-Harmonic 16:1	5th Sub-Harmonic 32:1
Grand Gallery	250 Hz	125 Hz	62.5 Hz	31.25 Hz	15.62 Hz	†7.81 Hz
King's Chamber	125 Hz	62.5 Hz	31.25 Hz	15.62 Hz	†7.81 Hz	3.9 Hz
Queen's Chamber	125 Hz	62.5 Hz	31.25 Hz	15.62 Hz	†7.81 Hz	3.9 Hz
Sarcophagus	125 Hz	62.5 Hz	31.25 Hz	15.62 Hz	†7.81 Hz	3.9 Hz

*Prime Resonance is to the nearest center frequency of analysis.

---

*Please note that although I do not agree with a lot of Tom Danley's conclusions, his work has provided corroboration of the 31.25 Hz subharmonic.

The 3:1 and 6:1 harmonics are also significant:

Area	Prime Resonance*	3:1	6:1
Grand Gallery	250 Hz	83.33 Hz	‡41.66 Hz
King's Chamber	125 Hz	‡41.66 Hz	20.83 Hz
Queen's Chamber	125 Hz	‡41.66 Hz	20.83 Hz
Sarcophagus	125 Hz	‡41.66 Hz	20.83 Hz

*Prime Resonance is to the nearest center frequency of analysis.

If we look at the frequencies marked [†] under 16:1 and 32:1 on page 307, we find 7.81 Hz. This is the frequency known as Schumann resonance. In her book *Sounding the Inner Landscape,* Kay Gardner cites the work of Michael Hutchinson.

> When a meditator reaches the rare point of oneness with all that is, the EEG reading is around 7.83 Hz. This is also the frequency rate of the electrical field resonating between Earth and the ionosphere (the Schumann resonance). Hutchinson writes that this frequency has the inner power of "seemingly integrating and harmonizing one's body and brain with the Earth's electromagnetic energy." This helps the meditator attune with Earth energy through the synchronization of vibrations.[8]

And here, firmly built into every room of Khufu's Great Pyramid is that very resonance. Coincidence?

There is also another potentially important inbuilt frequency: in the 3:1 and 6:1 columns of the chart above, 41.66 Hz is marked [‡]. There is evidence to suggest that frequencies in the region of 40 Hz are consciousness related. A device called a magnetoencephalograph (MEG), which can measure the brain's holistic magnetic activity, has enabled scientists to identify brain-wide oscillations in the region of 40 Hz and clearly associate them with the presence of consciousness. Research by Rodolfo Llinas and his team at the New York University

School of Medicine has also confirmed that these oscillations cover the whole cortex and move in waves from front to back. In *SQ, Spiritual Intelligence: The Ultimate Intelligence,* Danah Zohar and Ian Marshall further report that "these oscillations 'bind' individual perceptual and cognitive events in the brain into a larger, more meaningful whole and there may be a quantum dimension to the ion channel activity that generates the oscillations."[9]

This reference to ions is particularly interesting given that the granite in the King's Chamber exudes ions as part of the "radon package." The inbuilt harmonic of around 40 Hz could well have been used by chanters to stimulate and increase brain coherence, and the presence of ion-exuding stone may well have helped to facilitate this. As we can see from the chart, the frequencies are precise, moving in octaves of earth and brain wave frequency. The effect upon the mind and the body when these frequencies are intoned would be potent. The pyramid—the Mr—is a meeting place of earth resonance and humanity for purpose of interaction. The Earth resounds to the cosmos. The pyramid is a gateway to God; when individuals enter it, God may enter them—which is perhaps why so many people have reported life-changing experiences within its chambers. Further to these acoustical inquiries, John Reid was also the first person to carry out cymatic experiments inside the Great Pyramid. Many inquisitive visitors have noticed that the sarcophagus rings like a low-pitched bell when struck. He suspected that the ancient Egyptians had deliberately tuned it for ritualistic purposes, and he decided to set up an experiment to "see" the sound waves circulating within its interior, in the hope of gaining some insight into this theory.

However, its southeast corner has been badly damaged by generations of souvenir hunters, so he had first to carry out a temporary repair using a preformed aluminium corner. He then placed a small loudspeaker into the sarcophagus and stretched a plastic membrane over its open top, which was then weighed down with sixty-three small bags of sand to create a taut, drumlike, surface. Onto this John sprinkled a layer of fine quartz sand (using a cappuccino chocolate shaker!). The speaker was connected to a sine-wave oscillator to create a range of pure tones.* He turned on the equipment and slowly adjusted the control on the oscillator while closely observing the sand grains for movement.

Even though John had taken something of a leap of imagination to create such an experiment, he had no idea that the sound would reveal something quite extraordinary. As the images began to form, he couldn't believe what he was seeing. One by one, in an atmosphere of intense anticipation, archaic motifs appeared—*a whole series of ancient Egyptian symbols: hieroglyphs.* The nemes, the ritual headdress of the pharaohs; the ankh, or key of life; the hieroglyph of Ra, the sun god; the Wadjet, or sacred eye of Horus: they were all there and many more.

John rushed back to the hotel where we were both staying, in a state of great excitement, eager to share his revelations with me. Sitting in the hotel lobby, he was breathless as he told of his news—I shall never forget the look on his face. I listened intently and caught his excitement as he recounted his experiences. Of the eye of Horus, I remember quite vividly what he said.

> *The chamber was filled with the low-pitched tone of the oscillator and I was watching the sand intently; it sat lifeless and motionless on the membrane. I slowly adjusted the pitch and suddenly all the sand grains jumped into life and began to move. I began to see what looked like two eye images trying to break through within a few centimeters of each other, giving me the eerie feeling that I was being watched. One of them formed more fully and momentarily pulsated, almost as if blinking. I told myself to remain objective but I was so mesmerised it is a wonder that I had the presence of mind to take a photograph. Only a second or so after I did, the eyes melted into oblivion.*

He went on to tell me that almost every tiny adjustment of the oscillator brought new images to the sand in a seemingly neverending stream, but he lacked both the time and the film stock to record them all. It was, he said, one of those rare events, which may come only once in a lifetime, when you know with every cell of your being that what you

---

*A sinusoidal wave form enables energy to be transferred smoothly to the membrane.

are witnessing is far greater than yourself and belongs to humankind.

He likened the sarcophagus to Arthur C. Clarke's monolith in *2001: A Space Odyssey,* frozen from the dawn of time, waiting with infinite patience for someone to trigger the message that it had been programmed to deliver. The cymatic images had convinced John that the sarcophagus contains ancient data, locked into a matrix of quartz crystals, embedded within the living stone. But while the scientist in John struggled with the inference that the ancient magician priests had somehow encoded data within the stone, I knew without doubt that his work was an amazing confirmation of my own research.

It was his description of the image that formed at 190 Hz that excited me most. This tone is yet another significant harmonic, being exactly ten times 19 Hz, the frequency that triggered the visionary experience investigated by Lawrence and Tandy referred to in chapter 2. John told me that the image strongly resembled a human backbone, complete with curvature; he said that it slowly pulsated in a state of flux, as if it were a living organism.

Stunned, I asked him to repeat his description; I thought that perhaps my ears had deceived me. But there was no change in his description, and the shiver ran up my own backbone with the realization that my quest was finally over. After almost twenty years, after hundreds of hours of study and research, destiny had brought me to Egypt to meet the man who could unlock the secrets at the heart of the pyramid and provide answers to so many questions.

His 190 Hz image was a Djed pillar!

However, it represented much more than that. It was a spine, a pillar, the trunk of the World Tree. It was the image of Osiris and of Gesu, and it had imposed itself upon the sand for all to see. Even if it had been the physical remains of Jesus himself that had been found, the moment would not have been sweeter for me. This image, the image of the djed, had been derived through the action of sound, thus confirming my understanding that the Word was far more than a philosophical concept.

Standing back to take stock of this news, I realized that I had come to a milestone on my journey. I had sought and—it still seems to me a wonder—that also I had found. Yet there was the evidence. Not only the presence of consciousness-expanding frequencies but also acoustically obtained language—the icons of religion—all gained in a 4,500-year-old building that bore the name of the divine mother, Mary. I could imagine the priest stretching the animal-skin over the sarcophagus and sprinkling the sand over it. The pyramid choir, the Maids of the Mer, the Mermaids, would have chanted and the images have appeared.

These sacred images would, in time, influence the design of many of their symbols and hieroglyphs, some of them ultimately becoming stylized versions of the cymatic forms. The technology is so simple yet so profound. But the revelations do not end here. I remember an image, from an Egyptian papyrus, of the fifth division of the Duat, the House of Sokar.

The cosmic Osiris, Sokar-Osiris, is shown in his boat beneath a pyramid form that has as its apex a head. Above this there is what I can only describe as a wave form, and indeed, the pyramid seems to reflect this, too. And from where is this resonance being emitted? An

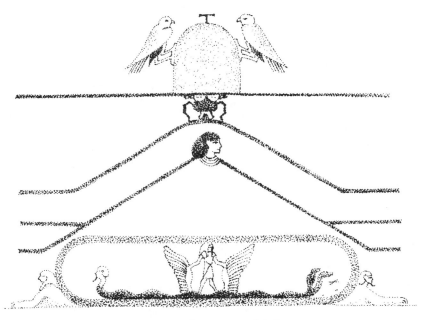

Fig. 8.4. The House of Sokar

omphalos-shaped bell, held high by two doves (the Christian symbol of the Holy Spirit). From out of the bell emerges the scarab, Khepera—the Egyptian verb *to be*. The resonance is "becoming," and upon it rides Sokar, who is within it.

The King's Chamber djed is a tower to the Lord, known to the Phoenicians as Baal: a Baal tower or bell tower, *bell* and *Baal* having the same root. Bells are the voice of the Lord, and if a plan of a church bell tower is overlaid on a plan of the King's Chamber complex, we get a match—both serve the same purpose.

Sitting with John and discussing the images, it occurred to me that the famous Egyptian Book of the Dead truly was the *Book of Coming Forth by Light*. John was blinking in the noonday sun after an extraordinary night's work. For the book was meant to be sung, in honor of the gods, and ultimately in honor of the One God. The word *oscillate* comes from the Latin *oscillum*—the mask of Bacchus. He was the Roman version of Osiris—Dionysus, the god who "trod the winepress alone."

Surely, the ceremony in Khufu's Great Pyramid was an acoustic Eucharist.

"I have eaten from the tambourine; I have drunk from the cymbal."[10]

These words, from the Mystery schools, suddenly made sense. The substance and the motion, the bread and the wine, the body and the blood of Jesus, Adonis, Aleyin, Apollo, Bacchus, Dionysus, Esus, Gesar, Osiris, and many more.

The power had manifested, in the very names of the gods.

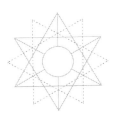

9

Acoustic Consciousness

In Egypt, when priests sing hymns to the gods, they sing the seven vowels in due succession and the sound of these vowels has such euphony that men listen to it instead of the flute and the lyre.

DEMETRIUS, CA. 200 BCE
(QUOTED IN REID, *EGYPTIAN SONICS*)

The import of John's analyses is remarkable, and it is paramount. Nothing like this had ever been undertaken in the Great Pyramid before.

However, it begged a question: What was it all for? Was it really only an ancient tomb? And if so, why had it, and the area immediately around it, been treated with such exceptional sanctity, even before the pyramid was constructed, from such a remarkably early date? And why such superlative construction, in so short a time?

If this truly was a place of initiation then we had encountered only some of its secrets, but there were others yet to be uncovered; and some of them were in plain sight. That the Great Pyramid was a place used for initiatory practices seems beyond doubt. Was it really a kind of spiritual birthing chamber?

The sheer accuracy of the acoustics inside the Great Pyramid are almost unearthly, almost impossible as is much else about the monument. But there is something else about the pyramid that stands out, as the Pyramid Texts make explicit, although not much attention has been paid to it. Could it be that what the acoustics of the Great Pyramid actually represent is the voice of Osiris himself, crying out, singing his voice as the voice of humanity into the vast great void of the cosmos? Was Osiris the first manifestation of the Anointed Christ, the eternal hero-king, *through* whom we must go to reach the stars?

The Pyramid Texts, from deep inside the pyramid of Unas, the ninth and final pharaoh of the Fifth Dynasty, make up the oldest body of religious literature in the world—dating to the twenty-fourth century BCE. They are all too frequently referred to as the Book of the Dead, but this is a euphemism. In Egyptian the texts are called, collectively, the *Pert em Heru*—the Book of Coming Forth by Day (or by Light).

As we have seen, these texts are divided into utterances, the most important of which, utterance number 600, states:

> *O Horus, this King is Osiris, this pyramid of the King*
> *   is Osiris,*
> *this construction of his is Osiris; betake yourself to it, do*
> *   not be*
> *far from it in its name of Mer . . . O Horus, as your*
> *   father Osiris in*
> *his name of "Mansion of the Monarch."*[1]

This is by far the most important textual quote. In other cultures, contemporary and post-contemporary, there are approximations of the text shown above regarding the ritual function and identity of the hero-god. The Babylonian Enuma Elish (third millennium BCE) quotes very much the same idea: as the most recent example, the catechism of the Roman Catholic Church states, *"The Church is the Body of Christ."* There are also numerous examples from other, more diverse areas of the world, as attested to by the ancient Greeks and the pre-Columbian cultures of Central and South America.

The pyramid is a resonator, and vowel sounds, which were known to have been chanted in succession in these places, charge them. The fertile process of rebirth could take place, the rebirth of life anew. I. E. S. Edwards has stated that Khufu's sarcophagus perhaps symbolized the womb of the goddess Nut. The sun god Ra apparently entered the mouth of the sky goddess every evening and emerged at dawn. When he died, the pharaoh was assimilated to Ra and was thought to undergo the same nightly journey as Ra, to be reborn as the sun god. The sarcophagus, it seems reasonable to speculate, may have been used as a part of the rebirth ritual. This is interesting, for the king in being the living Horus is also Set—his untamed wild side, so Osiris does not stand alone on the Giza Plateau: the king has to go *through* Osiris to ascend beyond Horus and Set, to escape earth and to become Ra. Such are the components of the spiritual technology. The spirit of the dead king, according to the Book of the Dead, ascends up, through the pattern of the seven musical notes, the seven sacred vowel sounds, just as in the East it is believed that energy rises through the seven chakra points of the backbone, the djed, where, upon reaching the crown, it is able to enlighten the spirit and release the soul. It is in this sense that the internal plan of the Great Pyramid is more or less a mechanism, a mechanism with resonance at its heart.

The spirit becomes reattuned, and the gods harken to it.

> *I purify my mouth.*
> *I adore the gods.*
> *I exalt Horus who is in the sky.*
> *I adore him. The Ennead listens, the inhabitants*
> *of the Underworld rejoice.*
> *They appear at my voice.*
>
> MAHU, EIGHTEENTH DYNASTY,
> CHIEF OF THE SINGERS OF AMUN[2]

The early church father Epiphanius (ca. 310/320–403 CE) quotes an early formula, the origin of which is Gnostic. It is a prayer designed to free the soul after death: "The Lord hath revealed to me the words the soul must use as it ascendeth up into heaven, and how it must make answer to each one of the Celestial virtues: I have known myself, I have collected myself from all parts, neither have I begotten sons to the Ruler of the World [the Demiurge]. I know thee who thou art, for thou art one from above."[3] This is very reminiscent of the same Osirian prayer from the Book of the Dead.

It is for these reasons that we must look at the Pyramid Texts as being not only an instruction manual of the spirit in the afterworld but also a songbook, a hymnal.

Music obviously had a more important role in ancient Egyptian society than has previously been realized. Perhaps it has been a mistake to see the Pyramid Texts purely as religious or liturgical literature, perhaps that is only half of the secret. Could this be the reason why gobble-dygook appears quite frequently throughout such texts? Not only that but doublings of names, oft-repeated names time and time again, and absurd images that seem to make little sense? The pharaonic texts are rich in such examples, almost all of them involved in one magical practice or another, a tradition that appears to have survived in the form of the *Hermetica*.

In a letter to King Ammon, Asclepius states: "As for us, we do not use simple words but sounds all filled with power." But what is the nature of this power? Why, nature itself, of course! *Nature* from Egyptian *neterw,* the essences of forces that comprised the divine. Life and its essential fertility.

The *w* in *neterw* is a quail hieroglyph. In hieratic (Greek, meaning "sacred") a cursive form of the hieroglyphic language, dating to the earliest times, this sign is abbreviated to ℓ, a strange form of proboscis that can be seen on the deshret crown of Lower Egypt and on the imperial crown of Japan.

The cry of the quail is a strange, low, resonant, guttural foreshortening of the *u* sound: it is a form of ululation. And ululations were generally a part of temple ritual. Today we know it as the hallelujah sung in praise of the Lord. This chanting would have helped to activate the interior resonance of the place. Was the proboscis a means of seeing resonance in action? A way of checking that the place was "switched on"?

The protrusion of the proboscis reminds me very much of the trails left by colliding subatomic particles after they have clashed and annihilated each other. Could this be symbolic of the spirit after death? Or of fertility and the power of life perhaps?

The same idea is portrayed at the Temple of Ramesses II at Abu Simbel in connection with

> the resurrection ceremony following the death of Osiris, at which Isis and her son Horus are officiating with initiatory tools. Horus holds to the back of Osiris's head a long rod with a tuning fork at its end, called the Rod of Resurrection, which Horus situates at the seventh chakra in the king's spinal system. Other tools are shown in the mural that further facilitate the king's spiritual metamorphosis and ascension to the light.[4]

In the Greek legend it is the head of Orpheus that sings and prophecies: another hero, one who had made the descent into the underworld in search of the feminine—Orpheus, who famously could sing the birds down from the sky. He was also an Argonaut in the famous quest for the Golden Fleece.

The first century BCE historian Diodorus of Sicily said that the rites of Dionysus were taken from the rites of Osiris by Orpheus. He states that it is only the names that have changed. The first century BCE statesman Cicero and the third century CE Neoplatonic philosopher Porphyry equated Dionysus with Osiris. The third century CE theologian Hippolytus equated Attis, Adonis, Osiris, and Bacchus-Dionysus with each other.

In the tomb of Ramesses VI (ca. 1141–1133 BCE) a serpent can be seen portrayed upon the walls, wearing the red crown; the serpent is symbolic of energy as is attested by many wild myths, the Coptic Saint George among them, overcoming the serpent or dragon, which almost invariably rises up from out of the earth. In the case of this representation the serpent was rising up as resonance within the coil. All of this is reminiscent of something else; it conjures up an image of living energy, vibrating at different frequencies—at every turn.

This raises another issue: all of this is pointing to the head of the god being somehow entombed in the pyramid. Pyramid Text, utterance number 600, underscores this: it says that the pyramid *is* Osiris. And yet in its original Egyptian pyramid is *Mer*—meaning literally the "womb of Isis." Isis who is the Queen of Heaven and shares all of the titles of another more familiar personage—the Blessed Virgin Mary. However, if the head of Osiris is in the pyramid, then surely the rest of his body is, too?

THE RITE OF THE MASS

The pharaoh, who was the living incarnation of Horus, the son of Isis and Osiris, would undergo at death a transformation from being the living Horus to being the dead and resurrected Osiris. Pharaoh was, by inheritance, the anointed king-god of his people: in essence he was the equivalent of the Greek Christos, a messiah. This is where it begins to get very interesting.

The word *pharaoh* means literally "great house." Throughout human history dynastic lineages have always been referred to as houses—the house of David is a case in point. This architectural term denotes the idea that the lineage descends from the gods—who reside in temples, and therefore, the origin of the line is divine. In the case of David and Solomon it was the tabernacle and then the Jerusalem temple. In the case of the pharaohs it can only mean the most revered monument in Egyptian history—and the most sacred place in its history: the Great Pyramid. The pyramid *is* Osiris; therefore the pharaoh, as Osiris's son Horus, has to be referencing this extraordinary monument as the meaning of his title; and as the anointed Son of God, he is also the master of its liturgy and ritual.

During his sojourn in the King's Chamber, John Reid discovered that the mysterious sarcophagus there shares the same resonant frequency as the heart of a newborn baby.[5] This remarkable fact immediately points us in only one direction.

Running in an outward direction from the interior of both the King's and Queen's Chambers in the Great Pyramid are four so-called ventilation shafts. There are two in each chamber. One of the shafts in the Great Pyramid was until very recently ovular in form. The two shafts do not run directly from the interior and outward; there is first a small entrance chamber at the end of which the shafts begin their ascent. When in the mid-1990s one of the shafts in the King's Chamber was earmarked for a new ventilation fan, in an attempt to control the humidity and keep the interior temperature constant, not much attention was paid to the symbolic nature of its form: its shape was duly destroyed after only a rudimentary survey was made and the form is forever lost. However, the shape remains intriguing.

The seed/ovule/egg symbolism ties in very nicely with the rebirth symbolism of the pyramid as a whole. The inference being that resonance plays a prime role in the atmosphere of both the King's and Queen's Chamber interiors. This is borne out by a curious phenomenon.

"Pyramid" in Greek is *pyramis*. This word literally means "wheat cake" (or barley cake). To the Greeks a pyramid was a giant wheat cake. The equivalent in ancient Egypt was the barley cake—also called, throughout historical tradition, *a sacrificial cake*. To the Romans it was known, revealingly, as *maza*—and these cakes were offered up to Osiris from time immemorial: they were unleavened. From *maza* we may derive *mass* (Hebrew *matzah*).

Reducing this further, *pur* is "fire" and *pyrrha* is "red." A field of wheat in the setting sun can look strangely red, as much as a single head of wheat looks like licking tongues of a flame. All over the Mediterranean, barley cakes and wheat cakes were, in ancient times, baked in honor of certain of the gods: from Attis and Adonis, to Dionysus and Jesus, whose communion wafers are consumed in sanctity to this day. These cakes were known as truth cakes—from wheat, the plant of truth. Those who ate the flesh of Osiris partook of his truth.*

In spring, with the coming of the inundation of the Nile, Osiris was associated with the verdancy of the coming season. Barley stalks, made

---

*The spires of many European churches resemble greatly the "spine of the god." How interesting that the word *spire*, meaning "steeple," comes from a Norse word meaning "spike of grass." It appears that the religion of the barley god has survived into modern times—and Osiris was heavily associated with the cultivation of barley. His Djed pillar symbol is also thought to have been a threshing pole for the sorting of the wheat from the chaff.

into corn dollies in the form of the hieroglyph for Osiris, with their heads protruding from it, were left at the entrance to the Great Pyramid. In the new season they would suddenly turn green with new growth.

Barley and Inspiration

Many years ago, as I walked through the green barley fields of northern Dorset, I noticed an interesting phenomenon. It was early morning and the sun was just rising above the horizon when I saw dewdrops caught in the whiskers of the barley. The effect in the early morning, as the sun was about to rise, was prismatic: gleaming with all seven colors of the rainbow. It also sings.

Sunrise and sunset are the times of the day when the pollen yield of these types of cereal crop are at the greatest. Furthermore, as any particular part of the agricultural landscape sees the sun go up or down it is actually emerging from out of the dense influence of the Earth's electromagnetic field, which itself is being pulled by the sun toward the sun. It is at this point that true earth resonance kicks in at roughly 8 Hz. It is this that stimulates the pollen overflow by vibrating the whiskers of the barley (or wheat) head, which in turn stimulates and excites the plant ovules—which upon fertilization become the seeds. The resonant god has worked in a mysterious way. Earth resonance is once again the key, for sunrise and sunset, called Matins and Evensong in church ritual, are the greatest periods of ELF activity.

As reported in *The Century Book of Facts,* "Adonis was an agricultural divinity and a vegetation spirit who, like Aleyin, was manifest in the seed of corn."[6] As Jesus states in the Gospel of John (6:35): "I am the bread of life." Barley's Latin classification is *Hordeum sativum,* literally "the Lord God cultivated." Barley and wheat fields in ancient Egypt were known as beds of Osiris. One of pharaoh's most ancient titles is his "Bee and Sedge" name; pharaoh was the Great Beekeeper. Bees, in Egypt, symbolized the tears of Ra—the prismatic qualities of dew again. When barley is ripe it weeps, its head sinks groundward as the hot summer sun brings the head to full ripeness.

In ancient Egypt the harvest falls in the months of late March to April and May—that is, the death of the god fell during what is spring-time in Western Europe, hence the date of Easter always being at this time. The pyramid/barley connection also leads us to other important links and observations: walking up the Grand Gallery deep in the interior of the Great Pyramid, you cannot but see the effect: seeing it in perspective it looks precisely like a stylized head of barley. The architect of this stunning monument is imparting an extraordinary secret. The so-called ventilation shafts are most probably whiskers of a head of barley—the barley god, Osiris—the original messiah who gave his life at Easter. However, if they are whiskers and Ra is the sun god stimulating them, then the base of the ventilation shaft tells us that the ova must at some stage be fertilized.*

We are inside the Mer—the Womb of Isis—somewhere inside there must be a child about to come forth, for the Mer is with child.

It is interesting in this regard to realize that the holy of holies in Solomon's temple in Jerusalem was built atop a threshing floor and was referred to as such—the place where the chaff was separated from the barley and wheat. (The halos we see around saints' heads are derived from the Greek *hálōs,* meaning "threshing floor.")

However, in the scheme of the Great Pyramid the ventilation shafts also represent an "escape hatch" for the departing pharaoh's soul. The king as a living, then dead but now spiritually resurrected, god, returns to the stars, his ka or double having mystically and mysteriously fertilized the seed by the resonant and powerful name of God. His ba or soul ascends, his job now done. The resonant earthly spirit of the king-god remains on earth and becomes incarnate in the king's son: an acoustic resurrection followed by an acoustic nativity.

Egyptologists and others have speculated that the body of the pharaoh was temporarily entombed within the sarcophagus of the King's Chamber and that vowels were intoned over the body. The granite sar-

---

*The ventilation shafts in the so-called Queen's Chamber (a later Arab appellation) were originally closed off from the rest of the chamber. It was Waynman Dixon who, upon investigating a crack in the masonry, realized that there was a void behind it and arranged for the shaft to be exposed.

cophagus, being rich in quartz, is highly resonant and would become excited by the rhythm and sound generated by the priestly chanting and singing. Could it be that the name of Osiris, in its original vowel form, the holy secret name of God, was used in this holiest and most secret of ceremonies?

A CHILD IS BORN

And they are Isis and Osiris, the divine intelligences. And when the story is written and the end is good and the soul of a man is perfected, with a shout they lift him into heaven.

THE BOOK OF THE DEAD,
TRANSLATION BY NORMANDI ELLIS

Jesus said heaven and earth will pass away, but my word will last forever.

MATTHEW 24:35; LUKE 21:33; MARK 13:31

The empty sarcophagus in the Kings Chamber, which has so baffled Egyptologists over the years, can be thus explained in much the same way that the story of the empty tomb in the Gospels is explained as a symbol of the resurrection.*

The ritual effect would have been greatly enhanced further still by the fact that the sarcophagus in the King's Chamber resonates at almost precisely 190 Hz; this is ten octaves above the movement of the eye (as we saw with Tandy and Lawrence's research in chapter 2). Therefore it is very likely that magical things would have been seen, thus adding to the overall atmosphere of wonder. This would have added to the overall myth and gossip associated with the monument that, most surely, would have passed into cultural myth and history, as well as language.

---

*Even the nativity can be explained within the context of another pyramid: Sneferu's pyramid at Dahshur. It has a very rough inner sanctum that has been hacked out and remains unfinished, unlike the rest of the pyramid, and gives the appearance of a cave or grotto. This is the grotto wherein the holy child/hero-god is born.

The use of granite is particularly interesting in that it is made up entirely of a complex mass of crystals brought together under the huge stresses of subterranean earth. Shamans in the ancient world saw crystals as the semen of the sky gods. They saw that the resonance of those gods was inside them, ready to seed the earth. In this respect the Aborigines of Australia see quartz as nothing more than frozen light, and light as the emanation of the gods.

Seed of the Gods

The most famous and prominent "seed of the gods" to be seen today is the Ka'aba at Mecca, which is annually kissed by millions while on the Haj pilgrimage. Another remarkable similarity is to be found in the same place, and it is this—the chief mosque at Mecca is called the Kufah and has been known as such since pre-Islamic times. This remarkable similarity to the name of King Khufu, builder of the Great Pyramid, is surely no coincidence. The body of the god was held to have been inside it—until the advent of Islam when the tradition was adjusted. It is nice to know that the tradition and reputation of Khufu's pyramid has altered little throughout the ages.

If utterance number 600 of the Pyramid Texts tells us that the pyramid is Osiris, then where is he? Just a little further on from utterance 600, the text tells us to look toward the Great Pyramid at dawn, as the celestial Isis in the form of the star Sothis/Sirius rises heliacally with the great sun god Ra.

This is the moment of the divine child, when his destiny is fulfilled and he emerges into the world to save it—and to save man from himself, by emerging from the womb of Wisdom herself, Isis.

This is the moment when the internal plan of the Great Pyramid's chambers can be observed. As we look east, they come together to reveal their secret scheme: they reveal the outline of a human body—

the form of the mummified Osiris.* Mortally he is dead: spiritually he is immortal, transformed into the divine child. By entering the pyramid, we go *through* him to the cosmos, to the Father, the God of all, of whom all the other gods are but aspects.

At the end of the Pyramid Texts the pharaoh cries: "I know myself, I know myself, I am at one with God!" It is the rebirth of awareness. Upon reflection, it is only too obvious that the internal plan should contain the mummified figure of Osiris.

From out of the womb comes the efflux of Osiris—Osiris has been placed in his tomb—here in the Great Pyramid; here where he has undergone the great shamanic rite of spiritual death, here where his bodily parts, once scattered, have been gathered together again, in this place where he now becomes immortal and where he rises again, resurrected, a judge of the dead.

Thus, the shafts of the King's and Queen's Chambers also act like the female fallopian tubes, which connect with the origin of the cosmic seed: pharaoh's soul has ascended to the stars and as the god he has reseeded the womb of Isis so that another Horus can be born, the new pharaoh. When one looks at the shafts that extend out from the King's Chamber it can be seen that they terminate at the exterior of the monument; whereas in the case of the Queen's Chamber they come to a halt somewhere deep inside the monument—within its body. In essence the Great Pyramid is a place of rebirth.

The symbolism of the interior goes very deep indeed. It has been suggested that the reason behind this is that the fallopian tubes in a woman's anatomy are closed off until the point of ovulation, whereupon they open, thus releasing the egg for fertilization. The sarcophagus itself may have been associated with the goddess Nut, whose domain was the sky.

A hymn to Osiris says that he "giveth birth unto men and women a second time,"[7] similar to the "born twice" motif of Dionysus and the Gospel of John.

---

*Recall that Monsieur Bovis, with his experiments in mummification, said the dead objects in preparation for the process had to be placed at a level *one-third up*—like the body of Osiris in the pyramid. (See chapter 3.)

Hildegard of Bingen, in *O Virga ac Diadema,* appears to be comparing Jesus to Osiris.

Unde, O salvatrix
Que novum lumen
Humano generi protulisti
Colligi membra filii tua
Ad celestum armoniam

Or:

O female savior who pours out
A new light
Upon the human race,
Gather the members of your son
Into heaven's harmony

The efflux of Osiris is the outpouring of the god, it is the act of creation-procreation: "This is the sealed thing, which is in darkness, with fire about it, which contains the efflux of Osiris, and it is put in Rostau."[8] Rostau is both the center of the underworld and the Giza necropolis—it is the meeting point of heaven and earth: at the center of it is Osiris in the womb of Isis, pouring out his efflux. This explains the extreme sanctity of the Giza Plateau.

THE ORIGINS OF THE MERMAID

The entrance to the Great Pyramid as originally constructed looks very akin to a hieroglyph for the word *hepet,* meaning "to proceed by boat, to a place." The Great Pyramid is the starting point for a journey—the journey of the soul into eternity, aided and abetted by certain ritual instruments, some of which were discovered in the northern shaft of the Queen's Chamber by Waynman Dixon in 1872. One of these instruments is a small bronze hook, then described as a kind of bronze grappling hook. This small double-hooked instrument is in fact a Pesh-en-kef knife. *Peshen* means "to split or divide" in the sense of separating. The Pesh-en-kef was a ritual instrument used in the famous Egyptian "opening of the mouth" ceremony.

In this ceremony the soul of the deceased king was given a spiritual reawakening into the afterlife, into which he was made resurrected. At the height of the ceremony the Pesh-en-kef, or "birthing knife," was used to cut the umbilicus of the newly dead pharaoh—for in everyday practice the knife was also used to sever the umbilicus of the newborn child. Now, the knife would have been used by a kind of "cosmic midwife." In the ritual of the dead pharaoh's body being separated from his soul the knife would have been an instrument for launching the newly resurrected king into the afterlife.

The Pesh-en-kef looks very much like the shape of the mythical mermaid's tail: it is the term *mermaid* that gives the game away, for as we have seen, the pyramid was termed *Mer* by the ancient Egyptians, and the cosmic midwife would have been a maiden in the royal court. Furthermore, the mermaid in legend inhabits the blue waters of the sea; space was held to have been a primeval watery abyss by the Egyptians, as well portraying its maidens as blue or dressed in blue, a sacred color in its day. How strange that, again in the legends, the mermaids were presented as luring poor unsuspecting sailors to their deaths—in the watery abyss.

This must be the last remaining vestigial rumor of what was once a very mysterious and enigmatic ritual, full of cosmic and mundane significance to the ordinary people of the day, who could only guess at what was taking place deep inside the Great Pyramid.

One last point is the fact that if one is able to see the long nave (recall *navy*) of the Gothic cathedral from high up, one gets the distinct sense that here is a picture of not only the cosmic boat of Ra, the boat of millions of years, but also of the death tunnel described in many myths and described by so many who have shared this extraordinary experience.[9] It is interesting that directly in front of the Great Pyramid is a pit that once contained a very sizable boat that now resides in its own dedicated museum.

Furthermore, *hepet* might be seen at the entrance, but it is not the key word to gain entry. Such a word has come down to us in the shape of fairy tale and legend. The word is *seshemu,* which in the mouth of Ali Baba (a pun referring to the ka or double of the god) is become the command "Open Sesame." The original of this, *seshemu* means "sexual intercourse." The reason for this is that it is a kind of pun; in entering the Great Pyramid one would most likely be committing a symbolic sexual act—due to the extraordinary symbolism of the internal plan of the pyramid. The Pyramid Texts of Teti (utterance 268) reinforce this: "Nephthys has reunited your members in the name of *Seshat,* the lady of the buildings through which you have passed."

THE OSIRIAN RITE: THE MERMAID RITUAL AND THE ETHMOID BONE

Osiris remains the epitome of myth, his birth is his death, his death his birth, and at the heart of it all is Isis, the great Maid of the Mer, the oldest of the old. And even here we can find echoes of the modern, echoes that have resonated down through all of the ages. Following are some further points.

In the process of mummification the removal of all the bodily internal organs was a matter of strict necessity in the preparation of the body for the journey to the underworld. However, in the case of the removal of the brain there seems little reason for it to have been removed, particularly in the arid meteorological circumstances of ancient Egypt.

As far as is known, the Egyptians did not regard the brain with quite the reverence that we give it today. Rather, they thought the seat of intelligence lay in the heart, so although all of the other removed body parts were stored in canopic jars, the brain would have been to a large extent hermetically sealed. In due course it would have broken down—while the mummy was in its tomb-enclosed recline—into the gruesome syrupy liquor beloved of Gothic novels, a liquid that eventually would have dried out; it almost certainly would not have penetrated other bodily areas of the mummy and thus have caused any terrible damage. So why remove it?

The answer to this riddle might lie somewhere in the area of the upper part of the nose: in the nostrils to be precise. In the process of mummification, when the skull was breached, to remove the brain a hook would have been used to break the nasal/skull membrane dividing the brain area from the rest of the skull. It was precisely this action that served to break the ethmoid bone leading into the brain cavity. The ethmoid bone contains a tiny piece of magnetite, which serves as a kind of digital recorder for the religious experience. It also acts as a gateway for certain stimuli to be detected and fed through to the pineal gland and the amygdala area of the brain. However, its main function is to give us our sense of direction; it is our anchor (recall the ankh) to this world. In severing this piece of bone-encrusted magnetite, thus making it possible to remove the brain, the officiating priests were symbolically releasing the soul of pharaoh into the cosmos—this was the point of the launch of his soul into the cosmos. Such was the expertise shown in the mummification of the human body in ancient Egyptian times, supported by textual evidence in the use of the healing arts and medicinal herbs, that, though it might be difficult to conceive of ancient Egyptians of having a working knowledge of the functioning of the brain, it would not be stretching a point to refer the reader to new breakthroughs in the field of historical shamanism. Research in the fields of earth resonance and human physiology have demonstrated an actual physical link that answers many unresolved questions about the subjective nature of human knowledge in the ancient world.

THE FINAL REVELATION

My quest was over, but the final end to it was to occur in the place I least expected—home. In late 2002, I found myself in discussion with my late friend Professor Christine el-Mahdy from the British School of Egyptology. Christine and I had known and encountered each other's work for years. However, it was only when I moved to the West Country that I got to know her better and to appreciate her work. She was a truly lovely, funny but serious lady, a scholar not afraid to look beyond the orthodox for answers to long-unsolved questions.

Her visit on this occasion was an eager one: she wanted to know more about my latest research, and, in the course of discussion, I brought up the work of John Reid inside the Great Pyramid. Christine was intrigued enough to want to see John's images. Unfortunately, the nature of the experiment that John undertook, in producing such surprising effects, was such that he had only room enough for his considerable assemblage of equipment—no one thought to take a video camera. The images from his still camera are good, but they don't do justice to what he saw as he scanned the King's Chamber frequencies and their effects upon the membrane now stretched over the sarcophagus.

It was when I produced the image of the double eye of Horus and related to her John's description of there being two eyes looking out at him from a Tutankhamun-like nemes headdress that, to my utter surprise, Christine fainted—she was out cold. I raced to the kitchen to get her a glass of water to revive her. She came around rapidly enough and dismissed my concerns. What had so transported her was something that I had not, until that moment, known.

And it was stunning.

Christine told me of the fact that, contrary to general perception, Egyptian religion was not polytheistic, that there was in fact an unspoken, ineffable concept of God. The gods, or neters, were merely aspects of the whole. The concept of this one god was so sacred that it was hardly ever mentioned in conversation or in texts, certainly not until the very late period of ancient Egyptian history. The name of this god was the *Neb-edjer*—the Lord of the Infinite—and the secret expression of his name, his holy, ineffable, unspeakable name was the double eye of Horus, the left eye and the right eye together, and this is what had appeared on the membrane atop the sarcophagus.

The name of God, the Lord of the Infinite had appeared in the midst of the greatest monument on Earth and in association with the body of Osiris, the original Anointed One.

This was the culmination of my quest.

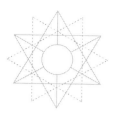

May the Force
Be with You

"This is absolutely amazing," said Robert Watts, the producer of the *Star Wars* and *Indiana Jones* films, when he had finished reading my thesis, ". . . we had no idea that this stuff was for real!"

Jesus said that "no one gets to the Father but *through* me." This statement has been taken to be metaphorical, but it is literally true. The ancient temple structure was designed to change our consciousness: we have to go through a seeming physical force to arrive at the point where we become the hero, to become at one with God.

Robert was uttering the word *djed* quietly to himself, using it as a mantra. I could see him arrayed in a brown cloak, the very image of a Jedi knight. Robert related to me some of the experiences he had had in the temple environment throughout his extraordinary career. He had definitely felt the force. "To think that there's an actual technology behind it all is, frankly, astonishing!"

Indeed, it is the temple that is critical to an understanding of Jesus as given in the Gospels—except that it has been written out of the story of early Christianity: it is the strangest thing about them.

The hero is Jesus, Gesar, Gesu, Asar—literally, the Prince of the Earth, in all cases but in a way that implies that he is not of this Earth. The original hero, Jesus was and is the bringer of harmony, his actual deeds are as the living incarnation of God, having achieved the hero state, having undergone the trials of the quest, and then having achieved the highest state of consciousness. This is "the pouring out of God," the pouring out of creation—and the sacred place is at the heart of it all. In fact, quantum physicists have speculated that 10 to 43 seconds after the big bang, the universe was shaped in the form of a tetrahedron: a pyramid.

Thus, the presence of the hero was an inevitability: all we have to realize is that he is in all of us all of the time.

The goddess Nut, the sky goddess, the symbolism of whose fallopian tubes in the Great Pyramid plays its part, gives us one more extraordinary correlation: EGGS—that is to say, **E**xtraneous **G**aseous **G**lobules. Deep inside the Orion nebula there are supermassive pillars of stellar gas and dust, some of which are many light-years in length. Inside these nebulae there is a miracle happening. In images of nebulae taken by the Hubble Space Telescope these giant pillars, looking much like the clutched fist of a cosmic giant—a Titan perhaps?—bundles of light, pockets of compressed dust and gas have fused and ignited; they are to be seen drifting off into the ether, giving light to the dark reaches of the universe. They are eggs in the birthing fields of the cosmos. This is the real nature of myth: it is gigantic in vision because it reflects the super-reality of who we are.

In conclusion, it could be argued that religion is an ancient *language of science* given expression by the world's sacred sites: places of profound power that have altered and directed the development of human society and civilization. *Religion* as a term means "to bind back to the source"; as a word, it is a fossil memory of the once profound belief in the Earth as our mother.

We stand on the cusp of great things; the hero is within us—and his mother, the figure of Wisdom, too. All we have to do is to use the gift given us by God.

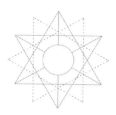

Human Wrongs and Human Rights in Scientific Inquiry

By Keith Hearne, Ph.D.

In evaluating ideas based on concepts and information coming from great antiquity, it is salutary to consider that current public opinion, and even "scientific" opinion, has very often resulted from considerable and deliberate political and/or religious distortions imposed in the past. Further, several natural psychological factors operate within us so that over time we tend to look uncritically (more than we would comfortably accept) upon the imposed view—so limiting our acceptance of a more truthful perception of the world. It is convenient to conform to mass thinking.

Within this framework we can begin to see how there may be an initial reluctance to countenance new ways of thinking—for instance, that the Ancients had a profound knowledge of acoustics. We assume too much and are cruelly disrespectful if we think them too "under-developed" to have known such things.

Ancient knowledge was staggering. For example, the ancient Greeks knew the circumference of the Earth and the distance to the moon. It was appreciated that the Earth was a globe, with the sun at the center of the planetary system—a concept that in the West was only rediscovered

333

by Copernicus in the sixteenth century. Unfortunately, many great truths that the Ancients knew have been ignored and no doubt many of them lost.

We should not automatically think that our social and technical advances over the past two thousand years also are the first that have occurred to humans or humanlike creatures on Earth. There is good evidence that humans exactly like you and I have been around for hundreds of thousands of years. There may have been several such progressions, followed by vast setbacks—such as mass destruction by super volcanoes (the last such event happened some 75,000 years ago in the Pacific region and reduced the human population, it is estimated, to a few thousand or so—hence the limited variance in mitochondrial DNA), or impacts from outer space, or even self-destruction. We all know of the rumored existence of Atlantis, a highly advanced society destroyed seemingly by natural forces.

It should not surprise us if we were to find here on Earth, or even on the moon and the planets, evidence of ancient technologically advanced inhabitation—not by distant alien beings but by earlier Earth civilizations of human beings like ourselves.

The hypothesis also exists that alien beings have visited this planet in the past and disclosed their insights to our ancestors.

Apart from the possibility that there have been repeated cycles of progression and regression in human development, we have in our own recorded history suffered other types of disaster.

We are familiar with catastrophic military conquest in history. However, there have been equally calamitous intellectual events with the resultant persisting thought biases acting on people like some long-lasting nuclear contamination having a half-life of centuries or millennia.

Until fairly recent times, a way of thinking could be readily and specifically imposed by powerful institutions, or a few powerful individuals, on whole areas of the world. In subtle ways, we are still manipulated. Unfortunately, once an attitude has been inculcated for centuries, the original, motivated, suppressive reason is lost, and people actually accept the idea as based on common sense. The attitude adopts the same status as folklore and wisdom. We find ourselves unwittingly giving lip service to someone else's self-motivated attitude from long ago.

This may be illustrated in two areas: the notion of reincarnation; and the area dealing with persistently reported anomalies—parapsychology. Both were recognized areas of understandings in the ancient world that became virtually forbidden topics in the West. The early Christians could accept repeated lifetimes. It was a natural notion, eminently sensible at a spiritual level, perhaps coming from ancient Egypt. It was favored in ancient China and was found strongly in ancient Greece. Pythagoras was a particular proponent of the idea. Metempsychosis (rebirth) was part of the teaching of the Greek Mystery schools, and Plato's Academy in Athens promulgated the idea. It was also present in the Gnostic teachings and Cathar thinking in the Middle Ages. Such groups were wiped out by the official church.

The concept of a cycle of life, death, and rebirth has, however, thrived elsewhere in the world. It was, and is, a perfectly acceptable and reasonable belief to Hindus and Buddhists.

One man, Emperor Justinian, was responsible for the willful destruction of the idea of reincarnation in the West. In May 553 CE, he convened the Second Counsel of Constantinople, also known as the Fifth Ecumenical Council. For his own reasons he wanted the council to ban certain popular ideas. The pope, Vigilius, refused to cooperate and was jailed. The Western bishops did not attend.

In one of the great "fixes" of history, the council "voted" to put an official curse, or anathema, on the topic. Anyone supporting the idea would be automatically excommunicated and dammed to eternal hellfire. There have been arguments about what really happened, and there is evidence that some documents were later altered, but, effectively, the council banned the idea of repeated existences.

At this same council, official anathemas were placed on the equally ancient concepts that the sun, moon, and the planets possessed a spirit. In one act, the accumulated wisdom and feeling about the nature of things that had sustained all cultures was eradicated from Western thinking. The many-lives notion remained taboo to the Western world for some 1,400 years, until it was reintroduced along with various

Eastern teachings. However, in the West many people consider it to be an unacceptable idea, their attitude directly repeating the dictates of Emperor Justinian.

The area is a seminal one, but conventional Western science, laboring under its own backward-looking restrictions, has no framework in which to encompass any kind of research in this field. Only a few brave scientists, like Ian Stevenson in the United States, have tried to gain understanding of this topic. He has, for instance, found strong links between birth defects or birthmarks in children reporting a previous life and injuries sustained during the previous reported incarnation.

In science, anomalous observations are extremely important and are usually seized upon because they indicate that the current theory is flawed in some way. There are actually many anomalous phenomena that science has studiously ignored because of a strict bias. The area includes paranormal effects such as telepathy, precognition, psychokinesis, and dreams.

That amazing block—like some gigantic neurosis—on the investigation of these fields was also engineered by the church. As a result of several papal bulls, notably of Pope Innocent VIII against witchcraft, such topics were considered to be forms of sorcery and therefore heresy. That suppression had consequences that still warps and stunts science today.

We like to think that we have complete freedom of thought and assume that science must be completely open-minded, but biases from centuries ago affect the thinking and attitude of all of us concerning many issues.

Censorious and authoritarian impositions on thinking from the past are, it could be said, kept alive by modern motivated and organized sceptics. Paradoxically operating with religious zeal, they unwittingly echo the restrictions on thinking imposed ages ago. Unfortunately, such skepticism never advances science; quite the opposite, it actively obstructs its natural progress.

Looking sensibly at counter-hypotheses to a new finding is a natural part of the scientific method. However, it is reasonable to assert that organized, motivated skepticism is an infringement on the freedom of those who seek an understanding of the curious situation of sentient existence in which we all find ourselves. At a time when far-reaching human rights legislation is becoming part of ordinary law in more and more countries (in Britain, for example), the same analogous concept of the human right to scientifically investigate any area, without undue and campaigned bias against that research, is bound to follow—just as racism and sexism have had their day. Organized skepticism is perhaps the last resort of the chauvinist.

This is not a book for the scientific chauvinist. I hope it marks the start of an awesome reevaluation of the wisdom of ancient peoples. The notion itself opens the doorway to vast mysteries and new vistas of possibilities concerning what happened on Earth long, long ago.

KEITH HEARNE is an internationally known innovator in science. He conducted the world's first research into lucid dreams for his Ph.D. and invented the "dream machine," now in the Science Museum, London.

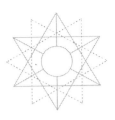

The Anatomy of the Hero

In referring to the "anatomy" of the hero I am of course speaking of the essential elements that link this archaic figure to us as humans: the ritual, the building of the sacred place, and the fact that, as in all mythologies, the hero is the arbiter between us and the divine. This list is by no means exhaustive, but it serves to reveal the many similarities on a worldwide level and to shine a light on the essential idea of humanity that emerged from deep within us at the outset of human history.

THE WORLDWIDE HERO: SYMBOLS, NAMES, MONUMENTS, AND LEGENDS

The etymologies given here are speculative, but I have taken great care to make them as accurate as I can in consulting many and varied sources: the important point is the syllabic bases—many of which are unconnected in world culture at the origin of their outset: this is an issue that I have not had time to look at in any kind of detail and must wait for a later time. The monuments given are noticeably pyramidal in form and are given largely as an example of the fact that they existed as such in those countries. I have placed the most prominent and important of the hero figures first in the list and then in descending order.

Hero: Osiris
Region: Egypt
Etymology: *Osiris* is the Greek translation of Egyptian As-ar. As a son of the earth god Geb, Osiris inherits the "earth" part of his father's name, *Ge* from *Geb,* his divine portion: so we have *Gesar* or *Gesa,* although the name never occurs, the intimation is there.
Legend: For details of the legend see chapter 6. Asar's coming was announced by three wise men. Angelic voices hailed his coming and time stood still. A star (Orion) rose in the east (called by the Tibetans Rishi-Agastya, after a holy king in very ancient times; in Persia this same star was called Messaeil; in other words, the Messiah.)

Osiris had a twin, Set, who was wild and uncultured, who in some versions of the legend is crucified. Osiris was killed by his jealous brother and harrowed hell before rising from the dead and ascending to "the Father" to become a judge of the dead. There are many variations of the legend. However, it is significant that in the boat of millions of years, in which the great god Ra sails, it is Set and not Osiris who overcomes the Serpent of Chaos so that the sun god may rise again. Set, the twin of his brother, becomes, in this act, his brother, overcoming his wild ignorant self.
Symbols: The Djed pillar; the ankh or ansate cross; the tamarisk tree
Monuments: The Great Pyramid of Giza, the Osireion at Abydos

Hero: Krishna
Region: India
Etymology: Called The Black, The Dark One
Legend: The deified hero of Indian legend, born of the goddess Devaki. His birth was announced by a star with the attendant voices of angels. Krishna was attended by three wise men and given gifts by shepherds. He was hailed as the Redeemer, Firstborn, Sin-Bearer, Savior, Liberator, and, significantly, he was called the Universal Word. Krishna's legend tells of a slaughter of the innocents and a tyrannical king who was closely related to him. Krishna was killed by being shot in the heel, his only vulnerable spot, by an arrow (as in the legend of Achilles). He was then hung between heaven and earth, and his blood fructified the soil.

Symbols: The elephant, called in the Hindu pantheon Ganesha, "man of the earth" in syllabic meaning; the Tree of Life.
Monuments: Various throughout India

Hero: Mithras
Region: Persia, first millennium BCE onward
Etymology: Latin *Missa* and *Mitre;* Mizd, the Mithraic sacramental meal; English *Mass,* compare *maize;* Indian *Mitra*
Legend: God of Light, born December 25—the birthday of the unconquered Sun. Born miraculously of a rock (*Petra Genetrix*), a female rock representing the Earth Mother. His birth was witnessed by shepherds and by magi bringing gifts. Mithra performed miracles, healed the blind, and cured the lame. He cast out devils. After his triumph he ascended to heaven, an event celebrated at Eastertime. Mithra's spouse was Anahita, the Mother of Waters, also known as Ma. The Mithraic high priest's name was the Pater Patrum, Papa, or Pope.
Symbols: The Cross of the Invincible Sun; the Chi-Rho
Monuments: Various. A temple to Mithras can be seen under the Vatican.

Hero: Adonis
Region: Syria ca. 3000 BCE
Etymology: From *Adon,* meaning "Lord" or "Master"
Legend: Born from a myrrh tree (Myrrh—Mary) at Bethlehem, Adonis was placed in an ark or coffin. His mother, called Myrrha, was a virgin and is identified with Mary. Myrrha, too, was a temple virgin. Adonis died at Eastertime. He was gored by a boar (the Egyptian god Set was also associated with this animal, called Typhon by the Greeks). Associated with the grain harvest, he rose again on the third day. He is associated with the sun.
Symbols: The cup or chalice; wine
Monuments: Various. See Tammuz/Dumuzi.

Hero: Attis
Region: Anatolia, Rome, Phrygia (ca. 3000 BCE–fifth century CE)

Etymology: Attis is very similar to Adonis in myth. It is a foreshortened form (*t* and *d* are cognate).

Legend: Born of the virgin Nana and lover of the great goddess Cybele. Born on December 25. A beautiful youth, he was killed by a boar while out hunting; in other versions he is crucified on a pine tree to bring salvation to mankind. A god of vegetation and healing, he was resurrected as the Most High God. His passion was on March 25, usually a Friday, called Black Friday, the Day of Blood. He descended into the underworld and on the third day he rose again from the dead. Attis's feast day is called the Hilaria.

Symbols: The pine tree

Hero: Aleyin/Baal

Region: Phoenician, western Semitic—Syria, Lebanon, Palestine, Turkey

Etymology: Baal, or "Lord." Aleyin, sometimes Aleyn, from el ayin–the pineal gland or letter *O*. Irish Bel, hence *Beltane*. Scandinavian Balder.

Legend: Aleyin/Baal was the son of El, father god of the Phoenicians (a general term meaning "god"). Baal had a twin brother, Mot, his mortal twin. In the version from Ras Shamra it is the goddess Anath who kills Mot, threshes him, and burns him. He revives seven years later. Mot then is the grain (compare the legend of Taliesin in the Mabinogion). Baal, having been tricked into the underworld, was under the thrall of Mot. Anath tried to persuade Mot to release him. Baal rose from the dead. In another version he killed Mot (who had been forsaken by his father, El) by driving a curved knife (the moon) under his fifth rib.

The throat of Mot, the underworld god, is huge and it is cavernous. Mot, the god of death, has no need to send messengers to the gods, he simply opens his mouth, resonating from his throat, his voice booming wide over land and sea, his words reverberating from the deep wide nothingness that was death's throat. Thus the hero emerging from the cave at birth is actually being resurrected.

Symbols: The moon; thunder (as Baal Hadad, Hadad-Rimmon, and Rimmon); pomegranate, in the case of Hadad-Rimmon and Rimmon (which name means "pomegranate" in Arabic, the symbol of resurrection—recall the legend of Demeter, Persephone, and Hades in Greek myth).

Monuments: Oak groves, Asherah pillars—stone circles and many sacred sites dotted throughout the region. See also Ras Shamra excavations.

Hero: Sarapis; Osiris-Apis
Region: Ptolemaic Egypt, ca. 300 BCE; Mesopotamia, 3000 BCE
Etymology: Sumerian Sar-Apsi, Easar Apsi—from *Esa* the bull (of the underworld), also known as Ptah-Hephaistos (or "the reception of Ptah," Ptah's strong point is his voice: resonance)
Legend: Sarapis is the fusion of Osiris with the Apis bull, lord of generative fertility, and a universal god, uniting within his person characteristics of Zeus and Hades.
Symbols: A kalathos—a basket-shaped headdress
Monuments: Many and various throughout Egypt

Hero: Tammuz/Dumuzi. Hebrew-Canaanite version of Dionysus or Adonis
Region: Mesopotamia, modern-day Iran-Iraq; later, throughout the Middle East
Etymology: Dumuzi—Sumerian, "True Son," "Good Son." Greek form is Thomas. Also called Esos in Greek, sometimes Ta-us, a grain god. Also called Eshmun.
Legend: Tammuz was the original Anointed King, or Christos. He was also the original Good Shepherd. Sometimes called Usir (a title of Osiris). Son of Ishtar-Mari—the virgin who gave birth to him. He was the lover of Inanna, who handed him over to the demons of the underworld. His twin was Azazel or Asar-el. He was killed on the Day of Atonement in the form of a lamb. At his death the Hallelujah went up, with ululations of despair. Tammuz was the only begotten Son of God, who rose from the dead on the third day. A god of healing and of vegetation.

Note that this mythology is very similar to the South American legend of the Calina people involving the twins Tamusi and Tamalu, respectively Light and Dark. The twins hated each other. They were the sons of Amana (Ana) the Creator, who sent them from her sea kingdom (Mer) in the Milky Way to protect earth. In the Sumerian myth, Dumuzi's mother is Ama'usumgal, whose mother is a heavenly dragon.

Symbols: Corn; cereal grain

Monuments: Various. Sumerian and Babylonian ziggurats, pyramid-type structures.

Hero: Esus, Hu-Esus, Hesus

Region: Celtic Europe, second millennium BCE

Etymology: Esar, Aesar. Phoenican Izr, Iasar. Called the earth spirit. God of the harvest, of wine and grapes.

Legend: There are many variants of this god throughout Europe. Esus was born of a virgin, was a healer god, and was crucified. Although the name Esus does not appear to occur on the British mainland, other variants do. On a carving from Roman Gaul, Esus appears in trinity, as Teutatis and Taranis. In Ireland he is Oengus or Angus, the earthly prince in the underworld, son of the Dagda. The Dagda's other name was Eochaidh the Hereman (Joseph the Herod-man). In another version of this myth, Eochaidh marries the Morrigan. In Britain she was called Eseye, another possible form of Isis.

Symbols: The cross; the bull; the crane—usually in triple form

Monuments: Newgrange, Silbury Hill, and various others

Hero: Quetzalcoatl, the Plumed, or Feathered, Serpent

Region: Central and South America

Etymology: An ancient Mexican god. The Aztec savior god. *Coatl* means "twin" or "earth." *Etza* is the central element of the name, *Qu* qualifies it. Also called Xototl (it is to be noted that the name begins with a cross, X).

Legend: Quetzalcoatl was born of a virgin. He represented corn or maize, and thus his life was symbolic of planting, growth, and the harvest. His other symbol, apart from the serpent, is the dove. He was the Lord of Knowledge, the god of the wind and of the zodiac.

Quetzalcoatl's twin is the typical red-haired, wild warrior twin that we have come to expect, called Tezcatlipoca (*Tezcatl,* or "mirror," *popoca,* or "smoking"). He tempted his brother, as Satan tempted Jesus in the wilderness. It was the representative of this wild twin on earth who was sacrificed by having his heart torn from his chest. Quetzalcoatl died, descended to hell, and rose again from the dead. His second coming was always expected imminently. He was crucified between two thieves.[1]

Symbols: The moon and the sun

Monuments: Represented by the pyramids at Teotihuacan

Hero: Jesus. Variants Gesu, Gsu, Esu, Esa, Isa, Issa. Greek Joshua. Hebrew Yeshua.

Region: Originally Palestine and Sinai

Etymology: Yehoshua, meaning "God saves," "God is salvation"

Legend: His birth is prophesied. He is born of a virgin, Mary, daughter of Anne. His birth is attended by signs and wonders, the visit of three kings or magi. Time stands still. The king orders a massacre of the innocents. The family goes into exile. The childhood of the hero is not spoken of in the main Gospels. Jesus appears as an adult and performs various miracles. He is put on trial and found guilty of an unspecified charge. He is mocked, made to wear a crown of thorns, processed, and crucified upon a hill. He harrows hell and rises from the dead on the third day before ascending to heaven.

Symbols: The cross; the fish—vesica piscis

Monuments: The Church of the Holy Sepulchre, Jerusalem; cathedrals and various worldwide

Hero: Gesar of Ling

Region: Western Tibet, China

Etymology: *Ga* or *Ge* (Greek, "earth"), *Sar* (Tibetan, "blood"). Gesar, Kesar (Caesar); Mongolian Kesar, Kyesar, Skye-gsar ("reborn"), from *kesa* ("hair"). Eshe or Esses ("the man").

Legend: The rescuer of Tibet from foreign invaders. A demigod who lies sleeping in a cave to return in the hour of crisis. He was fatherless (or his father is the sky god, while his mother is the goddess of water) and underwent a miraculous birth. He is a wonder-working king who had eighteen uncles and could render himself invisible. He led a quest for the Chiata Mani—a sacred stone, jewel, or cup, most often a stone in a tower in the city of Shambhala. He was an architect.

Symbols: Sun cross

Monuments: Potala Palace and the Pyramids of Xian

Hero: Gesar Khan

Region: Asia—Mongolia and Tibet

Etymology: See Gesar of Ling

Legend: A deity of the Mongols and Tibetans and the hero of a lengthy epic known in many different versions throughout an area that stretches from Lake Baikal to Bhutan and from Peking to Hunza in North Pakistan. Gesar's form is that of an equestrian deity, wearing a helmet and cuirass. He is rooted in pre-Buddhist religion. He was among the protective deities of Lama-ist Buddhism and entered the Mang cult circa 1600. In the nineteenth century Gesar became identified with the Chinese god of war and protector of the Manchu dynasty. Gesar's role was an oracular deity. In Tibetan belief, Gesar is the divine hero who descended in ancient times from heaven to the center of the world in order to restore order on Earth.

Hero: Esaugeteh Emissee

Region: Southeastern United States

Etymology: According to the Native American Creek people, name means "Master of Voice, of Breath"

Legend: The Creator, the First Principle. Very reminiscent of the legend of the Egyptian creator Atum, who, like Esaugeteh, hovered over the surface of the waters (compare the Hebrew creation story). Esaugeteh built a hill to reside in and in which he created people—out of mud. There are many variations upon the name and legend in ancient America.

Symbol: Cross

Monuments: The prime existing monuments are the pyramidal Cahokia mounds in Illinois.

Hero: Al Issa
Region: Arabia ca.1800 BCE–600 CE
Etymology: Arabic *Isa,* meaning "God saves," "Prince of God," "Son of God"
Legend: Son of Arabian Mara, sometimes Mari. A good spirit raised from death, a god of fertility and of the creation.
Symbols: Cross, phallic pillar
Monuments: Before the advent of Islam, the Ka'aba at Mecca and various monuments throughout Arabia

Hero: Ise, Jizu, sometimes Joshi
Region: Japan, Okinawa (the Ainu)
Etymology: "Earth kami" or "spirit"
Legend: Son of Maia, Maya, or Mary. A good spirit raised from death.
Symbols: Cross
Monuments: The Ashitake pyramid, the temple at Ise

Hero: Marduk, also Asarluhi, Esa
Region: Mesopotamia—ancient Sumer (modern Iraq)
Etymology: Also called Amar. Utu "Bull calf of the sun-god Utu." Like the Hebrew god, Marduk had many names, fifty in all according to the sources.
Legend: Grandson of An. Born of a virgin, Anu. Performed heroic deeds. Put on trial, dies, and harrows hell. Is resurrected three days later at Eastertime. He is the rising sun.
Symbols: Cross
Monuments: The Esa Gila, Babylon, a pyramidal ziggurat meaning "Esa's Song" or "Esa's Glory"

Hero: Ashur, Assur, Anshar—Asura (India)—Kansa;[2] Suriakos (Greek)
Region: Syria, northern Mesopotamia, southeastern Asia Minor
Etymology: *An.Sar,* Osiris—known as the Great Mountain
Legend: Ashur was a deified form of the city of Assur in Assyria. He was the equivalent of Enlil. The god of storms, of war.

Symbols: The winged sun; the Tree of Life; the solar disk; the bow
Monuments: Holy City of Harran, also known as Assur

Hero: Nimrod
Region: Babylon, Mesopotamia
Etymology: Greek Orion. Egyptian Asar (Osiris). The Hunter.
Legend: Patriarch of the Babylonians and Assyrians, called Assur, another form of Asar. A renowned king, portrayed as an oppressor in the Old Testament.
Monuments: Mount Nimrod, a tumulus of white limestone resembling a pyramid

Hero: Iasius, Triptolemus,
Region: Crete
Etymology: Triptolemus means "three plowings"
Legend: The young god with whom Demeter lay three times in the plowed fields of Crete[3]

Hero: Jason, Iasion
Region: Greece
Etymology: Greek Iason, from *iasthai,* "to heal." A form of Joshua.
Legend: The *Argo,* or ark, refers to a boat and its "crew" are, possibly, astronomical markers in the sky. Jason had his Argonauts, seventy-two of them, while Jesus, too, had his disciples, at first twelve and then seventy-two much later in the Gospels. Jason was killed in later life by one of the timbers of the *Argo* falling on his head; therefore, like Jesus, Jason's death is associated with a tree.
Symbols: The boat or ark, *Argo*

Hero: Eshbaal, Eshmun
Region: Ancient Israel—Palestine; Lebanon, Phoenicia
Etymology: "Baal's Man." Eshbaal was the son of King Saul and may have been Canaanite Phoenician. Eshmun was the tutelary god of Sidon, a Phoenician city-state. He was also worshipped in Tyre, Beirut, Cyprus, and Carthage. His name appears to mean "the Eighth," a

reference to completion. He was the eighth son of Sydyk—the righteous.

Legend: His legend is very similar to that of Adonis. He was fond of hunting and was amorously pursued by the goddess Astronoë (possibly Astarte). In desperation he castrated himself and died. He was restored to life by the goddess who changed him into a god. He is the equivalent of the Greek Asclepius.

Symbols: The moon

Monuments: The Temple to Eshmun, just outside Sidon; Asherim; oak groves; stone circles

Hero: Aiza

Region: Modern-day Benin (Dahomey)

Etymology: Asar, Ai-sar

Legend: One of the most important spirits in Dahomey religion, a protector of groups. Every village, region, compound, and marketplace has its *aiza,* or guardian spirit, and its mound or shrine. To'aiza is a spirit of the founder of a place. Xwe'aiza is the spirit of a compound and Ax'aiza of a marketplace.

Monuments: Mounds or shrines throughout the country

Hero: Woden, Odin

Region: Scandinavia

Etymology: Unknown

Legend: Wounded himself with his own spear and hung himself on the "Windy Tree"—the great ash Yggdrasil. For nine nights the god remained hanging and bleeding, until he observed runes on a stone in the abyss below him. He managed to lift the runes and was immediately released from his ordeal and filled with new life.

Symbols: His son Thor carries a T-shaped hammer or cross

Hero: Obassi Osaw and Obassi Nsi

Region: Southern Nigeria

Legend: Two gods who created everything together until Obassi Osaw decided to live in the sky and Obassi Nsi to live on the earth. Osaw gives light and moisture but brings storms. Nsi is a nurturer. Nsi taught man and woman about planting. The people learned to grow and eat the fruit of the palm tree.

Hero: Izanami and Izanagi
Region: Japan
Etymology: "She who invites you to enter" and "The Lord who invites you to enter." *Iza,* "Lord" or "Lady."
Legend: Male-female creators. One version says the male twin Izanagi reached down from heaven and stirred "the deep" with his spear.
Symbols: The sun and the moon
Monuments: Various throughout Japan. Caves and grottos.

Hero: Amaterasu
Region: Japan
Etymology: A form of Anath?
Legend: The sun goddess born from Izanagi's left eye

Hero: Susa No-Wo
Region: Japan
Etymology: Unknown
Legend: The dragon-slaying hero born from the nose of Izanagi

Hero: Ganesha and Buddha
Region: India
Etymology: "Lord of Hosts" (see Krishna)
Legend: Ganesh is the Hindu elephant-god, father of Buddha. Begot Buddha on the virgin Maya. At Elephantine in Egypt he appeared as a form of Yahweh, consort of Anath or "the Virgin Zion."[4]

HEROIC NAME VARIATIONS

As we have seen, the names of the hero have myriad variants worldwide and through the ages. These names have been adopted and incorporated in many contexts. They can take the form of toponyms (for instance, Gaza is a variation of the name Osiris), thus later countries are named after heroes. Modern ideas of these etymologies can seem a world away from the initial impulse to name after a hero. This is but the beginning of a longer, more in-depth inquiry into the fact that the ancient hero has left his imprint upon the world—and we have moved away from these origins . . . and into denial.

ABASSI: African deity in the myths of the Efik people of Nigeria. Associated with plowing, sowing, and harvesting. (Literally, "the heart of Issa.")

AESIR: A race of gods in Norse mythology.

AMATERASU: Daughter of Izanagi.

ANASAZI: A Native American people.

ANSAR: The Babylonian totality of heaven (compare to the Aesir).

ASAPH: A famous musician of King David's time. He is supposed to be the founder of the hereditary choir of Bene Asaph in the Second Temple. Psalms 50 and 73–83 are ascribed to him.

ASIA: Derives from the Hittite Assuwa.

ASIL: A Norse god.

ASOKA: Indian emperor (third century BCE).

ASSARIO: A tribe of South American Indians.

ASURA: The old gods of the Vedas.

AZAN: A son of Arcas.

AZAR: The name of the patriarch Abraham's father, a vizier to King Nimrod (Orion).

AZARIAH: A king of Judah, whose name gives away Osirian connections.

AZERBAIJAN: A place-name, Asar Baijan.

AZIRRU: A Phoenician king with seemingly Osirian connections.

AZURE: Blue, also lapis lazuli; Jesus was robed in blue.

BRIZO: A goddess of Delos.

DUSARIS: The chief deity of Petra in Edom, "the Red Land."

EDESSA: "Place of Issa"? A town (now called Urfa) in Eastern Turkey where there was a temple to the Mesopotamian god Marilaha—"the Lord of All." His chief ambassador was the moon god. If Abraham was born at Edessa/Urfa then this had to be the original home of the moon cult. For more information, see *Edessa* by Judah B. Segal.

ELISSA: "Issa is my god." An alternative name of Dido, queen of Carthage, ca. 800 BCE.

ESHU: Also Yoshi or Yoshu, a god of the Ninjas.

ESHU: Messenger god of the Yoruba people of Nigeria. He brought the gods' instructions to earth and could speak the languages of every living thing.

ESSA: An Arcadian word for an ear of corn.

ESSEN: Alternatively, hoshen. A word for the breastplate worn by Hebrew high priests.

EZANA: Ethiopian king of Axum, ca. fourth century CE. A convert to Christianity.

GAESA OR GEIS: A taboo or oath in Irish Celtic society.

GAESUM: A Celtic weapon. Its exact form is unknown.

GAGSISA: A Sumero-Babylonian term for the star we call Sirius.

GAZA: The temple demolished by Samson; one of the many places where the encoffined body of Osiris was said to have landed, to be found later by Isis.

GAZARII: A term used for witches in the Pyrenees.

GE-EZ: Ethiopian, a Semitic language with no known etymology. The word means "self-awareness" as in "I Am," the creative context of god.

GEEZER: "The man who came." Slang, but Wellington's soldiers picked it up from the Basques during the Spanish Peninsula Campaign of 1808–1814.

GEISHA: Japanese term for courtesans.

GERIZIM: The holy mountain in Samaria; site of an archaic temple of the Hebrew God.

GESHUR: A region directly north to the Sea of Galilee.

GESSO: A kind of plaster that is trodden with grapes in the making of wine in Spain.

GEZA: A king of Rumania, immediate post-Roman period.

GISAN: A province of Japan. Noted for the monastery of Shigatse.

GIZO: The spider/man of Hausa folklore; trickster hero (West African).

GUSI: A cult followed by the Dusan of Borneo incorporating the veneration of green porcelain jars that are inhabited by the spirits of ancestors. (*Vase* from Latin *vas,* earlier form of which was *vasus,* plural form *vasa.*)

HESI: A name of the heifer goddess.

HISSAR: A range of mountains north of Turkmenistan.

HZA: The serpent god of the Incas.

HZAMNA: Enlightener of Yucatan, recognized by the Maya as founder of their civilization and "Master of the Dawn." (Another variant of Itza or Issa.)

IASO: A son of Aesculapius.

IASUS: Greek. Son of Phoroneus, brother of Io (Isis).

IAZGES: A tribe of Sarmatians (fourth to sixth centuries CE).

IBIZA: The popular Mediterranean island.

IOSHEKA: "The good one"; One of a pair of twins in Native American lore, he is the "Good Twin" or the "Godly One." He serves the needs of nature and vanquishes his evil brother Tawiscara (*Tau*—Hebrew "cross").

ISA: A mountain in Queensland, Australia.

ISE: A Shinto shrine in Japan.

ISHA: A title of Shiva.

ISHNYA: A place in Northern Iceland.

ISSA: Collective term for the initiates of the Mysteries of Isis.

ISSAPOO: An island off Fernando Po. The people of Issapoo regard the cobra as their guardian deity, and it is hung tail downward from a branch of the highest tree in the public square. (The serpent being symbolic of Wisdom.)

ISSAS: Djibouti is the territory of the Affars and the Issas.

IXIAN: Crucified on a fiery wheel, a symbol of the sun (Greek).

IYASU: Emperor Lij Iyasu, emperor of Ethiopia in the early twentieth century.

IZZARRA: A Basque liqueur.

JEHOSAPHAT: A king of the Israelites.

JESI: Latin *Aesis,* a town on the Esino river in central Italy.

JESSE: The father of King David.

JESTER/GESTE: "Deeds" or "exploits" (*geste*) hence *jester,* "a trickster and truth-teller."

KOZI: A name for Nimrod, another variant of Orion—Egyptian Osiris in cosmic form. The Canaanites called him Baal-Hadad.

KRISHNA: A deity of India.

LHASA: Capital city of Tibet. The Potala (the palace that is a focus of

Tibetan Buddhism) is pyramidal in structure. Elements of its design, such as the stupas, reflect aspects of Egyptian architecture and religion, particularly the Osirian Djed pillar.

MASAR: A Hopi word meaning "the Creator."

MASSI: A Maori legend of a hero thrown into water (comparable to Moses).

MIZOS: A people of Mizaram in Burma.

NINGISHZIDA: Sumerian, "Lord of the Right Tree." An earth god. A healer god. Like Anubis he is portrayed as dog-headed, the spirit of the hero in the underworld.

NIZIR: The World Mountain of the Chaldaeans.

OISIN: An Irish hero, member of the Fianna, the warband of Fionn MacHumail (Finn MacCool).

OMPHALOESSA: The omphalos of Essa (Essa being the power within the stone).

ORISSA: A state on the western coast of India.

OSAIN: Yoruban (African) deity of leaves and herbs, conceived as living in the forest and giving medicines to man. He has been retained in the Afro-American cults of Brazil, Trinidad, and Cuba, and as Ossange in Haiti.

OSHOSSI: Yoruban deity of the forest and the chase. His devotees dance for him carrying the miniature bow that is his signature. (Compare to Jesus's dance in the Gnostic Acts of John.)

OSSA: A mountain mentioned by Herodotus.

PHRIXUS: Boeotian prototype of Isaac who was almost slain by his father as an offering to God. A ram with a golden fleece suddenly appeared to take his place.

SESA: An Indian snake demon who bears the Earth or enfolds it. He is king of the Nagas and under the name of Ananta is the symbol of eternity.

SHIGATSE: Buddhist monastery in Gisan province in Japan.

SIRE: Form of address for royalty, et cetera, in medieval English, meaning literally "prince" or even perhaps "messiah" (Me-sire, Assire-Asar).

SRAOSHA: Persian, *Saresha,* God manifest.

TESUP: A form of Atabyrius, a Kassite god, ca. second millennium BCE.

ULYSSES: Greek hero. The Latin form of Odysseus.

VISHNU: A principle deity of India.

YASU KUN: "Country at Peace." A Shinto shrine to and of the emperor in Tokyo.

YESHE: A king in Tibet.

ZAR: Ethiopian demon, may be related to an ancient name of Osiris, worshipped during the First Dynasty at Abydos as the god-king Zer, who became Lord of the Underworld.

Notes

INTRODUCTION

1. Barker, *The King of the Jews,* 43.
2. Rory Carrol, "Pisa Baptistery Is a Giant Musical Instrument, Computers Show," *Guardian* (London), December 2, 1999.
3. *The Times* (London), December 9, 1999.
4. Ouspensky, *Theology of the Icon* 1:94ff.

CHAPTER 1. SECRETS OF LIFE

1. Teilhard de Chardin, *The Phenomenon of Man.*
2. Cade and Coxhead, *The Awakened Mind,* describes details of research in this area.
3. Becker and Selden, *The Body Electric,* 249.
4. Rauscher, *The Power of Place,* 300.
5. Cousens, *Spiritual Nutrition,* 81–91.
6. Devereux, *Earthlights Revelation; Earth Memory.*
7. Roney-Dougal, *Where Science and Magic Meet,* 140.
8. Cousens, *Spiritual Nutrition,* 84.
9. Davenas, Beauvais, Amara, et al. "Human Basophil Degranulation," 816–18.
10. Quoted in Playfair and Hill, *The Cycles of Heaven,* 218.
11. Michael Ibison, personal communication, August 1995.
12. Jahn, Devereux, and Ibison, "Acoustical Resonances of Assorted Ancient Structures."

13. Jahn, Devereux, and Ibison, "Acoustical Resonances of Assorted Ancient Structures."

14. Jahn, Devereux, and Ibison, "Acoustical Resonances of Assorted Ancient Structures."

15. Jahn, Devereux, and Ibison, "Acoustical Resonances of Assorted Ancient Structures."

16. Hunt, *Infinite Mind,* 65–66.

17. Stirniman, "The Wallace Inventions," 44.

18. Watson, *Supernature,* 99.

19. Blair, *Rhythms of Vision,* 126–27.

20. Blair, *Rhythms of Vision,* 126–27.

21. *Science Now,* BBC Radio 4, March 1999.

22. Le Mee, *Chant,* 123–25.

23. Ash and Hewitt, *The Science of the Gods,* 123 (now retitled *The Vortex*).

24. Ash and Hewitt, *Science of the Gods,* 123.

25. Cousens, *Spiritual Nutrition,* 85.

26. Cousens, *Spiritual Nutrition,* 85; and Roney-Dougal, *Where Science and Magic Meet,* 129.

27. Persinger in Devereux et al., *Earthmind.*

28. Roney-Dougal, *Where Science and Magic Meet,* 140–44.

29. Aisling Irwin and Roger Highfield, *Daily Telegraph* (London), December 1998.

30. Irwin and Highfield; and Watson, *Supernature,* 94–95.

31. A. J. Scott-Morley, personal communication, July 1997.

32. Blair, *Rhythms of Vision,* 132–33.

33. Cited in Mann, *Sacred Architecture,* 92.

34. Mann, *Sacred Architecture,* 92; and A. J. Scott-Morley, personal communication July 1997.

35. Devereux, *Secrets of Ancient and Sacred Places,* 105ff.

36. Mann, *Sacred Architecture,* 92.

37. Mann, *Sacred Architecture,* 92.

38. Devereux, *Secrets of Ancient and Sacred Places,* 27.

39. Devereux, Steele, and Kubrin, *Earthmind,* 107.

40. Ashe, *The Ancient Wisdom,* 82.

41. Huxley, *Doors of Perception,* 16–17.

42. Huxley, *Doors of Perception,* 16–17.

43. Hearne, *Visions of the Future,* 132; and personal communication, May 1998.

44. Concar, "I Feel Therefore I Am."

45. Devereux, *Earth Memory,* 298.

46. From the *Aitareya Brahmana, Aranyaka II.*

47. Heath, *Sun, Moon and Stonehenge.*

48. Lockyer, *The Dawn of Astronomy.*

49. Watson, *Supernature,* 69; and Jenny, "Visualising Sound."

50. Watson, *Supernature;* and Crystal, *The Cambridge Encyclopedia of Language,* 128–30.

CHAPTER 2. THE ACOUSTIC DIVINE

1. R. Graves, *The White Goddess;* and Campbell, *The Hero with a Thousand Faces.*

2. "UK Museum Gives Up the Ghost," BBC News, February 2, 1999. Accessed on BBC Online Network.

3. Robert Matthews, "Ultra-Low Sound Waves Blamed for Visions, Feelings of Terror," *Sunday Telegraph* (London), June 28, 1998; and Tandy and Lawrence, "The Ghost in the Machine." See also, Andreas Sommer, "Reasons Not to Scoff at Ghosts, Visions and Near-Death Experiences," Aeon (website), January 6, 2020.

4. David Fontana, *Visions of the Queen of Heaven,* quoted in the *Daily Mail,* July 1998.

5. David Elkington, "Secrets of Sound," *Amateur Astronomy and Earth Science* magazine, May 1996.

6. Aurobindo, *On Yoga,* vol. 2.

7. Satprem, *Sri Aurobindo or the Adventure of Consciousness,* 79.

8. Rossetti, "The Blessed Damosel," 1870.

9. Results cited in Cornwell, *Powers of Darkness, Powers of Light,* 49–51.

10. *Daily Telegraph* (London), July 1, 1999.

11. Gerber, *Vibrational Medicine;* and A. J. Scott-Morley, personal communication, July 1997.

12. Tompkins and Bird, *Secrets of the Soil,* 104.

13. Tompkins and Bird, *Secrets of the Soil,* 104.

14. Devereux, *Places of Power;* Michell, *The Earth Spirit;* and A. J. Scott-Morley, personal communication, July 1997.

15. Devereux, Steele, and Kubrin, *Earthmind,* 177.

16. Devereux, Steele, and Kubrin, *Earthmind,* 177.

17. Emoto, *The Miracle of Water.*

18. Becker and Selden, *The Body Electric,* 249.

19. A. J. Scott-Morley, personal communication, July 1997.

20. Becker and Selden, *The Body Electric,* 249.

21. Campbell, *The Masks of God: Oriental Mythology;* Hawking, *A Brief History of Time;* Seymour, *Astrology;* and Wilber, *Quantum Questions.*

22. Targ and Puthoff, *Mind Reach;* and Roney-Dougal, *Where Science and Magic Meet,* 43–44.

23. Cornwell, *Powers of Darkness, Powers of Light,* 17, 161.

24. Richard Harries, *Sunday Times* (London), March 22, 1998.

25. *Enuma Elish.*

26. *Flightpaths to the Gods,* BBC TV, August 21, 1997.

27. R. Graves, *Greek Myths;* Graves and Podro, *The Nazarene Gospel Restored;* McKenna, *Food of the Gods;* and Reddish, *Spirits of the Jaguar.*

28. Heath, *Sun, Moon and Stonehenge;* and Thom, *Megalithic Sites in Britain.*

29. Pennick, *The Secret Lore of Runes and Other Ancient Alphabets;* Wallis-Budge, *Gods of the Egyptians;* and Spawforth, *The Oxford Classical Dictionary.*

30. Walker, *Woman's Encyclopedia of Myths and Legends,* 603.

31. R. Graves, *Greek Myths;* and Walker, *Woman's Encyclopedia of Myths and Legends,* 201–2.

32. Farmer, *The Oxford Dictionary of Saints;* Cross and Livingstone, *The Oxford Dictionary of the Christian Church;* and R. Graves, *The White Goddess,* 175n1.

33. Frazer, *The Golden Bough.*

34. Scholes, *The Oxford Companion to Music,* 949.

35. George, *Encyclopedia of Heresies and Heretics,* 218.

36. Seznec, *The Survival of the Pagan Gods,* 149.

37. Charpentier, *Mysteries of Chartres Cathedral;* Anderson, *The Rise of the Gothic;* and Fletcher, *A History of Architecture.*

38. Anderson, *The Rise of the Gothic,* 10.

39. Anderson, *The Rise of the Gothic,* 12.

40. Panofsky, *Abbot Suger on the Abbey Church,* 63.

41. Barrow, *The Artful Universe.*

42. Merz, *Points of Cosmic Energy,* 105.

43. Charpentier, *Mysteries of Chartres Cathedral,* 86.

44. Anderson, *The Rise of the Gothic,* 156.

45. Whone, *Church, Monastery, Cathedral;* West, *The Traveler's Key to Ancient Egypt;* and Schwaller de Lubitz, *Sacred Science.*

46. Cited in Blair, *Rhythms of Vision,* 136.

47. Blair, *Rhythms of Vision,* color insert.

48. Wulfstan of Winchester, *De Organis,* translated by Jamie Leader, Pembroke College, Cambridge University, private letter to the author, May 1998 (my emphasis).

49. Wulfstan of Winchester, *De Organis,* translated by Jamie Leader. Pembroke College, Cambridge University, private letter to the author, May 1998.

50. Owen Barfield, quoted in Carpenter, *The Inklings,* 36–39.

51. Damasio, "How the Brain Creates the Mind."

52. John 1:1.

CHAPTER 3. THE POWER AND THE PLACE

1. Saint Francis of Assisi, "Canticle of the Sun."

2. Hunt, *Infinite Mind,* 26, 32.

3. Hunt, *Infinite Mind;* A. J. Scott-Morley, personal communication, July 1997.

4. Hunt, *Infinite Mind,* 33.

5. U.S. Department of the Interior, *Field Studies of Radon in Rocks, Soils, and Water,* 39–50; Roney-Dougal, *Where Science and Magic Meet;* and Michell, *The Earth Spirit.*

6. Bigu, Hussein, and Hussein, "Radiation Measurements in Egyptian Pyramids and Tombs," 245–52.

7. Watson, *Supernature;* Roney-Dougal, *Where Science and Magic Meet.*

8. See, for example, Charpentier, *Mysteries of Chartres Cathedral;* Devereux, *Earth Memory;* Pennick, *Mysteries of Kings College Chapel;* and Watson, *Supernature.*

9. Roney-Dougal, *Where Science and Magic Meet,* 163.

10. Lynch, *Prehistoric Religion and Ritual,* 64.

11. DeMeo, *The Orgone Accumulator Handbook.*

12. "Man Made Diamonds—A Cut above Nature." *Daily Telegraph* (London), August 21, 1999.

13. R. Graves, *The White Goddess,* 33.

14. A. N. Wilson, *Daily Telegraph* (London), December 27, 1998; and Wilson, *Jesus.*

15. Robert Hardy, *Daily Telegraph* (London), November 20, 1998.

16. Grinsell, "Folklore of Prehistoric Sites in Britain," 23ff.

17. Bede, 86ff.

18. Christopher Somerville, *Daily Telegraph* (London), July 26, 1997.

19. Underwood, *The Pattern of the Past,* 161.

20. John Michell, personal communication, 1999.

21. Frayling, *Strange Landscape,* 57.

22. Hitching, *Earth Magic;* and Gerber, *Vibrational Medicine.*

23. T. Graves, *Needles of Stone Revisited,* 34–35.

24. Roney-Dougal, *Where Science and Magic Meet;* Smith and Best, *Electromagnetic Man;* and Cowan and Silk, *Ancient Energies of the Earth.*

25. Roney-Dougal, *Where Science and Magic Meet,* 147.

26. T. Graves, *Needles of Stone Revisited,* 34–35.

27. T. Graves, *Needles of Stone Revisited,* 34–35.

28. Grinsell, "Folklore of Prehistoric Sites in Britain," 23; and Hitching, *Earth Magic.*

29. T. Graves, *Needles of Stone Revisited,* 37.

30. David Whitehouse, "Prehistoric Moon Map Unearthed," BBC News Online, April 22, 1999.

31. Anderson, *The Rise of the Gothic,* 156.

32. von Pohl, *Earth Currents.*

33. G. Schneck, *British Society of Dowsers Journal,* 1994–1995.

34. Lawton and Herald, *Giza: The Truth,* 20.

35. Watson, *Supernature,* 101–2.

36. Smith and Best, *Electromagnetic Man,* 121–22.

37. Johnson, *The Ancient Magic of the Pyramids,* 45.

38. Oestrander and Schroeder, *PSI: Psychic Discoveries,* 362.

39. Morrell, *3rd Stone,* Summer 1999.

40. Faulkner, *The Ancient Pyramid Texts.*

41. Devereux, Steele, and Kubrin, *Earthmind,* 115.

42. Quoted in Trevelyan, *A Vision of the Aquarian Age,* 7.

43. Michell, "Interpretations," 16.

CHAPTER 4.
ENIGMA AT THE HEART OF MYSTERY

1. Herodotus (trans. Sélincourt), *Histories,* 21.

2. Freke and Gandy, *The Jesus Mysteries;* and Wallace-Hadrill, *Augustan Rome.*

3. Frazer, *The Golden Bough,* 347.

4. R. Graves, *King Jesus,* 11–12.

5. Eisler, *Orpheus the Fisher.*

6. Campbell, *The Masks of God: Occidental Mythology,* 350.

7. Freke and Gandy, *The Jesus Mysteries,* 42.

8. Freke and Gandy, *The Jesus Mysteries,* 42; and Apuleius, *The Golden Ass.*

9. Freke and Gandy, *The Jesus Mysteries,* 42.

10. Robertson, *Pagan Christs,* 97.

11. Robertson, *Pagan Christs,* 97.

12. Freke and Gandy, *The Jesus Mysteries,* 41.

13. R. Graves, *Greek Myths* and *King Jesus;* Walker, *Woman's Encyclopedia of Myths and Legends;* and Kerenyi, *Dionysos.*

14. Quoted in Godwin, *Mystery Religions in the Ancient World,* 28.

15. De Santillana and von Dechend, *Hamlet's Mill,* 33.

16. R. Graves, *Greek Myths,* 103; and Euripides, *The Bacchae.*

17. Bayley, *The Lost Language of Symbolism* 2:25–27.

18. Plutarch, *Isis and Osiris.*

19. De Santillana and von Dechend, *Hamlet's Mill,* 33.

20. Homer, *The Illiad.*

21. Wallis-Budge, *Gods of the Egyptians.*

22. De Santillana and von Dechend, *Hamlet's Mill,* 33.

23. De Santillana and von Dechend, *Hamlet's Mill,* 33.

24. Kerenyi, *Dionysos,* 251.

25. Quoted in Lemesurier, *The Great Pyramid Decoded,* 272.

26. Frazer, *The Golden Bough.*

27. Belostockaya, *Obashhenie so Slushatelyami.*

28. Birks and Gilbert, *The Treasure of Montsegur,* 113.

29. Birks and Gilbert, *The Treasure of Montsegur,* 113.

30. Augustine of Hippo, *Retractions,* 1:13.3.

31. Quoted in Freke and Gandy, *The Jesus Mysteries,* 41.

32. Quoted in Freke and Gandy, *The Jesus Mysteries,* 29.

33. Interview of John Strugnell, *Haaretz,* November 9, 1990; and Baigent and Leigh, *The Dead Sea Scrolls Deception,* 95–96.

34. Bernal, *Black Athena,* vols. 1–2; see also Baldick, *Black God.*

35. Eliade, *Encyclopedia of Religion* 12:331–33.

36. A. N. Wilson, *Jesus,* 239.

37. Massey, *The Logia of the Lord,* 1.

38. Massey, *The Logia of the Lord,* 1.

39. Quoted in Freke and Gandy, *The Jesus Mysteries,* 63.

40. Freke and Gandy, *The Jesus Mysteries,* 63.

41. R. Graves, *Greek Myths,* vol. 2, book 30.

42. Merry, *The Flaming Door,* 152.

43. Merry, *The Flaming Door,* 152.

44. *Brewer's Dictionary of Phrase and Fable,* s.v. "Hogmanay," 523.

45. See Bernal, *Black Athena* 2:106.

46. De Santillana and von Dechend, *Hamlet's Mill,* 114ff.

47. Massey, *The Logia of the Lord,* 5.

48. Massey, *The Logia of the Lord,* 11, and *The Historical Jesus and the Mythical Christ,* 156ff. (My emphasis.)

49. Frazer, *The Golden Bough;* Walker, *Woman's Encyclopedia of Myths and Legends,* 10–11; and Freke and Gandy, *The Jesus Mysteries,* 33.

50. Freke and Gandy, *The Jesus Mysteries,* 33; and Frazer, *The Golden Bough,* 347.

51. R. Graves, *The White Goddess,* 175, 327n.

52. R. Graves, *The White Goddess,* 175, 327n.

53. Angus, *Mystery Religions,* 136.

54. Walker, *Woman's Encyclopedia of Myths and Legends,* 78.

55. Shlain, *The Alphabet versus the Goddess,* 83.

56. Merry, *The Flaming Door,* 143; R. Graves, *The White Goddess;* and Morgan, *The Mabin of the Mabinogion,* 70.

57. Chadwick, *The Celts,* 170.

58. R. Graves, *The White Goddess;* and Walker, *Woman's Encyclopedia of Myths and Legends,* 602ff.

59. As compiled and translated by Lady Charlotte Guest between 1838 and 1845.

60. Cited in R. Graves, *The White Goddess,* 95.

61. Robin Heath, personal communication, June 1995.

62. De Santillana and von Dechend, *Hamlet's Mill,* 119–20.

63. Grimal, *The Churches of Rome.*

64. Dr. Philip Blair, personal communication, April 1995.

65. Salibi, *Who Was Jesus?*

66. Bailey, *The Caves of the Sun,* 35.

67. Ouaknin, *Mysteries of the Alphabet,* 178.

68. Ouaknin, *Mysteries of the Alphabet;* and Davenas, Beauvais, Amara, et al. "Human Basophil Degranulation," 816–18.

69. Ouaknin, *Mysteries of the Alphabet,* 200.

70. Christine and Abdul el Mahdy, personal communication, September 2000.

71. Bernal, *Black Athena.*

72. Lehner, *The Complete Pyramids,* 84.

73. De Santillana and von Dechend, *Hamlet's Mill,* 6–7.

74. Shlain, *The Alphabet versus the Goddess,* 258.

75. Mead, *Thrice Greatest Hermes* 1:286n8; and Plutarch, *Isis and Osiris.*

76. Mead, *Thrice Greatest Hermes* 1:286n8.

77. H. Sayce, the Hibbert lectures on the Religion of Ancient Babylonians, 1898.

78. Salibi, *Who Was Jesus?,* 51–52.

79. Graves and Podro, *The Nazarene Gospel Restored,* foreword, xi.

80. Graves and Podro, *The Nazarene Gospel Restored,* introduction.

81. Eusebius, *The History of the Church,* chapter 22; Eisenman, *James the Brother of Jesus,* xxxvi, 779–80.

82. Clement of Alexandria, *Stromateis,* book 1, chapter XV, page 70.

83. Epiphanius, *Panarion;* also see Cross and Livingstone, *The Oxford Dictionary of the Christian Church,* 553 for a breakdown of Epiphanius's life.

84. Quoted in Mead, *Thrice Greatest Hermes* 1:270n3.

85. Wallis-Budge, *Gods of the Egyptians;* and Collier and Manley, *How to Read Egyptian Hieroglyphs.*

86. See Ouaknin, *Mysteries of the Alphabet,* 252, regarding the letter *N.*

87. Clement of Alexandria, *Stromateis,* books 1–3.

88. McLeish, *Myth.*

89. R. Graves, *King Jesus,* chapter 1.

90. *Book of Jasher* 68:1–32.

91. Farmer, *The Oxford Dictionary of Saints,* 337.

92. Bible, in the books of Joshua and Ruth.

93. Salibi, *Who Was Jesus?,* 57ff.

94. Wallis-Budge, *The Mummy,* 2.

95. De Santillana and von Dechend, *Hamlet's Mill,* 221.

96. R. Graves, *The White Goddess,* 95.

97. Monier-Williams, *A Sanskrit-English Dictionary,* s.v. Murti.

98. Walker, *Woman's Encyclopedia of Myths and Legends,* 453; and Wallis-Budge, *Gods of the Egyptians.*

99. Edwards, *The Pyramids of Egypt,* 279ff.

100. Margaret Barker, personal communication, May 2010.

101. Harriet Pavey, "British scholar: 'Don't take the bible literally,'" *Evening Standard* (London), August, 23, 2017, quoting Hugh Houghton.

102. Davies et al., *The Case for the Jordan Lead Codices;* and Solignac, "Divine Revelation."

103. Hooke, *Myth, Ritual and Kingship.*

104. David Keys, *The Independent* (London), November 19, 1996.

CHAPTER 5. SALVATION AND THE SOUL

1. Weil, *Intimations of Christianity among the Ancient Greeks,* 27.

2. Osman, *In the House of the Messiah,* 171–72. (My emphasis.) See also Carter and Mace, *The Tomb of Tutankhamun.*

3. Osman, *In the House of the Messiah,* 171–72.

4. Faulkner, *The Ancient Egyptian Pyramid Texts.*

5. Wallis-Budge, *Gods of the Egyptians* 2:191.

6. Yates, *Giordano Bruno and the Hermetic Tradition,* 115ff.

7. Sattin, *The Pharaoh's Shadow,* 171.

8. Bernal, *Black Athena* 1:3.

9. Michael Wood, *Hitler's Search for the Holy Grail,* C4 TV, "Secret History" season 7, episode 3, aired August 19, 1999, director Kevin Sim; and Pennick, *Hitler's Secret Sciences.*

10. Baldick, *Black God.*

11. Luckert, *Egyptian Light and Hebrew Fire,* 241.

12. Luckert, *Egyptian Light and Hebrew Fire,* 241.

13. Bernal, *Black Athena* 1:26.

14. Bernal, *Black Athena* 1:26.

15. Dupuis, *Was Christ a Person or the Sun?,* 8.

16. Dupuis, *Was Christ a Person or the Sun?,* 8.

17. Letter of Montmorency-Laval, June 22, 1825, quoted in Bernal, *Black Athena* 1:252.

18. Bernal, *Black Athena* 1:184.

19. Ferrier, *Herald of the Cross,* 89.

20. Thory, *Acta Latomorum.*

21. Temple, *The Sirius Mystery,* 193.

22. Wallis-Budge, *Gods of the Egyptians.*

23. Lurker, *Gods and Goddesses,* 345.

24. Frankfort, *Kingship and the Gods,* 318.

25. Faulkner, *The Ancient Egyptian Pyramid Texts,* 247, utterance no. 600, "A prayer for the King and his pyramids."

26. Michell, *The Earth Spirit,* 29.

27. Frankfort, *Kingship and the Gods,* 318.

28. Wallis-Budge, *Legends of Our Lady.*

29. H. Wedeck, quoted in Walker, *Women's Encyclopedia of Myths and Secrets,* 614.

30. Scott, *Hermetica* 4:262.

31. Scott, *Hermetica* 4:261.

32. Newton, quoted in Devereux, Steele, and Kubrin, *Earthmind,* 55.

33. A. J. Scott-Morley, personal communication, July 1997.

34. Gerber, *Vibrational Medicine,* 130.

35. Gerber, *Vibrational Medicine,* 131.

36. Schwaller de Lubicz, *The Temple in Man,* 17.

37. Goldman, *Healing Sounds,* 17.

38. Tame, *The Secret Power of Music,* 23.

39. Tame, *The Secret Power of Music,* 24.

40. Tame, *The Secret Power of Music,* 24.

41. Cohen, "Echo of the Past," 7.

42. Quoted in "Birds Trigger Songs That Enlighten," *Times* (London), April 21, 1999.

43. Meyer, *The Ancient Mysteries,* 216.

44. Cohen, "Echo of the Past," 7.

45. Goldman, *Healing Sounds,* 42–43.

46. Goldman, *Healing Sounds,* 42–43.

47. Goldman, *Healing Sounds,* 42–43, quoting William Grey.

CHAPTER 6. IN THE REALM OF THE MER

1. Wallis-Budge, *The Mummy,* 2.

2. Gardiner, *Egyptian Grammar,* 495.

3. Wallis-Budge, *Gods of the Egyptians* 2:202.

4. Freke and Gandy, *The Jesus Mysteries,* 262n35.

5. Luckert, *Egyptian Light and Hebrew Fire,* 42–43.

6. De Santillana and von Dechend, *Hamlet's Mill,* 62–63.

7. De Santillana and von Dechend, *Hamlet's Mill,* 149.

8. H. Wilson, *Understanding Hieroglyphs*, 1.

9. *Daily Telegraph* (London), August 8, 2000.

10. Hooke, *Middle Eastern Mythology*, 73.

11. *Corpus Hermeticum*, chap. 16, 2. In Scott, *Hermetica*.

12. *Daily Telegraph*, January 26, 1999.

13. A. J.Scott-Morley, personal communication, May 1998.

14. De Santillana and von Dechend, *Hamlet's Mill*, 57.

15. Tompkins, *The Secret Life of Nature*, 153.

16. Tompkins, *The Secret Life of Nature*, 155, quoting Stephen M. Phillips.

17. Gary Stix, "Beam It Up," *Scientific American* 277, no. 3 (September 1997): 40–41.

18. Wallis-Budge, *Gods of the Egyptians* 2:114.

19. Cirlot, *A Dictionary of Symbols*, 341.

20. Hart, *A Dictionary of Egyptian Gods and Goddesses*, 151.

21. Emery, *Archaic Egypt*, 122.

22. Freke and Gandy, *The Jesus Mysteries*, 33.

23. Bernal, *Black Athena* 1:70ff.

24. Graves, *Greek Myths*, 196.

25. Faulkner, *The Ancient Egyptian Pyramid Texts*.

26. Wallis-Budge, *Book of the Dead*.

27. Wallis-Budge, *Gods of the Egyptians* 2:126, 2:141.

28. Freke and Gandy, *The Jesus Mysteries*, 32; and Murray, *Egyptian Religious Poetry*, 68.

29. Wallis-Budge, *Legends of Our Lady Mary*.

30. Stewart, *The Foreigner*, 77.

31. Churchward, *Signs and Symbols of Primordial Man*, 236.

32. Gardiner, *Egyptian Grammar*, 587.

CHAPTER 7. EPIPHANY

1. Godwin, *Mystery Religions in the Ancient World*, 28.

2. Schwaller de Lubicz, *The Temple in Man*; and West, *Serpent in the Sky*.

3. Teilhard de Chardin, *The Phenomenon of Man*, 218.

4. Brooke, *Towards Human Unity*.

5. Ouaknin, *Mysteries of the Alphabet*, 266ff.

6. Ouaknin, *Mysteries of the Alphabet*, 266ff.

7. Hall, *Man*, 209.

8. Hall, *Man,* 209.

9. *Shepherd of Hermas,* ninth similitude.

10. See Bauval and Gilbert, *The Orion Mystery.*

11. Lawton and Herald, *Giza: The Truth.*

12. Churchward, *Signs and Symbols of Primordial Man,* 381. Horus is Hu—in the eightieth chapter of the Ritual (the Pyramid Text), "I have seized upon Hu."

13. Bayley, *The Lost Language of Symbolism* 2:201.

14. Liddell and Scott, *Greek-English Lexicon,* 1006–1251.

15. See Gardiner, *Egyptian Grammar,* 489, sign N27.

16. Baines and Malek, *Atlas of Ancient Egypt,* 140–41.

17. Quoted in Scott, *Hermetica* 4:256.

18. Quoted in Scott, *Hermetica* 4:255–56.

19. Wallis-Budge, *The Egyptian Heaven and Hell,* 131.

20. For a detailed discussion of this see Lehner, *The Complete Pyramids;* Lawton and Herald, *Giza: The Truth;* and Alford, *The Phoenix Solution.*

21. See Hodges, *How the Pyramids Were Built.*

22. Tompkins, *Secrets of the Great Pyramid,* 278–79.

23. Tompkins, *Secrets of the Great Pyramid,* 278–79.

CHAPTER 8. COMMUNION

1. Josephus, *Antiquities of the Jews,* book 16, chapter 7.

2. Gardiner, "The Secret Chambers of the Sanctuary of Thoth," 2.

3. Douglas-Klotz, *Prayers of the Cosmos.*

4. Reid, *Egyptian Sonics.*

5. Reid, *Egyptian Sonics.*

6. Danley, "The Great Pyramid."

7. Danley, "The Great Pyramid."

8. Gardner, *Sounding the Inner Landscape,* 91.

9. Zohar and Marshall, *SQ, Spiritual Intelligence,* 87.

10. Godwin, *Mystery Religions in the Ancient World,* 28.

CHAPTER 9. ACOUSTIC CONSCIOUSNESS

1. Faulkner, *The Ancient Egyptian Pyramid Texts,* utterance no. 600.

2. Manniche, *Music and Musicians in Ancient Egypt,* 60.

3. Epiphanius, *Panarion*, 94.

4. LePage, *Mysteries of the Bridechamber*, 19.

5. Reid, *Egyptian Sonics*.

6. Ruoff, *The Century Book of Facts*, 234.

7. Wallis-Budge, *Egyptian Religion*, 71.

8. Faulkner, *The Ancient Egyptian Coffin Texts* 3:147, Spell 1080.

9. See also De Santillana and von Dechend, *Hamlet's Mill*, 221.

APPENDIX. THE ANATOMY OF THE HERO

1. *Codex Borgianus;* see Díaz and Rodgers, *The Codex Borgia*.

2. De Santillana and von Dechend, *Hamlet's Mill*, 78.

3. Walker, *Woman's Encyclopedia of Myths and Legends*, 1020.

4. Walker, *Woman's Encyclopedia of Myths and Legends*, 336.

Bibliography

Adams Leeming, David, and Margaret Adams Leeming. *Encyclopedia of Creation Myths*. Oxford, UK: ABC-Clio, 1994.

Aesop. *The Complete Fables*. Translated by Robert and Olivia Temple. London: Penguin, 1998.

Aghion, Irene, Claire Barbillon, and François Lissarrague. *Gods and Heroes of Classical Antiquity*. Paris: Flammarion, 1996.

Alford, Alan. *The Phoenix Solution*. London: Hodder and Stoughton, 1998.

Anderson, William. *The Rise of the Gothic*. London: Hutchinson, 1985.

Andreu, Guillemette. *Egypt in the Age of the Pyramids*. London: John Murray, 1997.

Angus. Samuel. *Mystery Religions*. Kent, UK: Dover Publications, 2003.

Apuleius, Lucius. *The Golden Ass*. Translated by Robert Graves. London: Penguin, 1990.

Ash, David, and Peter Hewitt. *The Science of the Gods*. Revised and retitled as *The Vortex* in 1995. Bath: Gateway, 1990.

Ashe, Geoffrey. *The Ancient Wisdom*. London: Macmillan, 1977.

Augustine of Hippo. *Retractions*. In *Fathers of the Church: Saint Augustine: The Retractions*. Washington, D.C.: Catholic University Press of America, 2002.

Aurobindo, Sri. *On Yoga*. Vol. 2. Pondicherry, India: Sri Aurobindo Ashram Trust, 1995.

Aveni, Anthony. *Stairways to the Stars*. Oxford, UK: Oxford University Press, 1998.

Baigent, Michael, and Richard Leigh. *The Dead Sea Scrolls Deception*. London: Jonathan Cape, 1991.

Bailey, Adrian. *The Caves of the Sun*. London: Jonathan Cape, 1997.

Baines, John, and Jaromir Malek. *Atlas of Ancient Egypt*. London: Phaidon Press, 1989.

Baldick, Julian. *Black God: The Afro-Asiatic Roots of the Jewish, Christian and Muslim Religions*. London: I. B. Tauris, 1997.

Baring, Anne, and Jules Cashford. *The Myth of the Goddess*. London: Arkana, 1993.

Barker, Margaret, *On Earth as It Is in Heaven*. Edinburgh: T & T Clark, 1995.
———. *The King of the Jews*. London: SPCK, 2014.

Barr, James. *The Variant Spellings of the Hebrew Bible*. Oxford, UK: British Academy/Oxford University Press, 1989.

Barrow, John D. *The Artful Universe*. Oxford, UK: Oxford University Press, 1995.

Bauval, Robert, and Adrian Gilbert. *The Orion Mystery*. London: Heinemann, 1994.

Bayley, Harold. *The Lost Language of Symbolism*. London: Williams and Norgate, 1912.

Becker, Robert O., and Gary Selden. *The Body Electric*. New York: Morrow, 1985.

Becker, Udo. *The Continuum Encyclopedia of Symbols*. London: Continuum, 1994.

Bede. *Ecclesiastical History of the English People*. Edited by D. Farmer. London: Penguin, 1990.

Belostockaya, N. G. *Obashhenie so Slushatelyami*. Moscow, 1984.

Bernal, Martin. *Black Athena*. Vols. 1–2. New Brunswick, N.J.: Rutgers University Press, 1991.

Betz, Otto, and Rainer Riesner. *Jesus, Qumran and the Vatican*. London: SCM Press, 1994.

Bienkowski, Piotr, and Alan Millard. *Dictionary of the Ancient Near East*. London: British Museum Press, 2000.

Bigu, Jaime, Mohamed I. Hussein, and A. Z. Hussein. "Radiation Measurements in Egyptian Pyramids and Tombs—Occupational Exposure of Workers and the Public." *Journal of Environmental Radioactivity* 47, no. 3 (2000): 245–52.

Birks, Walter, and R. A. Gilbert. *The Treasure of Montsegur*. Wellingborough, UK: Aquarian Press, 1987.

Black, Jeremy, and Anthony Green. *Gods, Demons and Symbols of Ancient Mesopotamia*. London: British Museum Press, 1992.

Blair, Lawrence. *Rhythms of Vision*. London: Paladin, 1976.

Book of Jasher. Reprint with foreword by Hilton Hotema. Hastings, UK: Society of Metaphysicians, 1965. Originally published by M. M. Noah & A. S. Gould in 1840.

Bowker, John, ed. *The Oxford Dictionary of World Religions*. Oxford, UK: Oxford University Press, 1997.

Brewer, Ebenezer Cobham. *Brewer's Dictionary of Phrase and Fable*. Edited by Andrian Room. London: Cassell, 1999.

Broadhurst, Paul. *Secret Shrines*. Cornwall, UK: Pendragon Press, 1991.

Brook, Elaine. *In Search of Shambhala*. London: Jonathan Cape, 1996.

Brooke, Anthony. *Towards Human Unity*. London: Mitre Press, 1976.

Brown, Guy. *The Energy of Life*. London: HarperCollins, 1999.

Brown, Mary-Ellen, and Rosenberg, Bruce A., eds. *Encyclopedia of Folklore and Literature*. ABC-Clio, Oxford, 1998.

Burl, Aubrey. *Circles of Stone*. London: Harvill Press, 1999.

———. *Great Stone Circles*. London: Yale University Press, 1999.

Cade, Maxwell, and Nona Coxhead. *The Awakened Mind*. Shaftesbury, UK: Element, 1987.

Campbell, Joseph. *The Hero with a Thousand Faces*. London: Fontana, 1993.

Campbell, Joseph. *The Masks of God*. Vol. 1, *Primitive Mythology*. London: Arkana, 1991.

———. *The Masks of God*. Vol. 2, *Oriental Mythology*. London: Arkana, 1991.

———. *The Masks of God*. Vol. 3, *Occidental Mythology*. London: Arkana, 1991.

———. *The Masks of God*. Vol. 4, *Creative Mythology*. London: Arkana, 1991.

Carpenter, Edward. *The Origins of Pagan and Christian Beliefs*. London: Senate Books, 1996.

Carpenter, Humphrey. *The Inklings*. London: HarperCollins, 2006.

Carter, Howard, and A. C. Mace. *The Tomb of Tutankhamun*. Bath, UK: Duckworth, 2000.

Chadwick, Nora. *The Celts*. London: Penguin, 1997.

Charpentier, Louis. *The Mysteries of Chartres Cathedral*. London: RILKO, 1972.

Chatwin, Bruce. *The Songlines*. London: Picador, 1987.

Churchward, Albert. *Signs and Symbols of Primordial Man*. Hastings, UK: Society of Metaphysicians, 1903.

Cirlot, J. E. *A Dictionary of Symbols*. Oxon, UK: Routledge, 1996.

Clement of Alexandria. *Stromateis*. Books 1–3. Translated by John Ferguson.

Fathers of the Church series. Washington, D.C.: Catholic University of America Press, 1991.

Cohen, Phillip. "Echo of the Past." *New Scientist,* October 10, 1998.

Collier, Mark, and Bill Manley. *How to Read Egyptian Heiroglyphs.* London: British Museum Press, 1998.

Concar, David. "I Feel Therefore I Am." *New Scientist,* March 11, 2000.

Cornwell, John. *Powers of Darkness, Powers of Light.* London: Viking, 1991.

Cousens, Gabriel. *Spiritual Nutrition and the Rainbow Diet.* Los Angeles: Cassandra Press, 1986.

Cowan, David, and Anne Silk. *Ancient Energies of the Earth.* London: Thorsons, 1999.

Cross, Frank L., and Elizabeth A. Livingstone. *The Oxford Dictionary of the Christian Church.* Oxford, UK: Oxford University Press, 1997.

Crystal, David, ed. *The Cambridge Encyclopedia of Language.* Cambridge, UK: Cambridge University Press, 1997.

Damasio, Antonio R. "How the Brain Creates the Mind." *Scientific American* 281, no. 6, (December 1999): 112–17.

Danielou, Alain. *The Gods of India, Hindu Polytheism.* Rochester, Vt.: Inner Traditions, 1985.

Danley, Tom. "The Great Pyramid: Early Reflections and Ancient Echoes." *Live Sound International* magazine, June/July, 2000.

Davenas, E., F. Beauvais, J. Amara et al. "Human Basophil Degranulation Triggered by Very Dilute Antiserum against IgE." *Nature* 333, no. 6176 (1988): 816–18.

Davidson, H. R. E. *Symbols of Power.* Woodbridge, Suffolk, UK: D. S. Brewer, 1977.

Davies, Phillip, Margaret Barker, David Elkington, Jennifer Elkington, and Keith Hearne. *The Case for the Jordan Lead Codices.* London: Watkins, 2014.

De Blois, Lukas, and Robartus van der Spek. *An Introduction to the Ancient World.* London: Routledge, 1997.

DeMeo, James. *The Orgone Accumulator Handbook.* Ashland, Ore.: Natural Energy Works, 1989.

De Santillana, Giorgio, and Hertha von Dechend. *Hamlet's Mill.* Boston: Godine, 1977.

Devereux, Paul. *Earth Lights Revelation.* New York: Sterling, 1989.

———. *Earth Memory.* London: Quantum, 1999.

———. *Places of Power*. London: Cassell, 1999.

———. *Secrets of Ancient and Sacred Places*. London: Cassell, 1995.

Devereux, Paul, John Steel, and David Kubrin. *Earthmind: Communicating with the Living World of Gaia*. Rochester, Vt.: Destiny Books, 1989.

The Dhammapada. Translated by Anne Bancroft. Shaftesbury, UK: Element, 1997.

Díaz, Gisele, and Alan Rodgers. *The Codex Borgia: A Full-Color Restoration of the Ancient Mexican Manuscript*. New York: Dover, 1993.

Douglas-Klotz, Neil, trans. *Prayers of the Cosmos: Meditations on the Aramaic Words of Jesus*. New York: HarperCollins, 1994.

Downer, John. *Supernatural: The Unseen Powers of Animals*. London: BBC Books, 1999.

Driver, Godfrey R. *Canaanite Myths and Legends*. Edinburgh: T & T Clark, 1956.

Drury, Nevill. *Shamanism*. Shaftesbury, UK: Element, 1996.

Dupuis, Charles F. *Memoire Explicatif du Zodiac*. Paris: Chez Courcier, 1806.

———. *Was Christ a Person or the Sun?* London: Holyoake, 1857.

Durrani, Saeed, and Ilic Radomir, eds. *Radon Measurements by Etched Track Detectors*. Singapore: World Scientific, 1997.

Dziemidko, Helen E. *Energy Medicine*. London: Gaia, 1999.

Edwards, I. E. S. *The Pyramids of Egypt*. London: Penguin, 1970.

Eisenman, Robert H. *James the Brother of Jesus*. London: Penguin, 1997.

Eisler, Robert. *Orpheus the Fisher*. London: Watkins, 1920.

Eliade, Mircea. *Encyclopedia of Religion*. London: MacMillan, 1986.

Emery, Walter B. *Archaic Egypt*. London: Penguin Books, 1991. Originally published by Penguin Books in 1961.

Emoto, Masaru. *The Miracle of Water*. New York: Atria Books, 2011.

Epiphanius. *The Panarion of Epiphanius of Salamis*. Translated by Frank Williams. Atlanta, Ga.: SBL Press, 2016.

Erdoes, Richard, and Alfonso Ortiz. *American Indian Myths and Legends*. London: Pimlico, 1997. Originally published by Pantheon Books in 1984.

Eusebius. *The History of the Church*. Edited by G. A. Williamson. London: Penguin, 1989.

Farmer, David. *The Oxford Dictionary of Saints*. Oxford, UK: Oxford University Press, 1997.

Faulkner, Raymond O. *The Ancient Egyptian Pyramid Texts*. Warminster, UK: Aris and Phillips, 1969..

————. *The Ancient Egyptian Coffin Texts.* Vol. 3. Warminster, UK: Aris and Phillips, 2004

Ferrier, John Todd. *Herald of the Cross.* Vol. 27. London: Order of the Cross, 1967.

Fletcher, Sir Bannister. *A History of Architecture.* London: Architectural Press, 1987.

Frankfort, Henri. *Kingship and the Gods.* Chicago: University of Chicago Press, 1978.

Frawley, David. *Gods, Sages and Kings.* New Delhi: Motilal Banarsidass, 1999.

Frayling, Christopher. *Strange Landscape: A Journey through the Middle Ages.* London: Penguin, 1996.

Frazer, James G. *The Golden Bough.* London: Batsford Press, 1996.

Freke, Timothy, and Peter Gandy. *The Complete Guide to World Mysticism.* London: Piatkus, 1997.

————. *The Jesus Mysteries.* London: Thorsons, 1999.

Gardiner, Sir Alan. "The Secret Chambers of the Sanctuary of Thoth." *Journal of Egyptian Archaeology* 11, no.1 (1925): 2–5.

————. *Egyptian Grammar.* Oxford, UK: Griffith Institute, 1957.

Gardner, Kay. *Sounding the Inner Landscape.* London: Thorsons, 1997.

Gaster, Theodor H. "'Ba'al Is Risen . . .': An Ancient Hebrew Passion-Play from Ras Shamra-Ugarit" *Iraq* 6, no. 2 (Autumn 1939): 109–43.

George, Leonard. *Encyclopedia of Heresies and Heretics.* London: Robson Books, 1995.

Gerber, Richard. *Vibrational Medicine.* Rochester, Vt.: Bear and Company, 1988.

Godwin, Joscelyn. *Mystery Religions in the Ancient World.* London: Thames and Hudson, 1981.

————. *The Mystery of the Seven Vowels in Theory and Practice.* Grand Rapids, Mich.: Phanes Press, 1991.

Goldman, Jonathan. *Healing Sounds.* Rochester, Vt.: Healing Arts Press, 2001.

Goodbody, G. B. H. *An Attestation to the Primary Resonances in the Great Pyramid of Khufu.* Vienna: Gebruder Spectre, 1921.

Grabsky, Phil. *The Lost Temple of Java.* London: Orion, 1999.

Graves, Robert. *The White Goddess.* London: Faber and Faber, 1961.

————. *King Jesus.* London: Hutchinson, 1983.

————. *Greek Myths.* London: Penguin, 1995.

Graves, Robert, and Joshua Podro. *The Nazarene Gospel Restored.* London: Cassell, 1955.

Graves, Tom. *Needles of Stone Revisited.* Glastonbury, UK: Gothic Image, 1986.

Green, Celia. *The Decline and Fall of Science.* London: Hamish Hamilton, 1976.

Grimal, Pierre. *The Churches of Rome.* London: Tauris Parke, 1997.

Grinsell, Leslie. "Folklore of Prehistoric Sites in Britain." *3rd Stone,* no. 32, (1998).

Grof, Stanislav. *Books of the Dead.* London: Thames and Hudson, 1994.

Hall, Manly P. *Man: The Grand Symbol of the Mysteries.* Whitefish, Mont.: Kessinger, 2010.

Hart, George. *A Dictionary of Egyptian Gods and Goddesses.* London: Routledge & Kegan Paul, 1986.

———. *Egyptian Myths.* London: British Museum Press, 1990.

Hawking, Stephen. *A Brief History of Time.* New York: Bantam 1998.

Hawkins, Gerald. *Stonehenge Decoded.* London: HarperCollins, 1982.

Hearne, Keith. *Visions of the Future.* London: Aquarian Press, 1989.

Heath, Robin. *Sun, Moon and Stonehenge: Proof of High Culture in Ancient Britain.* Cardigan, UK: Bluestone Press, 1995.

Herodotus. *The Histories.* Translated by Robin Waterfield. Oxford: Oxford University Press, 1998.

———. *The Histories.* Translated by Aubrey de Sélincourt. Revised with introduction and notes by John Marincola, London: Penguin, 2003.

Heselton, Philip. *The Elements of Earth Mysteries.* Shaftesbury, UK: Element, 1998.

Heyerdahl, Thor. *The Maldive Mystery.* London: George Allen, 1986.

Hitching, Francis. *Earth Magic.* London: Cassell, 1977.

Hodges, Peter. *How the Pyramids Were Built.* Shaftesbury, UK: Element Books, 1989.

Hooke, Samuel H. *The Origins of Early Semitic Ritual: The Schweich Lectures of the British Academy.* Oxford, UK: Oxford University Press, 1925.

———. *Myth, Ritual and Kingship.* Oxford, UK: Clarendon Press, 1958.

———. *Middle Eastern Mythology.* Harmondsworth, UK: Penguin, 1963.

Hornung, Erik. *Conceptions of God in Ancient Egypt: The One and the Many.* London: RKP, 1983.

Houston, Jean. *The Passion of Isis and Osiris.* New York: Ballantine/Wellspring, 1995.

Hunt, Valerie. *Infinite Mind.* Malibu, Calif.: Malibu Publishing, 1995.

Imel, Martha Ann, and Dorothy Myers Imel. *Goddesses in World Mythology.* Oxford, UK: ABC Clio, 1993.

Jahn, Robert G., Paul Devereux, and Michael Ibison. "Acoustical Resonances of Assorted Ancient Structures." *Journal of the Acoustical Society of America* 99, no. 2 (1996): 649–58.

James, Edwin O. *The Ancient Gods.* London: Weidenfeld and Nicholson, 1960.

James, Peter, and Nick Thorpe. *Ancient Mysteries.* New York: Ballantine, 1999.

Jaynes, Julian. *The Origins of Consciousness in the Breakdown of the Bicameral Mind.* London: Allen Lane, 1979.

Jeans, Sir James. *Science and Music.* Chatham, Kent: Dover Publications, 1998.

Jenny, Hans. "Visualising Sound." *Science* (June 1968).

———. *Cymatics.* 2 vols. Basel, Switzerland: Basilius Presse, 1974.

Johnson, Ken. *The Ancient Magic of the Pyramids.* London: Corgi, 1977.

Jones, Carleton. *Temples of Stone: Exploring the Megalithic Tombs of Ireland.* Cork, Ireland: Collins Press, 2007.

Jones, Prudence, and Nigel Pennick. *A History of Pagan Europe.* London: Routledge, 1995.

Jordan, Michael. *Plants of Mystery and Magic.* London: Blandford, 1997.

Josephus, Flavius. *Antiquities of the Jews.* Translated by H. St. J. Thackeray. Cambridge, Mass.: Loeb Classics, 1989.

Jung, Carl G. *Modern Man in Search of a Soul.* London: Ark, 1990.

Kerenyi, Carl. *Dionysos: Archetypal Image of an Indestructible Life.* Princeton, N.J.: Princeton University Press, 1976.

———. *The Gods of the Greeks.* London: Thames and Hudson, 1995.

———. *The Heroes of the Greeks.* London: Thames and Hudson, 1997.

Kharitidi, Olga. *Entering the Circle: Ancient Secrets of Siberian Wisdom.* London: Thorsons, 1996.

Knight, Vernon James, and Vincas P. Steponaitis. *Archaeology of the Moundville Chiefdom.* Washington, D.C.: Smithsonian Institution Press, 1998.

Lamie, Lucie. *Egyptian Mysteries.* London: Thames and Hudson, 1981.

Lang, Andrew. *Myth, Ritual and Religion.* 2 vols. London: Longmans, Green, 1906.

Laszlo, Ervin, Stanislav Grof, and Peter Russell. *The Consciousness Revolution.* Shaftesbury, UK: Element, 1999.

Lehner, Mark. *The Complete Pyramids.* London: Thames and Hudson, 1997.

Leick, Gwenolyn. *A Dictionary of Ancient Near Eastern Mythology.* London: Routledge, 1998.

Le Mee, Katherine. *Chant.* New York: Bell Tower, 1995.

Lemesurier, Peter. *The Great Pyramid Decoded*. Shaftesbury, UK: Element, 1996.

LePage, Victoria. *Mysteries of the Bridechamber*. Rochester, Vt: Inner Traditions, 2007.

Liddell, Henry, and Robert Scott. *Greek-English Lexicon: Revised Supplement*. Oxford, UK: Clarendon Press, 1996.

Lockyer, Sir Joseph Norman. *The Dawn of Astronomy*. London: Cassell, 1894.

Luckert, Karl. *Egyptian Light and Hebrew Fire*. Albany: State University of New York Press, 1991.

Lurker, Manfred. *Gods and Goddesses, Devils and Demons*. London: Routledge, 2004.

Lynch, Francis. *Prehistoric Religion and Ritual: Essays in Honour of Aubrey Burl*. Edited by Alex M. Gibson. Cheltenham, UK: Sutton, 1998.

MacCana, Proinsias. *Celtic Mythology*. London: Chancellor Press, 1996.

MacKenzie-Brown, C. *The Devi Gita*. New York: State University of New York Press, 1998.

Man, John. *Alpha, Beta*. London: Hodder Headline, 2000.

Mann, A. T. *Sacred Architecture*. Shaftesbury, UK: Element, 1993.

Manniche, Lise. *Music and Musicians in Ancient Egypt*. London: British Museum Press, 1991.

Massey, Gerald. *The Historical Jesus and the Mythical Christ*. Trenton, N.J.: A & B: 1997.

———. *The Logia of the Lord*. San Diego, Calif.: Book Tree, 2008.

Matthews, Caitlin. *Sophia, Goddess of Wisdom*. San Rafael, Calif.: Mandala, 1991.

McKenna, Terence. *Food of the Gods*. New York: Bantam, 1993.

McLeish, Kenneth. *Myth: Myths and Legends of the World Explored*. London: Bloomsbury, 1996.

Mead, G. R. S. *Thrice Greatest Hermes*. 3 vols. Wheaton, Ill.: Theosophical Publishing, 1906.

Merry, Eleanor C. *The Flaming Door*. East Grinstead, UK: New Knowledge Books, 1962.

Merz, Blanche. *Points of Cosmic Energy*. Saffron Walden, UK: C. W. Daniel, 1987.

Meyer, Marvin W. *The Ancient Mysteries*. Philadelphia: University of Pennsylvania Press, 1999.

Michell, John. *The Earth Spirit*. London: Thames and Hudson, 1975.

————. "Interpretations." *Cereologist* 7 (Winter 1993): 16.

Miller, Malcolm. *Chartres Cathedral*. Andover, UK: Pitkin, 1996.

Milner, George R. *The Cahokia Chiefdom*. Washington, D.C.: Smithsonian Institution Press, 1998.

Moier, Monier-Williams. *A Sanskrit-English Dictionary*. Oxford: Clarendon Press, 1963.

Morgan, Morien O. *The Mabin of the Mabinogion*. London: RILKO, 1984.

Morris, Craig, and Adriana von Hagen. *The Inka Empire*. New York: Abbeville Press, 1993.

Murray, Margaret A. *Egyptian Religious Poetry*. London: John Murray, 1949.

Narby, Jeremy. *The Cosmic Serpent: DNA and the Origins of Knowledge*. London: Gollancz, 1998.

Naydler, Jeremy. *Temple of the Cosmos*. Rochester, Vt.: Inner Traditions, 1996.

Oestrander, Sheila, and Lynn Schroeder. *PSI: Psychic Discoveries behind the Iron Curtain*. London: Abacus, 1973.

Osman, Ahmed. *In the House of the Messiah*. London: HarperCollins, 1992.

Ouaknin, Marc-Alain. *Mysteries of the Alphabet*. New York: Abbeville Press, 1999.

Ouspensky, Leonid. *The Theology of the Icon*. New York: St. Vladimir's Seminary Press, 1990.

Panati, Charles. *Sacred Origins of Profound Things*. London: Penguin, 1997.

Panofsky, Erwin, ed. *Abbot Suger on the Abbey Church of St. Denis and Its Art Treasures*. 2nd ed. Princeton, N.J.: Princeton University Press, 1979.

Patai, Raphael. *The Hebrew Goddess*. Detroit, Mich.: Wayne State University Press, 1990.

Peat, F. David. *Blackfoot Physics*. London: Fourth Estate, 1996.

Pennick, Nigel. *Hitler's Secret Sciences: His Quest for the Hidden Knowledge of the Ancients*. Suffolk, UK: Neville Spearman, 1981.

————. *The Secret Lore of Runes and Other Ancient Alphabets*. London: Rider, 1991.

————. *Mysteries of Kings College Chapel*. London: Aeon Books, 2012.

Playfair, Guy Lyon, and Scott Hill. *The Cycles of Heaven*. London: Souvenir Press, 1978.

Plutarch. *Isis and Osiris*. Translated by Frank Cole Babbitt. Vol 5 in *Moralia*. Loeb Classical Library 306. Cambridge, Mass.: Harvard University Press, 1936.

Prache, Anne. *Cathedrals of Europe*. New York: Cornell University Press, 1999.

Purce, Jill. *The Mystic Spiral.* London: Thames and Hudson, 1974.

Rauscher, Elizabeth. *The Power of Place.* Edinburgh: Floris, 1996.

Read, Howard. *In Search of the Immortals.* London: Hodder Headline, 1999.

Reddish, Paul. *Spirits of the Jaguar.* London: BBC Books, 1996.

Reid, John. *Egyptian Sonics.* Cumbria, UK: Sonic Age, 2001.

Robertson, John M. *Pagan Christs.* New York: Barnes and Noble, 1993.

Robinson, Andrew. *The Story of Writing.* London: Thames and Hudson, 2000.

Roney-Dougal, Serena. *Where Science and Magic Meet.* Shaftesbury, UK: Element, 1991.

Room, Adrian. *The Cassell Dictionary of Word Histories.* London: Cassell, 1999.

Ross, Anne. *Pagan Celtic Britain.* London: Routledge, Kegan and Paul, 1967.

Ruoff, Henry W. *The Century Book of Facts.* Springfield, Mass.: King-Richardson, 1907.

Salibi, Kamal. *Who Was Jesus?* New York: I. B. Tauris, 1998.

Satprem. *Sri Aurobindo or the Adventure of Consciousness.* Faridabad, India: Thomson Press, 2000.

Sattin, Anthony. *The Pharaoh's Shadow.* London: Gollancz, 2000.

Scarre, Chris. *Exploring Prehistoric Europe.* Oxford, UK: Oxford University Press, 1998.

Scholes, Percy A. *The Oxford Companion to Music.* Oxford, UK: Oxford University Press, 1992.

Schwaller de Lubitz, R. A. *Sacred Science.* Rochester, Vt.: Inner Traditions, 1981.

———. *The Temple in Man.* Rochester, Vt.: Inner Traditions, 1981.

Scott, Walter. *Hermetica.* 4 vols. Boston: Shambhala, 2001.

Segal, Judah B. *Edessa.* Piscataway, N.J.: Gorgias Press, 2001.

Sellers, Jane B. *The Death of the Gods in Ancient Egypt.* London: Penguin, 1992.

Settegast, Mary. *Plato Prehistorian.* Northumberland, UK: Lindisfarne Press, 1990.

Seymour, Percy. *Astrology: The Evidence of Science and the Paranormal.* Luton, UK: Lennard Publishing, 1988.

———. *The Paranormal: Beyond Sensory Science.* London: Arkana, 1992.

Seznec, Jean. *The Survival of the Pagan Gods.* Princeton, N.J.: Princeton University Press, 1953.

Shaw, Gregory. *Theurgy and the Soul.* Philadelphia: Pennsylvania State University Press, 1995.

Shlain, Leonard. *The Alphabet versus the Goddess.* London: Allen Lane, 1998.

Smith, Anthony. *The Human Body.* London: BBC Books, 1998.

Smith, Cyril W., and Simon Best. *Electromagnetic Man*. New York: Dent, 1973.

Solignac, Jennifer. "Divine Revelation." Unpublished manuscript.

Spawforth, Anthony. *The Oxford Classical Dictionary*. Oxford, UK: Oxford University Press, 1998.

Stevenson, Victor. *The World of Words*. New York: Sterling Publishing, 1999.

Stewart, Desmond. *The Foreigner*. London: Hamish Hamilton, 1981.

Stirniman, Robert. "The Wallace Inventions, Spin Aligned Nuclei, the Gravitomagnetic Field, and the Tampere 'Gravity-Shielding' Experiment: Is There a Connection?" *Frontier Perspectives* 8, no. 1 (1999).

Strachan, Gordon. *Christ and the Cosmos*. Edinburgh: Labarum, 1985.

Streit, Jakob. *Sun and Cross*. Edinburgh: Floris, 1977.

Sullivan, William. *The Secret of the Incas*. New York: Three Rivers, 1997.

Tame, David. *The Secret Power of Music*. Rochester, Vt.: Destiny Books, 1984.

Tandy, Vic, and Tony R. Lawrence. "The Ghost in the Machine." *Journal of the Society for Psychical Research* 62, no. 851 (April 1999): 360–64.

Targ, Russell, and Harold Puthoff. *Mind Reach*. New York: Delacorte Press, 1977.

Teilhard de Chardin, Pierre. *The Phenomenon of Man*, London: Harper Perennial, 1976.

Temple, Robert. *The Sirius Mystery*. London: Sidgwick and Jackson, 1981.

Thom, Alexander. *Megalithic Sites in Britain*. Oxford, UK: Oxford University Press, 1967.

Thomas, Keith. *Religion and the Decline of Magic*. London: Weidenfeld and Nicolson, 1997.

Thory, Claude-Antoine. *Acta Latomorum: ou Chronologie de L'Histoire de la Franche-Maconnerie Francaise et Etrangère*. Paris: Dufart, 1815.

Tompkins, Peter. *Secrets of the Great Pyramid*. New York: HarperCollins, 1978.

———. *The Secret Life of Nature*. London: Thorsons, 1997.

Tompkins, Peter, and Christopher Bird. *Secrets of the Soil*. London: Penguin, 1992.

Townsend, Richard F., ed. *Ancient West Mexico: Art and Archaeology of the Unknown Past*. London: Thames and Hudson, 1998.

Trevelyan, Sir George. *A Vision of the Aquarian Age*. Bath, UK: Gateway Books, 1994.

Troev, Theodor. *The Argonautica Expedition*. London: Ian Faulkner, 1991.

Underwood, Guy. *The Pattern of the Past*. London: Sphere, 1982.

U.S. Department of the Interior. *Field Studies of Radon in Rocks, Soils, and Water,* edited by L. C. S. Gundersen and R. B. Wanty. U.S. Geological Survey Bulletin, 1971.

von Pohl, Gustav Freiherr. *Earth Currents: Causative Factor of Cancer and Other Diseases,* Stuttgart, Germany: Frech-Verlag, 1983.

Walker, Barbara. *Woman's Encyclopedia of Myths and Legends.* New York: HarperCollins, 1983.

Wallace-Hadrill, Andrew. *Augustan Rome.* London: Bristol Classical Press, 1993.

Wallis-Budge, Sir Ernest A. *Legends of Our Lady Mary, the Perpetual Virgin, and her Mother Hanna.* London: Oxford University Press, 1933.

———. *The Mumm.* New York: Biblio and Tannen, 1964.

———. *Gods of the Egyptians.* Chatham, Kent, UK: Dover Publications, 1969. Originally published in 1904.

———. *Egyptian Religion.* London: Arkana, 1987.

———. *Book of the Dead.* London: Arkana, 1989.

———. *The Egyptian Heaven and Hell.* Chatham, Kent, UK: Dover Publications, 2003.

Watson, Lyall. *Supernature.* London: Hodder and Stoughton, 1973.

Watterson, Barbara. *Gods of Ancient Egypt.* Stroud, UK: Sutton Publishing, 1996.

Wayland Barber, Elizabeth. *The Mummies of Urumchi.* London: MacMillan, 1999.

Weil, Simone. *Intimations of Christianity among the Ancient Greeks.* London: Routledge, 1998.

West, John Anthony. *Serpent in the Sky.* Wheaton, Ill.: Quest Books, 1993.

———. *The Traveler's Key to Ancient Egypt.* Wheaton, Ill.: Quest Books, 1995.

Westwood, Jennifer. *Sacred Journeys.* London: Gaia, 1997.

Whone, Herbert. *Church, Monastery, Cathedral.* Shaftesbury, UK: Element, 1990.

Wilber, Ken. *Quantum Questions.* Boston: Shambhala, 2001.

Wilson, A. N. *Jesus.* London: HarperCollins, 1992.

Wilson, Colin. *Atlas of Holy Places and Sacred Sites.* London: Dorling Kindersley, 1996.

Wilson, Hilary. *Understanding Hieroglyphs.* London: Michael O'Mara Books, 1995.

Yahuda, Joseph. *Hebrew Is Greek.* Oxford, UK: Becket Publications, 1982.

Yates, Dame Frances. *Giordano Bruno and the Hermetic Tradition*. London: Routledge, 2002.

Zink, David. *The Ancient Stones Speak*. London: Paddington Press, 1979.

Zohar, Danah, and Ian Marshall. *SQ, Spiritual Intelligence: The Ultimate Intelligence*. London: Bloomsbury, 2000.

Index

Numbers in *italics* preceded by *pl.* indicate color insert plate numbers.